T0137189

Mathematical Advances Towards Sustainable Environmental Systems

James N. Furze • Kelly Swing • Anil K. Gupta
Richard H. McClatchey • Darren M. Reynolds
Editors

Mathematical Advances Towards Sustainable Environmental Systems

Springer

Editors
James N. Furze
Faculty of Environment and Technology
University of the West of England
Bristol, UK

Anil K. Gupta
Indian Institute of Management
Coordinator, Society for Research and
  Initiatives for Sustainable Technologies
  and Institutions
Ahmedabad, Gujarat
India

Darren M. Reynolds
Centre for Research in Biosciences
University of the West of England
Bristol, UK

Kelly Swing
Founding Director, Tiputini
  Biodiversity Station
College of Biological and Environmental
  Sciences
University of San Francisco de Quito
Quito, Ecuador

Richard H. McClatchey
Centre for Complex Cooperative Systems
University of the West of England
Bristol, UK

ISBN 978-3-319-82938-8   ISBN 978-3-319-43901-3   (eBook)
DOI 10.1007/978-3-319-43901-3

Printed on acid-free paper

This Springer imprint is published by Springer Nature
The registered company is Springer International Publishing AG
The registered company address is: Gewerbestrasse 11, 6330 Cham, Switzerland

This volume is dedicated...
To all who have contributed in the past, to those who see and are yet to see the beauty we rely on and can collectively renew— environmental systems hold bounty and sustenance for us all

(James N. Furze)

To all who depend upon biodiversity, whether they know it or not—with hopes of acting before it's too late

(Kelly Swing)

To all those creative, compassionate, and collaborative communities whose coping strategies with climatic risk require models that deal with their vulnerability, and yet don't underestimate their resourcefulness in a given socio-ecological and institutional context

(Anil K. Gupta)

To a future that binds us all

(Darren M. Reynolds)

To all who are concerned by, and driven to help guarantee, the sustainability of our wonderful worldwide ecosystem

(Richard H. McClatchey)

Foreword: The Vocabulary of Nature

Nature expresses herself through a broad array of options. Her immense vocabulary includes the majesty of rugged snow-capped mountain ranges, the terrifying power of tornados, the profound serenity of a golden sunset along the ocean's shore, or the countless species, animal and vegetable, great and small, assigned to every bit of land and sea imaginable. It is precisely this vocabulary that we wish to catalog, categorize, understand, exploit, and manage on a long-term basis. Her tremendous richness is precisely what presents the challenge, the opportunity, and the duty we have as the thinking agents of Earth.

As a writer hitches words together, one after another, to masterfully construct a story, so nature weaves species, along with their terrestrial and aquatic arenas into a tapestry of infinite functionality. When an author searches for just the right word, there exists only one real option due to its particular meaning, its sound, its history, and its connotations. Nature has likewise produced combinations of species meshing poetically together to compose the wondrous and seamless ecosystems that make our planet live and breathe.

And now, humankind is beset upon the task of eradicating the vocabulary of Nature herself by extinguishing species upon species and even entire panoramas from arctic extremes to the expanses of the sea, from the cloud forests of the cordilleras and the wettest of lowland rainforests to the driest of deserts, selfish, oblivious, and merciless in our advance. All for some "greater good," we push ahead never understanding that Nature's poetry can never be surpassed, and cannot be sustained without all her words, all the right words, all the integral elements of the fabric of life. Without all the fibers in their proper places, the tapestry becomes threadbare and begins to fall into tatters, becoming irreparable even for the most capable weaver, the last remnants serving essentially no function whatsoever.

Imagine the cruel punishment for the sculptor, the removal of an implement from his studio every day until left with no capacity to go on uncovering the spectacle within the material before him. Imagine eliminating pigmented oils from the painter's palette along with brushes and trowels to apply them. What could the carpenter accomplish if each day we took another tool from his workbench—hammer and saw, drill, chisel, and plane? And as to the novelist, the removal of a word from her treasure chest of emotive terms each day might eventually still her pen altogether. Such losses would not only affect the creators, but all those who might have appreciated their wonderful ideas as they are left intangible. Fortunately, true creativity, nevertheless, will ultimately triumph over such an onslaught. In spite of all the challenges, the driven artist or artisan will devise a way to overcome adversity; history confirms that evolution can also begin anew after a cataclysm, albeit with very different results.

Every culture has invented distinct words to represent each kind of place and resource, every sort of plant and animal. When we don't have a word of our own, usually because the phenomenon or species doesn't exist in places where people speak our language, we borrow the names from other tongues, because the standard human perspective is that every single thing simply must have a unique name. Part of basic human nature is the utilitarian desire to both quantify and qualify the things around us. Across the globe, in every ethnicity, we want a word, or combination of words, to match each and every thing we see, including every species. As an illustration, a basic repertoire of common names makes up an important portion of the first words learned by all children everywhere. And to avoid being confused by multiple common names of various origins, science has stepped in to apply a standard, universal name to every recognizable sort of organism, in the form of a genus combined with a qualifying specific epithet.

As species are driven to extinction, their word counterparts will be left orphans without any sense to their existence, or with only a perverse meaning parallel to current usage of "dodo," a species deemed too stupid to be allowed to share the planet with us any longer by a few myopic individuals blinded by hunger. We must rise above our overwhelming capacity to justify almost any loss or trade-off given momentary desperation situations. If we are to survive, there is no choice but to put our brains to earnestly resolving unsustainable scenarios.

How many organisms live in places where English isn't spoken? All those have names in other languages that we borrow to be able to talk or write about them. The enrichment of languages through incorporation from other languages is immense and enlightening. Just think of the marvelous examples like gorilla, narwhal and boomslang, chimpanzee, aye-aye and orang, panda, puma and piranha, take matamata, koala and kangaroo, condor, anaconda and caribou, wombat, wobbegong and wahoo. And there are so many more that leave us with no choice but to make use of their scientific names as common names. *Boa constrictor* and *Tyrannosaurus rex* are two of the few uttered every day in their complete forms. But it's much more typical to use the genus name alone; hardly anyone notices where they came from. How about all those ornamental plants and flowers? Geranium and chrysanthemum, ficus and philodendron, dieffenbachia and rhododendron just to name a few. And among the animals, we must consider rhinoceros and octopus, and also hippopotamus; contemplate alligator and python, iguana as well as mastodon never overlooking archaeopteryx and stegosaurus, triceratops, and brontosaurus!

The richness of language depends upon the richness of our surroundings—always has, always will. How many words exist in the English language?

What proportion of them do we use on a regular basis? Or ever? Do we have to use them frequently for them to have value? How many do we utilize in a lifetime? Does their disuse imply they're useless? If we don't use them, does that mean they're dispensable? If they have limited usage, say, by one peculiar erudite, does that indicate some potential for isolation or even discrimination or exclusion? In such a case, would we expect social pressure to avoid them? If such words disappear, would anyone notice or care?

How many languages exist? And how many species exist? How many species are useful? How many species are directly useful on a daily basis? What do we mean by "useful"? Who decides? Words come and go as far as popular usage is concerned, but they don't simply disappear. They continue to exist in dictionaries and in past publications and recordings no matter their most recent occurrences. And they can instantly be retrieved at any moment. But that cannot happen for species if every individual has been pushed off its proper habitat or immersed in sealed jars of embalming fluid on the dusty shelves of a museum. Words don't multiply when we ignore them, but populations of organisms, if they have sufficient numbers, can recover from neglect, abuse, or overuse and increase on their own, given the chance.

Can we only understand our planet and its biota, and consequently manage its resources, by cataloging all its actors and amassing information about their functioning? How much information is necessary for us to be able to effectively apply modelling strategies so that a rational balance between our wants and needs and the requirements for a planet in equilibrium may be struck?

"Not everything that can be counted counts, and not everything that counts can be counted." Often attributed to Albert Einstein but actually the words of William Bruce Cameron.

The more species and habitats we lose, the more I am left without words.

Tiputini Biodiversity Station, Yasuní,   Kelly Swing
UNESCO Biosphere Reserve
Quito, Ecuador
6 April 2016

Preface: From the Coordinating Editor

This volume focuses on diverse systems and sustainability. Included are component subjects of relevance with coverage of frontier research from subject specialists in 13 different areas. Following a coauthored introduction to establish balance and context, indication of the current state of research in each of the chapters will be marked. The volume unites multiple subject areas within sustainability, enabling the techniques applied in each chapter to be applied to other chapter areas in future research, giving a synergistic function for knowledge advancement, interdisciplinary cooperation, policy formation/governance, and subsequent areas.

The book is not of particular political focus; it is the scientific basis on which we can protect and enhance environmental sustainability within Earth Systems, faced with changes and pressures imposed by our expansive needs.

The target audience includes the "layman," graduate, postgraduate, doctoral, and postdoctoral researchers. Benefits are for national organizational structures, policy formation teams, and regional management bodies as well as the general public. This is a maturely written volume for the same audience.

- Each chapter describes frontier research which may be applied in different locations and groups as well as those that the authors quote.
- Together the chapters explain how we may proceed and progress in the subject disciplines with the use of systematic approaches.
- Each chapter provides a unique perspective of leading international authors, giving advancement and enrichment of knowledge and understanding of sustainability within diverse systems, while managing subject knowledge, development, and application for the benefit of multiple, expansive populations.

This book is an edited volume; the main purpose of the coordinating and other editors' work has been to locate and bring together the subject specialists, many of whom are editors and journal founders in their own disciplines.

Unique selling points of the volume are as follows:

- The volume gives pertinent mention of key areas which should be employed for generation of subject synergies, previously unpublished.
- The range of subjects considered has not been included in one complete volume before.
- Cooperation between high-standing authors across the range of subjects covered has value for policy makers and the general public, enhancing our progress within developmental pressures.

The book has a strong multidisciplinary nature at its core. This holds a wide range of interest to both generalist readers and to future subject specialists in information, mathematics, biology, physics, and chemistry as well as readers of the arts, history and political sciences. Upon reading the chapters together, the expressions of hope offered by each should allow great expansion in both separate chapters and the whole subject of sustainability/complex thinking. As knowledge expands in more descriptive chapters and contracts in more scientific chapters, great synergy may result; in mathematical thinking, the residual error between the approaches of each chapter offers a "functional resonance" which enables synergy beyond the volume. This statement becomes philosophical in terms of thinking and concrete knowledge advancement which we all keep in mind through modelling and societies; to create the abstract, one must have encountered the concrete and become able to ignore parts of it on purpose, and reaching rather than lessoning the very presentation of the concrete. What and if we should ignore of course is determined by our own individual, collective visions and perspectives of sustainability.

In addition to the 27 leading authors across 10 countries (including the United Kingdom, Belgium, Italy, Iran, Iraq, Egypt, India, China, Canada, and Ecuador), the editorial team was built of 5 leading members firmly uniting Eastern and Western ideologies in the name of knowledge advancement and sustainability for both natural and socioeconomic systems. We extend our gratitude to the publishers and authors who allowed the use of some figures/material in the volume, many reviewers (in additional countries including some in universities in Japan, Switzerland, Spain, Germany, Morocco, and Greece), and organizational units (including members of the Food and Agricultural Organization; United Nations Development Program and United Nations Environment Program) who gave support and assisted in the preparation of the volume. Finally, I appreciate all the authors and editors who sacrificed their time and effort to write the book.

The coordination of a sustainability effort represents a lifelong journey for all of us. It is hoped that this volume will represent a marker in the journey of the advancement of mathematics and individual research areas which will be followed by conference activities, journal platforms, and further editions of this volume as well as the use of increasingly diverse formats to enable our knowledge and ability to manage systems collectively to expand.

Bristol, UK   James N. Furze, Coordinating Editor

Bristol, UK   James N. Furze
Quito, Ecuador   Kelly Swing
Ahmedabad, Gujarat, India   Anil K. Gupta
Bristol, UK   Richard H. McClatchey
Bristol, UK   Darren M. Reynolds
17 May 2016

Contents

Contributors

J.F. Adamowski Faculty of Agricultural and Environmental Sciences, Department of Bioresource Engineering, McGill University, Ste Anne de Bellevue, QC, Canada

M.H. Bazrkar Department of Water Engineering, Isfahan University of Technology, Isfahan, Iran

S.A.F. Bonnett Department of Applied Sciences, University of the West of England, Bristol, UK

A. Dey Indian Institute of Management, Ahmedabad, India

M.D.F. Ellwood Department of Applied Sciences, University of the West of England, Bristol, UK

D. Ernst Systems and Modeling Research Unit, University of Liège, Institute Montefiore (B28, P32) Grande Traverse, Liège, Belgium

S. Eslamian Department of Water Engineering, Isfahan University of Technology, Isfahan, Iran

R. Fontenau Systems and Modeling Research Unit, University of Liège, Institute Montefiore (B28, P32) Grande Traverse, Liège, Belgium

James N. Furze Faculty of Environment and Technology, University of the West of England, Bristol, UK

A. Gohari Department of Water Engineering, Isfahan University of Technology, Isfahan, Iran

Anil K. Gupta Indian Institute of Management, Ahmedabad, Gujarat, India

Society for Research and Initiatives for Sustainable Technologies and Institutions, Ahmedabad, Gujarat, India

H.A. Hashem Faculty of Science, Botany Department, Ain Shams University, Cairo, Egypt

R.A. Hassanein Faculty of Science, Botany Department, Ain Shams University, Cairo, Egypt

A.A. Hill Department of Applied Sciences, University of the West of England, Bristol, UK

J. Hill Faculty of Environment and Technology, University of the West of England, Bristol, UK

P.J. Jena Department of Botany, Ravenshaw University, Cuttack, Odisha, India

S. Kumar Department of Botany, Ravenshaw University, Cuttack, Odisha, India

M. De Marchi Department of Civil, Environmental and Architectural Engineering, University of Padova, Padova, Italy

P.J. Maxfield Department of Applied Sciences, University of the West of England, Bristol, UK

F. Qiao Faculty of Information and Control Engineering, Shenyang JianZhu University (SJZU), Shenyang, Liaoning, China

S.M. Raafat Automation and Robotics Research Unit, Control and System Engineering Department, University of Technology, Baghdad, Iraq

F.A. Raheem Automation and Robotics Research Unit, Control and System Engineering Department, University of Technology, Baghdad, Iraq

B. Sengar Department of History and Ancient Indian Culture School of Social Sciences, Dr. Babasaheb Ambedkar Marathwada University, Aurangabad, India

Gurdeep Singh Vinoba Bhave University, Hazaribagh, Jharkhand, India

Kelly Swing Tiputini Biodiversity Station, College of Biological and Environmental Sciences, University of San Francisco de Quito, Quito, Ecuador

M.J. Zareian Department of Water Engineering, Isfahan University of Technology, Isfahan, Iran

Q. Zhu Faculty of Environment and Technology, University of the West of England, Bristol, UK

Chapter 1
Mathematical Advances Towards Sustainable Environmental Systems: Context and Perspectives

S.M. Raafat and Kelly Swing

Abstract This volume focuses on diverse systems/sustainability. Included are component subjects of relevance with coverage of frontier research from subject specialists in different areas. Following a co-authored introduction to establish balance and context, indication of the current state of research in component chapters will be marked. The volume unites multiple subject areas within sustainability, enabling the techniques applied in each chapter to be applied to other chapter areas in future research, giving a synergistic function for knowledge advancement, interdisciplinary cooperation as well as policy formation/governance and subsequent areas.

Each of the chapters of this book provides description of frontier research which may be applied in different locations and groups as well as those specifically mentioned. Together the chapters give indication of how we may proceed to advance related disciplines with use of systematic approaches. Each chapter provides a unique perspective of leading international authorities, encouraging advancement and enrichment of knowledge and understanding of sustainability within diverse systems, whilst managing subject knowledge, development and application for the benefit of multiple, expansive populations.

Keywords Frontier research • Balance • Context • Multiple subjects • Synergistic function • Cooperation • Policy formation • Governance • Knowledge enrichment • Sustainability • Expansive populations

> Learning to live sustainably on Earth is going to require enormous advances in our understanding of the natural world and our relationship with it. To acquire that understanding progress in the mathematical sciences is Essential
>
> (Rehmeyer 2011)

S.M. Raafat (✉)
Automation and Robotics Research Unit, Control and System Engineering Department, University of Technology, Baghdad, Iraq
e-mail: 60154@uotechnology.edu.iq

K. Swing
Founding Director, Tiputini Biodiversity Station, College of Biological and Environmental Sciences, University of San Francisco de Quito, Quito, Ecuador
e-mail: kswing@usfq.edu.ec

© Springer International Publishing Switzerland 2017
J.N. Furze et al. (eds.), *Mathematical Advances Towards Sustainable Environmental Systems*, DOI 10.1007/978-3-319-43901-3_1

1.1 Introduction

Sustainability science is dedicated to understanding human–environment interactions—patterns that play out over periods of decades to centuries and over significant expanses of space (Levin and Clark 2010).

In the face of global climate change and the ongoing mass extinction event, the urgency for cataloguing, understanding and managing resources at the planetary scale increases, thereby making the discovery and implementation of functional solutions an ever greater priority.

Constructing this science of sustainability requires a multi-disciplinary method that combines practical understanding with knowledge and proficiency deduced from across the natural and social sciences, mathematics and computation, medicine and engineering. Moreover, the developed research on sustainability affords a magnificent challenge to mathematical disciplines and computer science in collaboration with natural, physical and social sciences (Levin and Clark 2010).

Sustainability issues are extremely complicated, demanding more sophisticated scientific, mathematical and statistical tools than those traditionally available. Biological systems, climate, water, renewable energy production capacities models, for example, are extraordinarily complex. Created by scientists from many disciplines, and requiring unusually powerful supercomputers to run, these strategies provide only crude and imprecise approximations of the true processes affecting climate. They are, in fact, raising mathematical and statistical questions that have never before been faced, and right now, there still remain tremendous research demand to obtain suitable answers (Rehmeyer 2011).

Almost every sustainability challenge needs new mathematical tools. Economic issues, which are deeply interlinked with sustainability issues, promote their own mathematical and statistical challenges. The only approach to realise the real impacts is to integrate economic models with mathematical models of climate, energy, biodiversity, etc. a task that presents dramatic new challenges. Additionally, with the increase in world population, we may have to revisit accepted definitions of a healthy economy. Complex issues arise, including issues concerning the carbon market, concepts of equity between nations and intergenerational equity, etc. Addressing these issues requires new partnerships between mathematical scientists and social scientists (Rehmeyer 2011).

Reaching the goal of a sustainability system is affected by generating new knowledge about the current state of the system under study and by considering all the influences of management and natural processes. This can be accomplished by acquiring vast amounts of new information. To handle new data types, classifier decision rules based on fuzzy logic, evidential reasoning and neural networks have been developed or modified for local and/or remote sensing applications. Decisions can be made either based on probabilistic rules or by using different mathematical theory and logic (Franklin 2001).

It has been recognised that mathematical models are an essential method to establish effective responses to critical situations like disease outbreaks (Rehmeyer 2011).

An Epidemiological mathematical model, for instance, could provide guidance to help make more informed decisions. Specific attention is required to ensure that solving interesting mathematical problems is useful for public health in the real world. To deal with public health challenges, it is recommended that foster collaborations between mathematical scientists and researchers from the economic, social and behavioural sciences be established. Agent-based models are one promising approach. They create individuals inside the computer and model their movement along with the movement of any pathogens they carry; still, there are limited applications in this context (Levin and Clark 2010).

Sustainable Environmental Systems (SES) are complex and adaptive. In spite of much study of such phenomena, however, we still have only the beginnings of an understanding of the vulnerability and flexibility of SES. Research is needed to understand—in both spatial scales and organisational levels—the impact of advances in mathematics and modern technologies on such systems (Levin and Clark 2010).

Many properties of SES can be adequately captured with conventional statistical or system-dynamic models. But the complex dynamics, inter-sectoral and multi-scale interactions, emergent properties and uncertainty that characterise many of the SES most relevant to sustainability concerns have proven highly difficult to deal with using such approaches. Advances in agent-based and network approaches to the modelling of complex adaptive systems offer promise of doing better, as do several approaches to the qualitative analysis of non-linear systems and the development of interdisciplinary, multi-scale scenarios. Research in this field has accomplished some goals (Levin and Clark 2010).

Humans and highly developed robots are not necessarily competitors in terms of sustainability. Indeed, it is an understanding of automisation in terms of the criteria for sustainability—economic, ecological and social—that ultimately leads to a significantly closer combination of humans and automisation systems. Consequently, these considerations call for a new paradigm in automation technology, i.e. one of human-centred automation. The objective of human-centred automation is not to copy human capabilities but to assist and support them. Examples of such human-oriented automation devices are cooperative robots, e.g. cobots, where robots do not need to be programmed with any particular movement in advance. The human worker controls the complex process, the robot supports him/her and at the same time learns, i.e. memorises the process. With the human-centred automation approach, sustainability in production is maintained in several ways: economically by an improved adaptability of semi-automated systems to change production processes, ecologically by a more long-term usability and energy efficiency of human-centred systems, and socially by an improved integration of humans into the production process and the possibility to avoid difficult and dangerous work tasks (Krüger 2008).

An important application of sustainability in automation technology can be found in the field of medical technology, where several successful examples already exist. Robot-assisted systems for supporting humans in the household, in care or in rehabilitation are still undergoing investigation (Krüger 2008).

Industrial robotics and automation currently play a quantitative role in increasing the productivity of human workers, which could actually have a destabilising effect on economics. However, robots also make a qualitative contribution to production. Sustainable productivity management does not, in principle, exclude the use of robots. First, robots have skills such as strength, precision and sensing often superior to those of humans. These skills facilitate the production of advantageous outcomes and it would not make much sense to abandon such qualitative benefits. Second, robots can be redesigned to exploit sustainable energy and material resources. Third, new robotic system application models can be conceived to fit in a human–robot corporation establishing a production unit, thereby making use of the capabilities of both partners. Fourth, new applications of robotics can appear, that support a sustainable economic model, like energy and resources production, the food chain and recycling. However, the most useful applications are still to emerge (Bugmann et al. 2011).

1.2 Chapter Outlines

This volume focuses on diverse systems/sustainability. Included are component subjects of relevance with coverage of frontier research from subject specialists in different areas. Following a co-authored introduction to establish balance and context, indication of the current state of research in component chapters will be marked. The volume unites multiple subject areas within sustainability, enabling the techniques applied in each chapter to be applied to other chapter areas in future research, giving a synergistic function for knowledge advancement, interdisciplinary cooperation as well as policy formation/governance and subsequent areas.

Each of the chapters of this book provides description of frontier research which may be applied in different locations and groups as well as those specifically mentioned. Together the chapters give indication of how we may proceed to advance related disciplines with use of systematic approaches. Each chapter provides a unique perspective of leading international authorities, encouraging advancement and enrichment of knowledge and understanding of sustainability within diverse systems, whilst managing subject knowledge, development and application for the benefit of multiple, expansive populations.

Chapter 2 introduces dimensions in which authors consider biological diversity of plant species (life history strategies, life forms and metabolic process). Maintaining and monitoring current levels of diversity in these categories are essential for understanding trophic systems. Diversity within the primary producing level of trophic systems enables subsequent levels of trophic organisation to attain stability and growth. Socio-economic consequences are relevant to sustainability and stability of our populations.

Details of the modelling process are shared using computationally efficient processes. Basic algorithmic structures, advanced mathematic dispersal of populations and functional approximation are shown with reference to case studies with contrasting levels of species richness. Alternative functions are considered in vulnerable areas in order that the use of plants within those areas can be maximised. Appropriate levels of protection status are recommended to national organisations and governing bodies.

Mathematic methods/modelling generates tools within research providing us with structures by which we can expand knowledge of related fields and systems, allowing us to apply appropriate safeguarding of diversity and communities in the face of developmental pressures.

About 80 % of world energy consumption is currently from non-renewable origin. Chapter 3 addresses the problem of the "energy transition", i.e. the switch to an energy production system using renewable resources only. One of the main challenges of this transition comes from the fact that switching to an energy system that would not depend on non-renewable resources is a process that needs itself—at least for the moment—to use non-renewable energy. The originality of this approach is to consider the deployment of renewable energy production units as an energy investment. This point of view is motivated by the fact that the Energy Return Over Investment (EROI) parameters characterising the two main rising renewable technologies—wind turbines and photovoltaic panels—are currently too low to be viable. The transition is formalised into a discrete-time optimal control problem constrained by a finite budget of non-renewable resources. The objective is to maximise over a given time horizon (in the order of 50 years) the level of renewable energy production while minimising the quantity of non-renewable resources used during the transition phase. We show that this problem can be reformulated as a reinforcement learning problem, opening the doors to a large class of resolution techniques.

Modelling plays a key role in the simulation, planning and management of water systems. Effective simulation allows for more sustainable decision making in water resources management. There are many different types of water resources modelling approaches, each with its own advantages and disadvantages. Chapter 4 will provide readers with a thorough overview of modelling approaches, as well as ways to decide which type of model to use in a given situation.

Biodiversity has finally entered conversations among scientists and the populace at large as a common topic. Chapter 5 brings to our attention high rates of species loss; getting an idea of the wealth of species on our planet takes on an air of urgency. In about 250 years of applying scientific names, we have managed to work through only a fraction of what is believed to be out there. Part of human nature involves collecting and categorising the things around us. Our primordial purpose can vary from simple curiosity to economic interests. Only recently has biodiversity, in and of itself, come to be regarded as a valuable resource. Undoubtedly, every species plays a role in nature and as such, has potential to provide humankind rewards at some level.

Chapter 6 discusses the fact that management and maintenance of nature are complicated because both nature and human nature are themselves quite complex. While traditional explanations based on human population growth combined with increasing resource demands offer some perspective on an ever worsening situation, they seem somewhat separated from root causes as well as potential solutions. Unfortunately, we live in a time of wilful ignorance of scientific facts. For some inexplicable reason, certain sectors of society have taken to rejecting the teachings of science as though they were nothing more than political rhetoric based on opinions and popularity polls. We have literally taken a step back in time to such pivotal moments as the ostracising of imminent figures such as Galileo and Copernicus, purposely choosing to toss aside well-established facts because their consequences imply the need for great expenditures. In earlier times, opposition mostly came from religious factions, but now it comes from politicians and their powerful self-interested business affiliates that will be impacted by transitions to different technologies.

One simplified model frequently used to depict relationships of humans and our world involves a triangle with the three corners, respectively, occupied by our species (the exploiters), another specific species (the exploited) and its habitat (the stage). Assuming that ecosystems were in some state of equilibrium before humans so recently appeared on the scene, it becomes evident that our addition to this geometric figure is precisely what has upset the balance. We know that humans have been increasingly responsible for impacts on their surroundings for millennia. We have repeatedly proven our capacity to modify nearly any habitat for our own benefit and concomitantly to the detriment of nature.

Biogeochemical cycling (Chap. 7) is fundamental to Earth systems. Carbon and nutrients are recycled primarily within ecosystems by the microbial processes of decomposition and mineralisation, with decomposition being one of the world's most important ecosystem services. The stability of such ecosystem services is being undermined by climate change, habitat loss and nitrogen loading.

Model systems are widely used for testing complex problems. However, accurately parameterising models for predicting intricate real-world scenarios is highly challenging. An effective solution is to use natural, controllable microcosms to generate representative models. We will review current biogeochemical deterministic and probabilistic models, focusing primarily on microcosms. Primary areas of interest will include ecophysiology, nutrient and enzyme stoichiometry, cycling of carbon and nitrogen, and isotope partitioning using tracers and kinetic isotope fractionations.

Having defined the biogeochemical parameters, we will review possible modelling approaches. The principle aim of Chap. 7 is to explore the potential destabilisation of the cycles to environmental perturbations through instability theory. By highlighting and predicting areas of unstable behaviour, we will show how this can facilitate wider studies into the effects of global change and the fate of natural ecosystems.

Case studies to be reviewed will include but not be restricted to conventional carbon and nutrient cycling models such as DNDC, CENTURY and FAEWE. Microbial-based models will be reviewed in the context of ecosystem function owing to their sensitivity to environmental change and their control over critical biogeochemical processes. Finally, we will examine and evaluate the use of oceanic islands and epiphytes in tropical rainforests as example model systems. The ultimate goal nonetheless is to accurately model the inputs and outputs of entire ecosystems rather than as a series of individual interactions.

Chapter 8 considers a complete review of the recent research of plant metabolite expression. It aims to illustrate the link between different elements in the metabolism regulation process with special reference to the environmental factors affecting metabolite expression in plants. This chapter covers both primary and secondary metabolism in plants. The complexity of transcriptional regulation of secondary metabolite biosynthesis including flavonoids, alkaloids and terpenoids will be highlighted. Recent advances in mitochondrial metabolism control as the power station of plant cells will be also summarised.

Plant metabolic engineering is an emerging discipline which promises to create new opportunities in agriculture, environmental applications, production of chemicals and even medicine. The key to improving the success rate of this technology is the enhanced understanding of the systems that are subjected to engineering, and this can be achieved by wide and comprehensive study of the expression and over-expression of plant metabolites. Potential in this field is driven by the diversity of physiological reactions within cellular and organismal biochemical arenas as well as the overall diversity of life forms. Specific examples within the Dioscorea family are discussed in Chap. 9, together with community and social uses of the plants.

Chapter 10 discusses Climate change, referring to a statistically long-term shift in the pattern (mean state and variability) of a regional or global climate. This phenomenon is attributed directly or indirectly to human activities that have resulted in an increased concentration of greenhouse gases in the global atmosphere. Climate change is already having major effects on the physical environment as well as societies. The aim of Chap. 10 is to investigate the effects of climate change on the sustainability of water resources at the watershed scale.

Chapter 11 considers how resilience may be formed in climatic and community terms and outlines the pertinent elements which must be considered in the maintenance of a homeostatic system. The chapter highlights the strengths and weaknesses of modelling approaches, outlines different dimensions of social systems and indicates the consequences of the changing climate on socio-economics in terms of the patterns of coping strategies.

Chapter 12 focuses on Robotics systems and how they play a crucial role in the world; their presence and our dependence on them are progressively growing. Qualitative benefits result from several of their inherent characters including computing power. The application of robotics for sustainable development is an exciting arena whose challenges may be met with assistance from scientific research and industry. Chapter 12 brings together mathematical developments in important fields of robotics: kinematics, dynamics, path planning, control and vision. An introduction to the development of robotics in different areas of applications (types of robots and applications) is provided. The kinematics of a robot manipulator is briefly described. The formulation of dynamics for the manipulator has been obtained based on Lagrange's energy function. Linear Segments with Parabolic Blends and Third-Order Polynomial Trajectory Planning have been described in detail. Different classical control strategies are presented. Basic concepts of Robot Vision are presented.

Intelligent control has a great influence in improving the performance of robotic systems. Similarly robust control proves to enhance the precision and accuracy of positioning in robotic applications. Chapter 13 describes some powerful intelligent and robust control methods for robots. The development of intelligent methods of path planning and obstacle avoidance is illustrated in detail. An efficient decision-making model for collective search behaviour for a swarm of robots in a risky environment is also demonstrated. In addition, an intelligent variable structure control for a robot manipulator has also been shown. Finally, discussion and further directions are given.

Chapter 14 acts as a summary for the volume and attempts to include emphasis from all chapter areas whilst giving an indication of the prospects for sustainability with our 'dynamic management of common resources'. Models formed by policy makers and sustainability scientists cooperatively produce tools with desired practical implications. Models presented in the volume are summarised, with existing policy methods for resource management and Global perspective for resource planning methods in context. Our holistic management vision for sustainability is emphasised with diverse digital, practical and wholly pragmatic imperatives. Additionally authors give details of case studies in two contrasting yet highly vulnerable areas and show how each area has different issues with similar priorities of land and ecosystem management being in the best long-term interest of each system. The two case studies, although complex may be given new hope through application and more efficient use of models and methods covered in this volume. A European example is further given to show how a complex demographic can be stabilised and sustained together with the ecosystem with use of modelling approaches. Finally, the authors concede that although sustainability offers many challenges, use of dimensional and alternative perspectives presented in each chapter will result in synergy which will advance our sustainability into the future.

Collaboration towards sustainable environmental systems must have key participation from the communities of vulnerable ecosystems, by scientists and from national and international governance bodies. We hope to see this impetus sustain the World for many years to come.

References

Bugmann G, Siegel M, Burcin R (2011) A role for robotics in sustainable development? Proc. of IEEE Africon '2011, 13–15 Sept., Livingstone, Zambia

Franklin SE (2001) Remote sensing for sustainable forest management. CRC, Boca Raton

Krüger J (2008) Sustainability of automated systems in production. Future. https://www.ipk. fraunhofer.de/fileadmin/user_upload/_imported/fileadmin/user_upload/IPK_FHG/ publikationen/futur/Futur_1_3_2008_englisch/sustainability_of_automated_systems_in_pro duction.pdf. Accessed 10 January, 2016

Levin SA, Clark WC (2010) Toward a science of sustainability, report from toward a science of sustainability conference Airlie Center—Warrenton, Virginia November 29, 2009—December 2, 2009, March 30

Rehmeyer J (2011) Mathematical and Statistical Challenges for Sustainability. In: Cozzens, M, Roberts FS (eds) Report of a Workshop held November 15-17, 2010

Chapter 2
Biological Modelling for Sustainable Ecosystems

James N. Furze, Q. Zhu, J. Hill, and F. Qiao

Abstract Modelling of biological systems is discussed in terms of the primary producers of trophic levels of organisation of life. Richness of plant communities enables sustainability to be reached within subsequent trophic levels. Plant dimensions of life history strategy, primary metabolic type and life form are defined and discussed with respect to the water–energy topography dynamic of climatic variables. The role of biogeography is given and classical approaches to species distributions and both antecedent and consequent variables within plants differentiated modelling frameworks are discussed. The algorithmic modelling frameworks applied in plant systems are justified statistically and with reference to established models. Mechanisms of dispersing discrete distributes are covered in consideration of genetic programming techniques. Further analysis expanding discrete approaches through functional transformations is considered and detail of a Gaussian Process Model is shown. Case study data of global locations is considered by means of discrete approaches of strategy and photosynthetic type and a continual approach of life form distribution. Algorithms for all processes and techniques are shown and graphical distributions made in illustration of the techniques. Synergy which may take place between the techniques is elaborated and flow diagrams of multiple benefits in pattern identification and further analysis are given. Recommendations for planting policies and policy implementation methods are covered in the final section, which also gives further direction on which we can base investigations which allow rational truth of distributes to be maximised and hence modulation of future modelling frameworks. Value is provided in terms of empirical, cause–effect and combinatorial approaches allowing us to process information effectively, structuring trophic levels and our own communities' expansive needs.

J.N. Furze (✉) • Q. Zhu • J. Hill
Faculty of Environment and Technology, University of the West of England, Frenchay Campus, Cold Harbour Lane, Bristol BS16 1QY, UK
e-mail: james.n.furze@gmail.com; quan.zhu@uwe.ac.uk; jennifer.hill@uwe.ac.uk

F. Qiao
Faculty of Information and Control Engineering, Shenyang JianZhu University (SJZU), 9 Hunnan East Road, Hunnan New District, Shenyang, Liaoning 110168, China
e-mail: fengqiao@sjzu.edu.cn

J.N. Furze et al. (eds.), *Mathematical Advances Towards Sustainable Environmental Systems*, DOI 10.1007/978-3-319-43901-3_2

Keywords Modelling • Life • Sustainability • Dimensions • Synergy • Combinatorial • Information • Trophic levels

2.1 Introduction

In characterising biological systems we may apply a modelling approach to species characteristics and ecological functions in order to provide policy makers with informative descriptions of life on Earth. In this introduction of biological modelling applications, the authors focus on different dimensions of those species which are the primary producers within trophic levels of organisation. In order to show the application of the dynamic conditions which exist on Earth, we may characterise plants in terms of life history strategies, metabolic (photosynthetic) type and life form. We characterise species within the main elements of climatic and topographic factors. In brief, habitats dominated by certain life history strategies, or with a major element of one of the three photosynthetic types, determine richness and hence variation in life forms. By considering these different dimensions of plant species, detail is covered of use in both empirical (Furze et al. 2011) and cause–effect-based modelling approaches (Best et al. 2011), which may lead to a combinatorial approach. Such an approach has minimum error in its consequential result (Fourcade et al. 2014; Zhu et al. 2013). Inclusion of different dimensions of form enables sufficient detail on which cause–effect modelling platforms can be based (Cuddington et al. 2013). Each of the primary groups considered here is fundamental to secondary and tertiary biological and biochemical processes and even abundance of species coverage. Each of the groups considered has implications in both ecological and socio-economic systems (given that we have many uses, for example for metabolites and plant materials). Thus, an appreciation of the diversity within the main sets equates to a resource in itself, notwithstanding the multiple benefits in ecosystem sustainability. This section provides a brief account of each of the three dimensions.

Globally characterised groups of plant life history strategies (Furze et al. 2013a, b; Grime et al. 1995; Grime and Pierce 2012) are as follows: Competitive species (C) are fast growing, often aggressive species, with rapid nutrient absorption and rapid root and leaf growth. They develop a consolidated growth form with vigorous lateral spread above and below ground, thriving in high nutrient soils. Stress-tolerating species (S) are slow growing, capturing and retaining scarce resources in a continuously hostile environment. Their leaves are long-lived and often heavily defended against predation. Ruderal species (R) have a potentially high growth rate within the seedling phase and display early onset of the reproductive phase. The allocation of resources to flowers and seeds is suited neither to development of extensive root and shoot systems needed for dominance of habitats, nor to highly stressed environments dependent on conservative patterns of resource use. Species combine the earlier described strategies (e.g. C–R, S–R, S–C and C–S–R), integrating different growth forms to suit the environment.

Kp

0	0.071	0.214	0.357	0.500	0.643	0.786	0.929	1
r	r-3	r-2	r-2	r↔K	K-1	K-2	K-3	K
	R	S-R	C-R	C	C-S-R	C-S	S	

Fig. 2.1 Separation of the rK and C–S–R theories in two dimensions (Barreto 2008)

In a study of the evolution of model plant populations in computer-simulated environments, where nitrogen availability and disturbance frequency alone were used as variables, the evolution of expected plant strategies and patterns proved consistent with described theoretical and field evidence (Grime 1979; Mustard et al. 2003). The illustration of plant strategies in computer-simulated systems supports the existence of patterns on all scales, which may be modelled in real space and time. Barreto (2008) linearly spaced plant species using simulation techniques to obtain the rK ('Kp') continuum, with seven definite partitions as shown in Fig. 2.1.

Barreto's (2008) 2D rK and C–S–R reconciliation shows low to high Kp value: plants of ruderal (R) strategy were isolated in places of high disturbance and productivity; stress-tolerant ruderal plants (S–R) were seen in lightly disturbed habitats with low productivity; competitive ruderal plants (C–R) were present in habitats where disturbance brought moderate competition by a relatively low level of stress; competitive plants (C or $r = K$) were found in environments with low disturbance and high productivity; competitive-stress-tolerant ruderal plants (C–S–R) were found in environments where there was a moderate intensity of stress and disturbance; competitive stress tolerating plants (C–S) were found in environments where a moderate intensity of stress and a situation of relative non-disturbance existed; stress tolerating plants (S) were found in environments where there was low productivity and low disturbance.

Figure 2.2 shows the extrapolation of plant life history strategies and rK strategies to a linear 3D continuum. The triangular distribution is effectively expressed within membership categories of x and y and strategy. The former axis categories may be used to express climatic/topographic conditions (as elaborated in later sections).

Photosynthesis is the primary metabolic process by which plants grow (Furze et al. 2013c). There are three main groups of photosynthesis in plants, C3, C4 and Crassulacean Acid Metabolism (CAM). C3 plants have three carbon compounds in the first step of photosynthesis, C4 have four carbon compounds in the first step of the process and CAM plants store carbon in the form of an acid before photosynthesis.

C3 plants' stomata are open during the day, allowing gaseous and water exchange via photorespiration (breakdown of sugars formed in photosynthesis, releasing CO_2 and H_2O). Ribulose bisphosphate-carboxylase oxygenase (RUBISCO) is the enzyme involved in uptake of CO_2 in C3 photosynthesis.

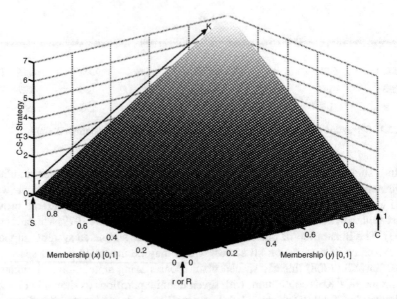

Fig. 2.2 C–S–R and rK Strategies in three dimensions

Photosynthesis takes place throughout the leaves of the plants. C3 plants represent the largest group of plant species, the process being highly efficient under cool and moist conditions (Niu et al. 2005).

C4 plants' stomata are open during the day. Phosphoenol pruvate (PEP) carboxylase is the enzyme involved in uptake of CO_2 with RUBISCO processing CO_2 in photosynthesis. Photosynthesis takes place in specialised Kranz cells, compartmentalised inner layers of the leaf. C4 plants photosynthesise faster than C3 plants under circumstances of high energy (e.g. light, temperature) and have much lower rates of photorespiration as the enzyme RUBISCO is more saturated with CO_2 for photosynthesis due to PEP activity. CO_2 uptake is more efficient in C4 plants, coupled with highly efficient water use in photosynthesis due to spatial separation. Stomatal closure effects less water-loss from plants and hence greater efficiency under warm and drier conditions. They are mainly summer annual species occurring in over 19 plant families (Keeley and Rundel 2003; Salisbury and Ross 1992; Wang et al. 2012).

CAM plants keep their stomata closed during the day and during both day and night in periods when water must be conserved (known as CAM-idle). During CAM-idling photosynthesis and photorespiration couple, the oxygen given off in photosynthesis is used during respiration and CO_2 given off in respiration is used in photosynthesis. CAM idling leads to a build up of toxic compounds over very dry periods. When moisture is available, the stomata reopen and CAM occurs as before. Opening of stomata at night enables more efficient use of water as temperatures and wind speeds are lower than during the day. CAM is an adaptation to very hot, dry conditions. Most cacti and succulent plants use this metabolism. It is also found in orchids and epiphytic bromeliads (Lüttge 2003). Intermediates occur between

C3-CAM and C4-CAM (Silvera et al. 2014). C3 metabolism evolved primarily, with C4 photosynthesis serving as an adaptation to warmer temperatures and enabling CAM plants to cover arid zones and extreme environments. Although the three main photosynthetic types lend themselves to the triangular ordination of Fig. 2.2, the large proportion of intermediates and subsequent levels of secondary and tertiary metabolism underline the requirement for the use of sophisticated mathematic systems for discrete space–time modelling.

There are five main groups of life forms: phanerophytes, chamaephytes, hemicryptophytes, cryptophytes and therophytes. Within the main groups there exist a total of 18 subgroups, with the following characteristics: Phanerophytes are of three types: (1) evergreen phanerophytes with bud scales, (2) evergreen phanerophytes without bud scales, (3) deciduous phanerophytes with bud scales. Phanerophytes are further divided according to height: Mega-(>30 m), Meso-(8–30 m), Micro-(2–8 m), Nano-(<2 m).

Chamaephytes have woody and herbaceous types. Chamaephytes are broken into (1) suffruticose (after the main growth period upper shoots die, only lower parts of the plant remain in "unfavourable" periods); (2) passive (in unfavourable conditions upper shoots become procumbent, protecting them from environmental stresses); (3) active (shoots only produced along the ground and remain so); (4) cushion (similar to passive type but shoots are so closely packed together so they form a "cushion").

Hemicryptophytes are further divided into (1) proto (leaves are well developed up the stem of the plant, partially developed leaves protect growing buds); (2) partial rosette (developed leaves form a rosette at the base of the plant, the following year a long aerial shoot may grow); (3) rosette (leaves restricted to a basal rosette, long exclusively flower bearing aerial shoot forms).

Cryptophytes are divided into (1) geophyte (underground organs such as bulbs, rhizomes, tubers, shoots emerge in growing season); (2) helophyte (growing buds are in soil or mud under water producing shoots above water); (3) hydrophyte (buds lie under water, unfavourable period spent completely below water).

Therophytes are annual plants, which survive the unfavourable period as seeds, completing their life cycle in the summer months.

Life forms are seen in differing proportions or spectra; plants show chaotic patterns of evolution in terms of their individual growth processes and numbers (Cui et al. 2012; Furze et al. 2011; Su et al. 2009). Life form differences are often associated with variable gradients in topographical and climatic conditions (Bhattarai and Vetaas 2003; Furze et al. 2013f; Schmidt et al. 2005).

In order to model the groups of life history strategies, photosynthetic type and life forms, we propose the use of a modelling framework, constructed with use of efficient antecedent (inputted) climatic and topographical variables and consequent species numbers. The remainder of the chapter is as follows: Section 2.2 gives detail of a digital elevation model (DEM) for the sourcing of topographical information, climatic variables and biological records. Section 2.3 provides mathematic detail of algorithmic structures. Section 2.4 shows a candidate dispersal method used in biological simulations. Section 2.5 proceeds to detail the use of both continual and discrete data in informative functional approximated ecological/ mathematic methods. Section 2.6 elaborates case studies of life history strategies, photosynthetic type and life form characteristics. Section 2.7 develops the use of functional approximation methods further and describes the proposed use of Lyapunov stability in combined use of algorithmic blocks. Finally, Section 2.8 concludes by summarising some of the advantages of and challenges for biological modelling and proposes further directions and applications.

2.2 Biogeographic studies, Digital Elevation Models, Climatic data and Biological Records

Plant characterisation originates in plant exploration. Alexander von Humboldt (1769–1859) published on botany and geography, founding biogeography itself (Wallace 1878). Humboldt was one of the first to describe the increase in species richness towards the equator (Humboldt 1806), the 'latitudinal gradient' as it later became known (Humboldt 1808; Rosenzweig 1995). Humboldt, 1845–1858, was the first to identify the Chocó region and Andean forests of Columbia, South America as one of the mega centres of plant diversity:

> Die dem Äquator nahe Gebirgsgegend … von Neugranada [today: Columbia] … ist der Teil der Oberfläche unseres Planeten, wo im engsten Raum die Mannigfaltigkeit der Natureindrücke [today: biodiversity] ihr Maximum erreicht—[Humboldt (1845), p. 12].
>
> (English translation by Otté (1860, p. 10):… The countries bordering on the equator [meant is the present-day country of Colombia] possess another advantage … This portion of the surface of the globe affords in the smallest space the greatest possible variety of impressions from the contemplation of nature [today: biodiversity] (Barthlott et al. 2005).

Humboldt (1808) hypothesised explanations for the diversity including complex topography and the variety of suitable climatic conditions in the Chocó region. He made the statement that plant richness declines at higher latitudes due to the fact that many species are frost intolerant and may not survive in the comparatively cooler temperatures of temperate zone winters, substantiating the water–energy dynamic (Wright 1983). Wright continues that plant productivity is limited primarily by energy from the sun and water availability. He adds however that the solar energy that transfers through each trophic level is what constrains richness as opposed to the total energy within a geographic area (Hawkins et al. 2003; Jetz et al. 2009; Wright 1983).

The relationship between species occurrence and area is given by Arrhenius (1921) as follows:

$$S = cA^z \tag{2.1}$$

Where S refers to the number of species, c is an environmental constant specific to the area, A refers to the area and z is a taxon-specific constant. To estimate what the values of c and z are from numbers of species we take logarithms of both sides of the equation and express the equation as a straight line. Hence the question of whether the number of species correlates significantly with area can be ascertained with use of t-testing between different areas. It has been found that unless the units of the earlier equation are standardised/distorted in preparatory Boolean statistical methods, the amount of species increase with area is not significant at 5 % probability (Furze et al. 2011; Hill et al. 1994). If we wish to see a true relationship of the complex of species presence and hence predict species occurrences and their conditions across different locations it is necessary to differentiate the conditions further than through the generalised expressions given in (2.1). Input data to be considered in plant species orientation is of altitude, water and energy variables, given that the effect of the water–energy dynamic alters at different altitudes (Bhattarai and Vetaas 2003; Gentry and Dodson 1987; Hawkins et al. 2003; Sommer et al. 2010).

Altitude data is most efficiently shown in digital elevation models (DEM), of which there are many candidate sources at different spatial resolutions and monitoring methods. We propose the use of the 1 km resolution dataset, GTOPO30 (Gesch and Larson 1996) which may be freely accessed online via the United States Geological Survey (USGS) from satellite data. GTOPO30 can be easily downloaded in the form of 22 tiles which may be uploaded into the technical computing platform Matlab. Display is simple via the use of a graphical user interface (GUI) and coding within the platform. Figure 2.3 shows examples of DEM data.

After processing GTOPO30 tiles through the GUI of Matlab, one may extract precise details of any location's coordinates and the range of elevation present. Ecuador has northernmost latitude of 1.464504°, southern most latitude of −5.02194°, easternmost latitude −75.19056°, westernmost latitude −81.023841°. Macedonia is located between latitude 43.000° North, 40.000° North and longitude

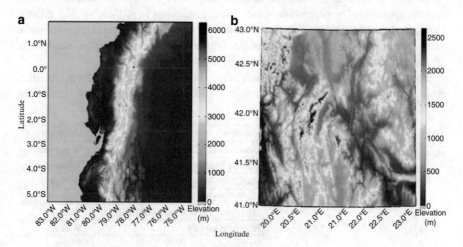

Fig. 2.3 Digital Elevation Maps of (**a**) Ecuador and (**b**) Macedonia

20.000° East to 23.000° East. Coordinates for Ecuador and Macedonia were validated by Bass et al. (2010) and Bilsborrow et al. (2012), respectively. Ecuador shows an elevation range from 0 to 6300 m above sea level, whereas Macedonia shows an elevation of 0 m to just over 2520 m above sea level. The range of elevation present is quantified and partitioned for the purposes of an algorithmic modelling framework in Sect. 2.3.

High resolution climatic data may be sourced from the Climatic Research Unit, East Anglia, UK, the use of which is validated by the Intergovernmental Panel on Climate Change (IPCC). The data originated from studies made by New et al. (1999) and are of great use in stochastic weather generator models by the IPCC under constantly developing scenarios of different variables (e.g. different CO_2 levels) reverting from natural processes. The use of scenario TS2.1 data in the time period 1961–1990 is seen as sufficient to encompass variation in today's climate (Evans et al. 2005; Kreft and Jetz 2007; Sommer et al. 2010). The data are available in the data distribution centre of the IPCC. Variables considered in TS2.1 are presented in Table 2.1 in terms of their water and energy qualities.

Consideration of the variables in terms of energy and water is pertinent as these categories also affect edaphic and atmospheric variable classes such as soil nutrients (e.g. nitrogen, phosphorus and potassium) and concentration of gaseous compounds (e.g. CO_2), via chemical relations determined in the former by the presence of hydrogen in water. Use of the water–energy dynamic is efficient in strategic modelling of plant characteristics and processes. The variables may be further reduced to avoid redundancy/covariance. In the approach shown in this chapter, the main variables considered are mean temperature, precipitation and ground-frost frequency (chosen for its disturbance effect on plant growth). Each variable has been considered in low resolution (30 arc minute or 50 km) which is of assistance in empirical Mamdani type consequential modelling (Furze et al. 2012). Increasing resolution (in this case to 10 min or 18.5 km) allows Takagi-Sugeno-Kang type modelling, which enables further analysis, elucidating numerical methods which may be applied in the process of cause-effect modelling (e.g. Genetic Algorithms; Functional Approximation). Such techniques as the latter are hybrid combinatorial and represent a significant advance in empirical and cause–effect modelling. (Further advances are noted in Furze et al. 2013b.)

Species data for biological modelling should be acquired from reliable, established data sources. Universities and research institutes undertake

Table 2.1 Water–Energy dynamic variables of IPCC TS.2.1 (New et al. 1999)

Variable	Category
Cloud cover (CC)	Water–energy
Ground frost frequency (GFF)	Water–energy
Maximum Temperature (MaxTemp)	Energy
Mean Temperature (MeanTemp)	Energy
Minimum Temperature (MinTemp)	Energy
Precipitation (P)	Water
Vapour pressure (VP)	Water
Wet day frequency (WDF)	Water

investigations of species presence and densities individually and may be used as sources of species data. However, when considered in isolation from other collecting individuals and bodies, these often provide distorted views of species presence as collections are made according to priorities of the organisation or individuals themselves. In order to establish an unbiased record of species presence and distribution patterns, meta-portals are used in which multiple users store data in collaborative efforts, many of which are available online. The Global Biodiversity Information Facility (GBIF) is one such meta-portal which makes the records stored available through its publically available online platform. The GBIF is seen as representative of global biodiversity (Yesson et al. 2007). The remit of the GBIF is to act as a source for over 110 million species records originating from field expeditions, herbarium and museum collections. Researchers have made an effort to link mapping platforms with species distributions in order to aid informative mapping studies (Flemons 2007; Beck et al. 2014). In order to cater for a sophisticated range of factors and variables and overcome technical challenges of processing the vast amount of data present, the authors of the current chapter suggest linking the GBIF with additional platforms such as the IPCC and USGS through an integrative computing platform with coding application of imaging, mapping and high level mathematics such as Matlab. The following section details a breakdown of the variables to be considered in algorithmic modelling structures for biological modelling.

2.3 Mathematic Detail of Algorithmic Structures

Variables are described by their probability of occurring. Distributions are defined by continuous (as in quantitative/constitutive classes of measurement) and discrete classes (such as absence and presence, 0, 1). It should be noted that in probability theory the spread of normal, continuous (Gaussian) variation is defined by:

$$f(x) = \frac{1}{\sigma\sqrt{2\pi}} e^{-\frac{(x-\mu)^2}{2\sigma^2}} \qquad (2.2)$$

where μ is the mean of the distribution, σ is the standard deviation, σ^2 is the variance of the distribution and e is the base of the natural logarithm (2.71828). If, for example, a particular plant's distribution is defined continuously in the interval a, b, the following probability density function integral applies:

$$\Pr[a \leq X \leq b] = \int_a^b f(x)dx \qquad (2.3)$$

Hence continuous variables may be plotted over a bell shaped curve or parabola. This category of distribution is made use of in this chapter in patterning of biotic and abiotic variables used for the modelling framework of this study. The derivative of Gaussian or normal deviation is the central sample of the variation, which is proportional to the standard deviation or greatest incline of the population range of variables under consideration. Gaussian patterns are also central to the inception of fuzzy techniques (Zadeh 1965). These techniques give causal relationships which are made use of in cause–effect (e.g. JULES, Huntingford et al. 2013), empirical process (e.g. Maxent, Fithian et al. (2014)) and combinatorial modelling (Furze et al. 2013f) approaches.

Discrete variables are a further category of distribution, each vector in a discrete class is independent. Discontinuous variables were used to classify defined classes of variation. Commonly used discrete distributions include the Poisson distribution, binomial and Pareto distributions, which show stochastic organisation qualities.

Absence/presence (0, 1) can be expressed within the continual range of variables of choice such as topographic and climatic factors. Water and energy variables display Poisson distribution in their global patterning (Schölzel and Friedrichs 2008; Sommer et al. 2010). In order to illustrate its suitability for expression of presence/absence, we show Poisson distribution in the following form:

$$f(k; \lambda) = \Pr(X = k) = \frac{\lambda^k e^{-\lambda}}{k!} \tag{2.4}$$

Where e is the base of the natural logarithm, $k!$ is the factorial of constant k, λ is λT when the number of events observed in a specified time interval is 1.

Further categories of estimation of distribution algorithms are given in copulas. Copula theory separates a joint probability distribution function into the component univariate margins of the distribution and a copula which represents the domain of the random variables. There are copulas that represent all classes of distribution in probability theory, there being Elliptical (Gaussian) copulas and Archimedean (representing classes of discontinuous variation). Essentially, the copula distribution may be used to generate the distribution of a random variable and identify dependencies between the variables used in modelling frameworks via differentiation of the terms contained in their expressions.

Essential notation for definition of copulas (Nelson 2006) is as follows: interval **I** [0,1] is an n-dimensional copula, a function C from **I** to **I**^{n} with the following property: for every $u = (u_1, u_2, \ldots, u_n)$ in **I**^{n}, $C(u) = 0$ if at least one vector of u is 0, if all vectors of u are 1 except u_k then $C(u) = u_k$.

Gaussian copula is therefore defined:

$$C_r(u_1, u_2, \ldots u_n) = \phi_r\left(\phi^{-1}(u_1), \phi^{-1}(u_2), \ldots, \phi^{-1}(u_n)\right) \tag{2.5}$$

Where ϕ_r is the multivariate Gaussian distribution with correlation $_r$, and ϕ^{-1} is the inverse function of the standard 1D Gaussian distribution function.

Gumbel copula (Gumbel, 1960) is:

$$\phi_G(u) = (-\log u)^{\phi_G} \tag{2.6}$$

where ϕ_G is the generator of the distribution, related to the Laplace transform which may be used in differential/inferential equations.

Sklar's theorem (1959) states that multivariate distribution functions can be expressed as a copula function evaluated as a 2D distribution function. Univariate distribution functions (e.g. $F_1, \ldots F_n$) can be linked to a multivariate distribution function, H as shown:

$$H(x_1, \ldots, x_n) = C(F_1(x_1), \ldots, F_n(x_n)) \qquad (2.7)$$

Elements of the water–energy dynamic, with bivariate dependency structures (Schölzel and Friedrichs 2008) are pertinently described using copula theory (Gumbel type, (2.6)) as they may show stochastic distribution in a Pareto Type I/Poisson domain.

In order to give informative, concisely stated relations in algorithmic form and make use of both continuous and discrete variation seen in the antecedent variables described in Sect. 2.2, we propose implementation of quintile partitions.

The partitions presented in Table 2.2 should be used to quantify the constituent parts of inference A, to create the union which, in combination with variables membership weight, express the consequent B (Zadeh 1965). The framework for the partitioned variables and Laplace transform union is as follows:

$$\text{If } A1 \ldots n_{(n)} - A1 \ldots n_{(n)} \quad \text{THEN} \quad B1 \ldots n_{(n, \ldots n)} \qquad (2.8)$$

Here, the linguistic connectors IF and THEN are used to unify inferential variables (numerically defined from partitioned data) to give the consequent range (numerically defined from the knowledge base of antecedent variables).

$$A = \int_U \mu_A(y)/y \qquad (2.9)$$

In (2.9) the Laplace transform $\int_U$ stands for the union of fuzzy singletons, and μ is the grade of membership in A (or y).

For strategy-based environments, the numerical data for each of the variables is considered in each of seven example environments, each representing the dominance of one of the seven strategies indicated in Section 2.1. Using the maximum and minimum inference of each variable's linguistic definition ($A1, \ldots, n$), the fuzzy rule-based algorithms were constructed so that each variable was expressed in

Table 2.2 Separation of variables into quintile partitions

Percentage range	Linguistic category	Notation
0–20	Low	1
20–40	Low–medium	2
40–60	Medium	3
60–80	Medium–high	4
80–100	High	5

terms of the number of species $(B1,\ldots,n)$ of each geographic location $(E1,\ldots,E7)$. Mean temperature was noted as $A1_{(1,\ldots,n)}$, precipitation was noted as $A2_{(1,\ldots n)}$, ground frost frequency is given as $A3_{(1,\ldots n)}$, altitude was noted as $A4_{(1,\ldots n)}$, the number of species was noted as $B_{(1,\ldots,n)}$. The linguistic connections 'If', 'AND' ($\wedge$) and 'THEN' ($\rightarrow$) were used to construct the conditional fuzzy rule base. The basic framework of the fuzzy algorithms is:

$$If\ A1_{(n)} \prec A1_{(n)} \wedge A2_{(n)} \prec A2_{(n)} \wedge A3_{(n)} \prec A3_{(n)} \wedge A4_{(n)} \prec A4_{(n)}$$

$$\rightarrow B(n) = E1, \ldots, E7 \tag{2.10}$$

Fundamental groups of plant life history strategies, metabolic (photosynthetic) type and life form, necessary for modelling of plant species, are outlined in the introduction, and further detail is given in the example case studies shown in Sect. 2.6. Using the root framework of (2.9) with defined resolution of antecedent variables enables the formation of Takagi–Sugeno–Kang fuzzy algorithms (2.10) which may be further analysed in stochastic terms, defined by example of genetic dispersal in Sect. 2.4.

2.4   Genetic Dispersal/Stochastic Methods

The higher mathematic theory of logic-based systems progresses to illustrate how plant strategies may be further differentiated, split into their rudimentary elements, each specifically optimal in different water–energy conditions present in ecological systems. A combination of random dispersal methods is elucidated in this section together with the stochastic evolutionary algorithm programming of strategies. This introduction serves as a brief discussion of stochastic methods in the context of biological systems.

The principal area of set theory underpinning plant species modelling with relation to development of evolutionary networks is species stochasticity. A stochastic process is generation of random variables, the key point being that evolution of variables is not univariate, but may potentially develop in many different shapes. Stochastic networks are well suited in the field of evolutionary algorithms and have extensive use in non-linear system modelling, computer technology and biological systems (Silvera et al. 2009). Just as any stochastic group, plants may be said to be functions of one or several deterministic arguments. To apply stochastic processes, the key point is that the variables determining the measured characteristics or vectors must share the same functional domain. In other words, they may be seen to show certain probability distributions such as Poisson, Gaussian or other continuous or discrete pattern, as discussed in Sect. 2.3 and hence they often share complex statistical relationships (Zhang et al. 2012).

Plants show chaotic patterns of evolution in terms of their individual growth processes and plant species numbers (Cui et al. 2012; Furze et al. 2011; Su et al.

2009). Patterning of plant species may be determined by key factors of the water–energy dynamic (Hawkins et al. 2003; Kreft and Jetz 2007; Sommer et al. 2010). Climatic variables such as rainfall and temperature often show discrete patterns across timescales, so are often made use of within fixed time ranges. As such, they can be said to be discrete stochastic patterns, which facilitate macro-level modelling of plants (Grime and Pierce 2012; New et al. 1999; Silvert 2000).

Stochastic vectors are numbers which distribute in a defined domain in a random fashion. As such, a stochastic matrix can be used to show a functional relationship between species and their surroundings (Cáceres and Saez 2011; Davison et al. 2010; Drenovsky et al. 2012; Furze et al. 2013e). Methods used to form such matrices involve combinatorial techniques considering both population dynamics of the species in question and the conditions in which they are located. In this section we show how mathematic genetic programming techniques can be used as a method of dispersal of field-based plant occurrence records.

Genetic algorithms (GAs) use a representation of the components of variation as vectors (chromosomes) within strings (populations). The chromosomes recombine in an iterative process under specified conditions involving the same elements of genetic recombination as in natural systems. As in biological systems, operators are selection, crossover and mutation. In combination with fuzzy set theory, GAs are robust, stochastic evolutionary computational algorithms (Su et al. 2009). GAs are adaptive algorithms for finding the best (global) solution to optimisation problems. Expansion of a hybrid multi-objective genetic algorithm is shown by application to biological data in Sect. 2.6. In combination with a GA approach, the rule base may be successfully used to generate the distribution present in natural systems (Elith et al. 2011). The process of generating multi-objective genetic algorithm (MOGA) involved the following main steps:

1. Define each vector for plant strategies.
2. Randomly generate an initial population of 20 solutions (chromosomes).
3. Evaluate each solution according to how well it fits into the desired environment (as defined in equation (2.10), examples given in Table 2.3).
4. Select chromosomes randomly (tournament selection). Keep those with the highest fitness function to improve the population and discard those with too low (value may be previously calculated) fitness.
5. Create new chromosomes by crossing selected solutions using crossover of proportions of the individual strings of solutions.
6. Mutate a previously determined proportion of the population's chromosomes.
7. Go back to step 3 until final population number is reached, then stop.

A natural progression of GA and logic systems is the estimation of a population's resultant distribution; an additional method of approximation: Estimation of Distribution Algorithms (EDAs) are algorithms with applications in computing, industry and natural systems. Copula theory is a concise, robust form of EDAs (Nelson 2006). Copulas join multivariate distributions to 1D marginal distribution functions, which make them ideal candidates for the analysis of patterns of genetic variation such as that which is produced by the multi-objective optimisation (MOO) or Pareto front of a multi-objective genetic algorithm (MOGA) process (Wang et al. 2012). Identifying the copula distribution with intuitive knowledge assists in recognition of the functional relationship within ecological systems. This is further discussed in Sect. 2.5.

2.5 Functional Approximation Algorithms/Using Continual and Discrete Data for Informative Expansion

Approximation of the functional distribution of any discrete biological character may be seen to be within the field of evolutionary algorithms and estimation of distribution algorithms. In this section, a novel approach is formulated by making use of process models to summarise the distribution of a characteristic such as plant life form seen within individuals of any set population of plants. Which function 'fits' the plants must be determined by the researcher's intuition of the system in question. Thus, the choice of function is formed by knowledge guidance. Continual functions such as De Jong's Function and Schwefel's Function are unimodal and can be used in conjunction with discrete categories of single variable modes. An advantage of initially using a Gaussian process model in estimation of multimodal rudimentary distribution is the estimation of the character's fitness function, dependant on two or more objective modes, which must be known when programming a MOGA such as that detailed in the previous section. The fitness function is a numerical estimation of chromosomal 'fitness'. One may assume a continual distribution on a wide scale before investigating whether a discrete relationship (e.g. Sinusoidal) may be applied or alternatively apply subject knowledge (e.g. related to plant strategy or photosynthetic type) to determine a discrete relationship with surroundings. Such discrete relationships may also 'fit' into functional types, though it is suggested that these should be determined following detailed analysis of habitats (species and characteristic components, relationships with climatic factors, topography) and their settings. Following this method, a functional approximation algorithm (FAA) for plant variation consists of the following four steps:

- *Step 1*. Geographic and climatic study to establish the framework for modelling species of plant life forms in candidate areas.
- *Step 2*. Adaptive fuzzy neural inference system (ANFIS) using identified variables based on consequent primary nodal number.
- *Step 3*. Multi-objective genetic algorithm dispersing expanded secondary nodal number.
- *Step 4*. Functional approximation of characters within secondary nodal number using a continuous/discrete surrogate process model.

The assumption that individuals within the populations of plant species in each of the studied areas are normally distributed in their life form characteristics is made, furthered by elaboration of the discrete ANFIS in Sect. 2.6. Additionally, standardisation of the dispersed strength Pareto population to zero mean and unit variance allows the population to be expressed across a bell-shaped (Gaussian) curve (Guo 2011).

The Gaussian model selected is Rastrigin's function, which serves the dual purpose of expanding the dispersal of the characteristic (18 sub-groups of Raunkier's life forms) simultaneously and verifying the validity of the GA process described in Sect. 2.4. The function is elaborated as follows.

$$f(x) = 10 \cdot n \sum_{i=1}^{n} \left(x_i^2 - 10 \cdot \cos\left(2 \cdot \pi \cdot x_i\right) \right) \tag{2.11}$$

where $i = 1/n$, x_i is an element of the interval $[-5.12, 5.12]$. In order to make use of Rastrigin's function as a surrogate Gaussian process model in the current case, the number of life form sub-categories ($n = 18$) occurs within the interval shown. This is further expanded by example in Sect. 2.6.

2.6   Case Studies: Plant Strategies, Life Forms and Metabolism

In this section, the modelled presence of the discrete dimensions of plant life history strategy and primary metabolic (photosynthetic) type is presented within the continual spectrum of plant life forms present in case study areas. The areas were selected at random following literature searches informing high diversity levels, shown in diversity zones (DZ8-10) in which 3000 species/km^2 have been reported (Barthlott et al., 2005). The candidate areas are presented in Table 2.3.

In Table 2.3, the number of individuals are those that were reported by multiple collecting institutes and logged into the Global Biodiversity Information Facility (GBIF, www:gbif.org, accessed December 2010). The dominance of each strategy is shown due to each location containing the approximate proportions presented in Table 2.4.

Table 2.3 Categorisation of environments and plant life history strategies

Environment	Plant life history strategy	Example location/number of individuals
1	R	Ecuador/51857–65535
2	S–R	Guyana/50700–51847
3	S–R/C–R	Cuba/33356–50700
4	C–R/C	Democratic Republic of the Congo/11355–33366
5	C–S–R/C–S	Georgia/8805–11355
6	C–S	Guinea/2203–8805
7	S	Macedonia/0–2203

Table 2.4 Plant species C–S–R balance in numerical form

Strategy/Environment	Competitive	Stress tolerant	Ruderal
R/E1	0	0	1
S–R/E2	0	0.5	0.5
C–R/E3	0.5	0	0.5
C/E4	1	0	0
C–S–R/E5	0.33	0.33	0.33
C–S/E6	0.5	0.5	0
S/E7	0	1	0

In reality, the existence of a pure strategy without any elements of the others very rarely exists due to the polyploidy level of plant species. Extending an ecosystem approach means that in the plant-strategy environments, levels of competitive and of stress tolerant species do exist in ruderal environments, the limit after which the strategy element is detected as equal or greater than 1/3 of the species. The latter leads to the statement that each environment $E1,\ldots, E7$ is dominated by the strategy elements presented in Table 2.4, following knowledge of the conditions in which each strategy element occurs (Grime et al. 1995). Proportionality (probability) of the three strategy elements may be investigated via the construction of a 3D linear mesh of the species elements as seen earlier (Fig. 2.2). Ascertaining the dominance of each strategy in different locations due to the presence of the water–energy dynamic enables several advantages for policy formation bodies on an international scale, principal of which are planting programmes, investigation of communication between trophic levels of life and further exploration of metabolic and life form groups which have substantial benefits for human and ecological sustainability. Regarding modelling application, the presence of the global water–energy dynamic allows ecological investigators to form cause–effect and empirical modelling frameworks. Section 2.3 gives detail as to how discrete relationships can be concisely stated; in brief the modeller works with the fact that plant species have a finite relationship with their surroundings and conditions. Hence we may develop T–S–K algorithm structures in order to predict the occurrence of plant strategies with the conditions found in each of the candidate areas given in Table 2.3. Once we have identified the climatic conditions and topographical ranges found in the areas we may proceed to quantify the conditions within percentage ranges as presented in Table 2.2. The quintile partitions indicated in Table 2.5 are translated into membership functions for each strategy. Thus, each environment in which the strategy is found operates according to a rule base with a weighted structure, with weights (0.25, 0.5, 0.75, 1) allocated following the occurrence of the conditions on a quarterly basis across a year. This is implemented according to the following stepwise process:

1. Define fuzzy inference system type (Sugeno for defined output type) variable names
2. Define membership functions of each variable
3. Define rules, weights according to algorithm

Table 2.5 Variable partitioning for T–S–K modelling of plant strategies of Guyana

Ling exp	% Quant/Not'n	Range			
		MT°C	MP (kg m^2)	MGFF (days)	Alt (m)
Low	0–20/$_1$	−75 to −51	0–100	0–6	0–600
Low–medium	20–40/$_2$	−51 to −27	100–200	6–12	600–1200
Medium	40–60/$_3$	−27 to −3	200–300	12–18	1200–1800
Medium–high	60–80/$_4$	−3 to 21	300–400	18–24	1800–2400
High	80–100/$_5$	21–45	500–500	24–30	2400–3000

Ling exp linguistic expression, *Quant'* quantification, *Not'n* notation, *%* percentage, *MT* mean temperature, *°C* degrees Celsius, *MP* mean precipitation, *kg m^2* kilogram per square metre, *MGFF* mean ground frost frequency, *Alt* altitude, *m* metre

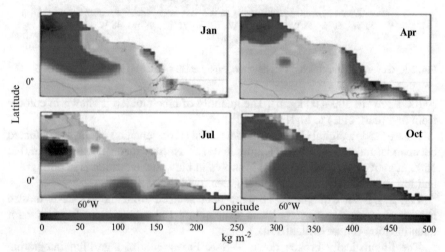

Fig. 2.4 Guyana quarterly mean 1961–1990 precipitation at 10 min (18.5 km) resolution (New et al. 1999)

4. Examine (adaptive neuro fuzzy/Sugeno) logic structure, modify as necessary
5. Test rules through data input
6. Display 2D and 3D views of resultant surface in order to graphically display the algorithm (efficiency)

Enhanced resolution of antecedent variables, in turn, enables greater accuracy in consequential statements to be obtained. This is an imperative when using a T–S–K modelling basis as any errors received during the modelling process are borne out during later stages of analysis (Sivanandam et al. 2007). Higher logic-based mathematics tolerate the error due to the fact that they are based on continuous (Gaussian) process models. The error detected in the methods is minimised by use of a triangular membership function (Fig. 2.7), which gives a discrete perspective of continuous variation.

Figure 2.4 shows example data, where Guyana mean precipitation is 0.75 (January, April, July) 0–100 kg m^2 to 200–300 kg m^2, and 0.25 (October)

Fig. 2.5 Guyana digital elevation model/topography at 30 s (1 km) resolution (GTOPO30)

0–100 kg m^2 to 300–400 kg m^2. The quantity of precipitation is shown in colours from low (dark blue) to high (dark red).

Figure 2.5 is a digital elevation model (DEM) representation of Guyana, situated between latitude 60–55° West, longitude 0–7.5° North with an elevation from 0 to 1500 m above sea level. Sea level is shown in blue, low elevation is in dark green and low–medium elevation in lighter green to white.

Fig. 2.4 is an example of climatic data sourced from the IPCC. It is also necessary to source quarterly mean temperature and ground frost frequency for quantification (Furze et al. 2013d).

The linguistically broken down T–S–K Fuzzy control algorithm integrating climate variables and DEM data for Guyana is written as follows:

$$IFA1_{(5)} \prec A1_{(5)} \wedge 0.75A2_{(1)} \prec A2_{(3)} 0.25A2_{(1)} \prec A2_{(4)}$$
$$\wedge A3_{(1)} \prec A3_{(1)} \wedge A4_{(1)} \prec A4_{(2)} \rightarrow B_{(51847)} = E2 \qquad (2.12)$$

Numerical quantification of the algorithm is as follows:

IF Variables A =

- Mean temperature = 80–100 % to 80–100 % ($A1_{(5)}$)
- Mean precipitation = 0.75 0–100 kg m^2 ($A2_{(1)}$) to 200–300 kg m^2 ($A2_{(3)}$), 0.25 0–100 kg m^2 ($A2_{(1)}$) to 300–400 kg m^2 ($A2_{(4)}$)
- Mean ground Frost frequency = 0–6 days to 0–6 days ($A3_{(1)}$)
- Altitude = –30–1366 m ($A4_{(1)}$) to 1366–1500 m ($A4_{(2)}$)

THEN $B_{(51847)} = E2$

Example locations of environments $E1$, $E2$, $E3$, $E5$, $E6$, and $E7$ were defined using the algorithmic control structure of (2.10), as given in Table 2.3.

Membership functions of each of the variables were defined as shown in Fig. 2.7, making use of triangular functions to discretely define each partition. Each vector function operated together under the instruction of the root algorithm of (2.12), which was expressed with the use of ten separately weighted rules as follows:

1. If (temperature is high) and (GFF is low) then (strategy is S–R) (1)
2. If (precipitation is low) then (strategy is S–R) (0.75)
3. If (precipitation is low–medium) then (strategy is S–R) (0.75)
4. If (precipitation is medium) then (strategy is S–R) (0.75)
5. If (precipitation is low) then (strategy is S–R) (0.25)
6. If (precipitation is low–medium) then (strategy is S–R) (0.25)
7. If (precipitation is medium) then (strategy is S–R) (0.25)
8. If (precipitation is medium–high) then (strategy is S–R) (0.25)
9. If (altitude is low) then (strategy is S–R) (1)
10. If (altitude is low–medium) then (strategy is S–R) (1).

The rules fed into the fuzzy inference system are shown in Fig. 2.6.

FIS is Fuzzy Inference System, Sugeno refers to the modelling type Takagi-Sugeno-Kang, f(u) is the fuzzy union consequent.

The discrete values for 'feet' (a, c) and 'peak' (b) of the triangular membership function are defined by

$$f(x; a, b, c) = \max\left(\min\left(\frac{x - a}{b - a}, \frac{c - x}{c - b}\right), 0\right) \qquad (2.13)$$

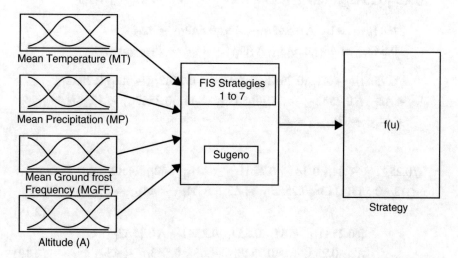

Fig. 2.6 Design of the fuzzy inference engine to differentiate plant strategies from the water–energy dynamic

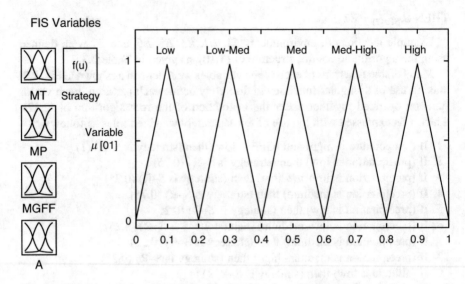

Fig. 2.7 Definition of triangular membership functions for ordination of plant strategies

Following the structure given in (2.10), implemented in the example environment of Guyana shown in (2.12), we may form the following location-dependent concise statements for the other locations presented in Table 2.3 as follows:

$$\text{If } A1_{(3)} \prec A1_{(4)} \wedge 0.75A2_{(1)} \prec A2_{(5)}0.25A2_{(1)} \prec$$
$$A2_{(4)} \wedge A3_{(1)} \prec A3_{(2)} \wedge A4_{(1)} \prec A4_{(5)} \rightarrow B_{(65535)} = E1 \tag{2.14}$$

$$\text{If } 0.25A1_{(4)} \prec A1_{(5)}0.75A1_{(5)} \wedge 0.5A2_{(1)}0.25A2_{(1)} \prec$$
$$A2_{(3)}0.25A2_{(2)} \prec A2_{(3)} \wedge A3_{(1)} \wedge A4_{(1)} \prec A4_{(2)} \rightarrow B_{(50700)} = E3 \tag{2.15}$$

$$\text{If } A1_{(4)} \prec A1_{(5)} \wedge 0.5A2_{(1)} \prec A2_{(4)}0.5A2_{(1)} \prec A2_{(3)} \wedge$$
$$0.5A3_{(1)} \prec A3_{(2)}0.5A3_{(1)} \wedge A4_{(1)} \prec A4_{(5)} \rightarrow B_{(33356)} = E4 \tag{2.16}$$

$$\text{If } 0.75A1_{(3)} \prec A1_{(4)}0.25A1_{(4)} \prec A1_{(5)} \wedge 0.75A2_{(1)} \prec A2_{(2)}0.25A2_{(1)}$$
$$\prec A2_{(3)} \wedge 0.25A3_{(2)} \prec A3_{(5)}0.5A3_{(1)} \prec A3_{(4)}0.25A3_{(1)} \prec A3_{(2)} \wedge A4_{(1)}$$
$$\prec A4_{(5)} \rightarrow B_{(11355)} = E5$$

$$\tag{2.17}$$

$$\text{If } 0.25A1_{(3)} \prec A1_{(4)}0.5A1_{(4)}0.25A1_{(4)} \prec A1_{(5)} \wedge A2_{(1)} \prec A2_{(2)} \wedge 0.25A3_{(3)}$$
$$\prec A3_{(5)}0.5A3(1)A3_{(3)}0.25A3_{(1)} \prec A3_{(2)} \wedge A4_{(1)} \prec A4_{(4)} \rightarrow B_{(8805)} = E6$$

$$\tag{2.18}$$

$$\text{If } 0.25A1_{(4)} \prec A1_{(5)}0.5A1_{(4)}0.25A1_{(5)} \wedge 0.75A2_{(1)} \prec$$
$$A2_{(2)}0.25A2_{(1)} \wedge 0.25A3_{(1)} \prec A3_{(5)}0.5A3_{(1)} \prec A3_{(2)} \tag{2.19}$$
$$\wedge 0.25A3_{(1)} \wedge A4_{(1)} \prec A4_{(3)} \rightarrow B_{(2023)} = E7$$

Algorithms for plant photosynthetic type environments share the same basis as earlier, there being three main types as detailed in Sect. 2.2. These provide the following concise statements:

$$C3 \geq C4 \geq \text{CAM} \forall E1; E2; E3 \tag{2.20}$$

$$C4 \geq C3 \geq \text{CAM} \forall E4; E5 \tag{2.21}$$

$$\text{CAM} \geq C4 \geq C3 \forall E6; E7 \tag{2.22}$$

The photosynthetic types given in (2.20), (2.21) and (2.22) are present in the orders shown according to the conditions in which each is best suited in terms of water and energy, as stated in Sect. 2.2. In order to disperse the rudiments of photosynthesis within candidate environmental types and show relations with the respective water and energy conditions present, we propose the dissection of each photosynthetic character as given in Table 2.6 in order to make use of genetic programming method.

Table 2.6 Photosynthetic characters and quantification solutions for multi-objective genetic algorithm cycle

Character/Chromosome	C3 (1,…,5)	C4(1,…,5)	CAM (1,…,5)
ws	(3,…,5)	(0,…,2.5)	(1,…,2)
fs	(1,…,2.5)	(2.5,…,5)	(2.5,…,5)
hs	(4,…,5)	(2.5,…,4)	(0,…,1)
fl	(0,…,2.5)	(2.5,…,5)	(2.5,…,5)
tl	(2.5,…,5)	(2.5,…4.5)	(0,…,2.5)
nll	(0,…,2.5)	(0,…,1)	(0)
tr	(0,…,2)	(2,…,5)	(0,…,2.5)
crb	(0,…,3)	(0,…,3)	(0,…,2.5)
drb	(3.5,…,5)	(0,…,2.5)	(0,…,2.5)
gpspod	(5)	(5)	(0)
gpspon	(0)	(0)	(4,…,5)
pka	(0)	(5)	(1)
ppep	(0)	(5)	(0,…,2.5)
sspc	(0,…,2)	(2,…,5)	(2,…,5)
tspc	(0)	(0,…,2)	(5)
sac	(0,…,2.5)	(2.5,…,5)	(5)
soc	(5)	(3)	(0,…,2.5)
C	(0,…,2.5)	(3)	(3)
S	(0,…,2.5)	(2.5)	(5)
R	(5)	(2.5)	(2)

Plant characters in Table 2.6 are: *ws* woody stem, *fs* fleshy stem, *hs* hairy stem, *fl* fleshy leaves, *tl* thin leaves, *nll* needle-like leaves, *tr* tap root, *crb* compact root ball, *drb* dispersed root ball, *gpsod* greater proportion of stomatal pores open during day, *gpspon* greater proportion of stomatal pores open during night, *pka* presence of Kranz anatomy, *ppep* presence of phosphoenol pyruvate, *sspc* spatial separation of photosynthetic compounds, *tspc* temporal separation of photosynthetic compounds, *sac* storage of acidic compounds, *soc* storage of carbohydrate, *C* competitor, *S* stress tolerant, *R* ruderal. Ideal quantification is shown in brackets in the table

The characters displayed in Table 2.6 represent a chromosomal population and are used to form a MOGA, in which the chromosomes are cycled (via roulette wheel selection) through 0–5, resulting in a Pareto front. The process is carried out in order that the distribution of each metabolism type and intermediate types is approximated within the previously algorithmically defined environment.

The variables should be directed through two main objective types (water and energy) and expressed as a double vector population. In the present case, we programmed the algorithm to stop when the total number of individuals had reached that of C–R/S–R, E3. The number of iterations required was recorded and the distribution of ideal solutions for each of the 20 chromosomes across objective space was noted. Linear and quadratic lines were fitted to the resulting Pareto in order that an approximation of the distribution is made. A plot was carried out in order that the position of the linear fit (representing the Utopia line) and the shape of the curve are visualised in order that the evolution of the 20 characters could be estimated through future transgressions.

The evolutionary strength Pareto plot of elements of photosynthesis present in the species of E3, Cuba, enabled estimation of the distribution of photosynthetic characteristics to be carried out, in accordance with the W–E dynamic as summarised in the utopia line and utopia curve expressions given in Fig. 2.8. These expressions are subject to the residual variance of each point as shown in Fig. 2.9.

Plotting the residual error of utopian space enables the formation of rules for the distribution of photosynthetic elements to be formed as given in Table 2.7.

In rule 1 of Table 2.7, y (objective 2, mean precipitation) is equal to an element of Z (given that x (objective 1, mean temperature) and y axis form the Z space), $\partial 1$ is -0.11175 and $\partial 2$ is -0.15059 following a weighted least squares structure subject to the error ($\varepsilon = 0.47457$). In rule 2, y (objective 2) is equal to an element

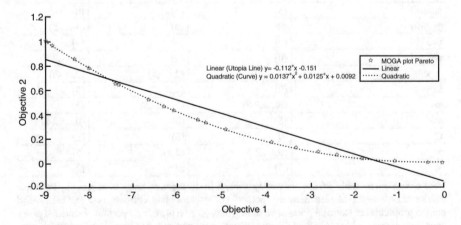

Fig. 2.8 Plant photosynthetic evolutionary strength Pareto of Cuba (E3)

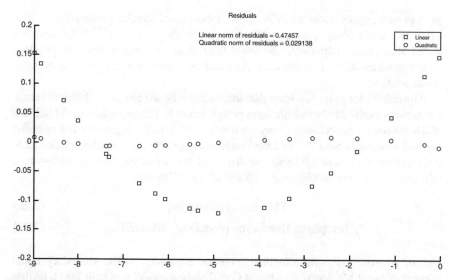

Fig. 2.9 Residual error of the utopian space of plant photosynthetic characters

Table 2.7 Utopia rules of photosynthetic character rudimental dispersal

Rule	Variables (3 significant figures)			
	$\partial 1$	$\partial 2$	$\partial 3$	ε
1. $Z = \partial 1x + \partial 2 \pm \varepsilon$	0.112	−0.151		0.475
2. $Z = \partial 1x^2 + \partial 2x + \partial 3 \pm \varepsilon$	0.0137	0.0125	0.00920	0.0291

of Z (given that x (objective 1, mean temperature) and y axis form the Z space), $\partial 1$ is 0.013653, $\partial 2$ is 0.012531, $\partial 3$ is 0.0092027, subject to the error ($\varepsilon = 0.029138$).

Construction of rules from linear and quadratic expressions enables estimation of each objective to be made given a fixed population number, as identified for each T–S–K algorithmically described location and character; hence the hybrid MOGA techniques enable us to predict climatic data and additionally species numbers subject to errors such as those detailed in Table 2.7. This point not only enables us to remove uncertainty in prediction of individual plant species occurrences, but also to enhance the knowledge bases on which the modelling system is founded (Furze et al. 2013a). Expression of the W–E dynamic in quintile terms together with quantification of plant characteristics in the same way allows the cyclical nature of the MOGA to produce optimal distribution of individual traits within the dynamic conditions. The benefit of carrying out the technique is that the likelihood of individual characteristics occurring in ecology/field conditions may be approximated within concise conditions. Knowledge of the characters being dispersed is very important, as it is this which supplies the root of intuition to infer which elements of a character are present given conditions of the dynamic in which they are dispersed. A major advantage of mathematics in this chapter is that it can be used to give highly accurate indications of both plant species presence/character-istics and, given plant species occurrence, the conditions of the W–E dynamic are stated with greater accuracy. The novel application of these techniques provides time and financial savings for ecologists and biologists in otherwise purely field-based studies.

Algorithms for plant life form dominance must be expressed in different terms, the primary nodal number of life form groups being 5. The secondary nodal number of 18 combined probabilities may be obtained using the dispersive optimisation method detailed in Sect. 2.3 and the functional approximation method of Sect. 2.5. Although the spectra of life forms are dispersed in a non-linear fashion the two ends of a continuum may be obtained with use of the following:

$$PhanerophyteDom \forall E1 \tag{2.23}$$

$$Therophyte; HemicryptophyteDom \forall E5 lim E7_{(max)} \tag{2.24}$$

Use of genetic mathematic methods enables expansion to the secondary nodal number of plant life forms. Further, a C–S–R-based model was split into 6 further categories as 18/3 returns 6. However, unless the investigators have field-based knowledge of the life forms within each location it is difficult to combine proba-bilities. It is suggested that the localised use of the Gaussian process model formed in Sect. 2.5 is implemented in combination with field data. The standardised mean of individual populations' life form categories will enable the statements of the water–energy dynamic and elevational conditions within which each life form category occurs. From the latter statements of objective dispersal we may proceed to make statements linked with further character dispersal. Figure 2.10 shows the implementation of the secondary number of plant life forms to the surrogate Rastrigin's function.

The multimodal continual distribution of Fig. 2.10 enables intuition of life form presence to be estimated with use of standardised objectives (e.g. precipitation, temperature) and additionally validates the genetic programmed dispersal shown in Fig. 2.8. It may also be used to estimate the fitness of each of the rudiments of the primary number of life forms chromosomes. Life form distribution spectra were seen between candidate environments of Ecuador and Macedonia in recent publi-cations (Furze 2014; Furze et al. 2013f). Field-based data may subsequently be used to estimate the relationships between individuals on a local scale according to dynamics which exist due to the communication between trophic levels at such a scale. However, use of functional distributions is a developing field of great application in highly vulnerable locations of the highest diversity (Kraft and Ackerly 2010; McCracken and Forstner 2014; Zytynska et al. 2012) .

Mechanisms of aligning and linking distributes that we observe in ecology through the mathematic methods presented allow rational interpolation in biolog-ical systems to be made. These are developed in the following section.

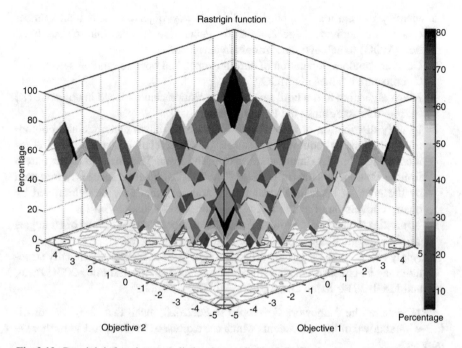

Fig. 2.10 Rastrigin's function as applied to the standardised distribution [0,. . .,5] of 18 sub-groups of plant life forms (Furze 2014; Furze et al. 2013f)

2.7   Heuristic and Optimal Search Capability and Application to Biological, Chemical and Physical Data

Heuristic searches are random searches made within defined parameters. Hence the power of a stochastic relationship, such as that which may be identified within species patterning of ecosystems, may be optimised by a secondary process such as MOGA (as identified in the previous section). In order to form a combinatorial approach between fuzzy and genetic methods, a ten-point summary for the combined fuzzy-MOGA methods is included here:

1. Determine structure via identification of complete sets of background/ environmental data.
2. Derive minimal TSK IF–THEN rule base for strategical nodes of network organisation.
3. Identify elements of strategies (node structure).
4. Use expert-based intuition to rank ideal solutions of each element, represent within chromosome population structure.

5. Identify constraints (stopping conditions), key objectives and total number (maximum) individuals/generations for generation of ideal solution combination (MOO) to achieve the Z utopia hyperplane.
6. Generate random population of chromosomes and operate random selection of chromosomal values via MOGA.
7. Allow algorithm to run until sets of ideal solutions have spaced in objective (Z) space—this determines the Pareto Frontier.
8. Identify distribution of Ideal Pareto Frontier, fit linear, quadratic and polynomial functions as required to precisely describe relative distribution of chromosomes within maximum number of individuals. The linear fit of the Pareto frontier is defined as the Utopia Line (The Utopia objective space is represented by the error of each individual Pareto solution; this may be determined by binary conversion of each Pareto (P) generating set, equal to 0-1).
9. Approximation of the distribution may determine real proportions of strategies within strategical nodes (stochastic distribution in this case).
10. Apply solutions to ranges identified through T–S–K FRBS, additionally structure may be checked via pattern identification (Angelov and Filev 2004; Juang and Hsieh 2012; Salah and Abdalla 2011).

Referencing the framework developed in Sect. 2.3, recall that (2.8) is a constitutive breakdown of the antecedents which are expressed in (2.9), and so further we may state:

$$X = \{x_1, \ldots, x_n\} \tag{2.25}$$

Where $x_1, \ldots, x_n$ are set values of generic set X.

The following definitions using multiple elements in T–S–K systems are used:

$$\bar{x} = [x_1, x_2, \ldots x_n] \tag{2.26}$$

$$\bar{y} = [y_1, y_2, \ldots y_n] \tag{2.27}$$

$$f_A(x; y) = X_\mu; Y\mu \tag{2.28}$$

In (2.26) the mean of set x is represented by normal set elements $x_1, \ldots, x_n$. Accordingly, in (2.27) the mean of set y is represented by normal set elements $y_1, \ldots, y_n$. In (2.28) f_A refers to function A in x and X_μ is the grade of membership of X and may also be used to express function A in y and Y_μ is the grade of membership of Y. Both X and Y are used in the antecedent linguistic A matrix in order to form:

$$\bar{y} = [\bar{x}^*R] \tag{2.29}$$

Where R is the relational index matrix as a function of the combined arguments of $\bar{x}$ and $\bar{y}$. In this it may also be stated to be the result of the Laplace transform union as in (2.8), or succinctly defined as '*IF A... THEN B*' (Zadeh 1965).

By developing ANFIS and subsequent MOGA we can see that the key variables are expressed in instructive algorithms of natural systems and can also be referred to as holomorphic or analytic functions. Holomorphic functions are subject to the Taylor series, when subjected to differentiation the degree of polynomial reflects the rational truth of the ecosystem with its determinants. Ecological systems may be iteratively differentiated as the complete set of variables ascertaining all relations cannot be reached for any single individual plant occurrence falling within a set population. Prediction may be made within natural systems using a high ordered polynomial, with only a finite residual variance occurring (Table 2.7).

$$F_i \leq \sum_{j=1}^{n} M_j B_{ji}, (i = 1, \ldots n) \tag{2.30}$$

Following the expression of ANFIS through (2.28), the relational index matrix function is shown as (2.30), where F represents the function of the data i in the space of Fig. 2.7, M represents the midpoint of the utopia front and B represents the Z matrix space. Developing concise system blocks of (2.14) through (2.22) reverting from field data within defined parameters and limits we have shown how numerical optimisation of these Gumbel copulas can be shown in combined objective space.

Application of the same combined objective can be shown on a local scale within species in individual locations. Should extreme anomalies occur in the proportions of strategies, photosynthetic type or indeed life forms in the field, given that the category has the majority of its contrasting (opposite) class, the domain of (2.30) is 'flipped'. To see the opposite/orthogonal population we simply change the upper bound of each function to its lower bound as follows:

$$F_i \geq \sum_{j=1}^{n} M_j B_{ji}, (i = 1, \ldots n) \tag{2.31}$$

Domain flipping allows identification of water–energy variables in field conditions at the highest resolutions, where the harshest environmental conditions occur within the richest populations of plants (McCracken and Forstner 2014).

Another use for the domain flipping function shown in (2.31) is to show the convergence points between contrasting distributes of system blocks, such as that which may be seen between areas of high concentrations of diversity and development potential of renewable energy systems in a local area, though this is a currently developing field. The authors recommend application of fuzzy sliding mode to identify concise relationships between the convergences of orthogonal distribution (Furze, Zhu 2014).

Combining novel fuzzy and MOGA techniques remains a method by which we can determine both species and species group concentrations as well as climatic variables within highly rational limits. The final section concludes the chapter and remarks on further tools that are available within biological modelling.

2.8 Conclusions/Further Directions

This chapter has summarised and justified categorical antecedent and consequent data which should be used in consideration of biological/plant system modelling. Although the nature of observation-based (empirical) modelling is often of a reflective basis, the case studies have been provided which enhance intuitive learning of the parameters in which plant species occurrence manifests in the most diverse areas of the planet. As such, the information provided in each section has applications which are related to the use of plants (in terms of material form, metabolic products and indeed within ecosystems) by our developing societies. As the methods make use of climatic and topographic data, the parameters within which these are considered in discrete time periods are clarified, with a stringency that is underlined by the relationship of over 300,000 primary producing species with their environmental conditions and surroundings. Biological modelling holds the promise of being a key tool by which viable recommendations of conservation policies may be made. As plants act as producers in trophic systems, these policies are imperative for formation of community governance in isolated and vulnerable (highly diverse) regions of the planet as well as in urbanised communities. Engaging with researchers and institutes remains a starting point from which planning/protection/rejuvenation of ecosystem policies can be implemented.

Further recommendations include formulating the difference (residual) between flipped distributions to form a harmonic function and applied in the context of a Gumbel copula for enhancement. The authors propose application of methods with variable data types (Mascaro et al. 2014) and integration of field data to build models for comparison. The models/instructive algorithms shown in this chapter should be used as a marker of the state of biological systems and climatic variables as they hold the basis for a stochastic weather generator with data from living (plant) systems. Use of sophisticated models with increased rationality as is shown in this chapter proves greater validity of the cause–effect, empirical models (Best et al. 2011; Zhu et al. 2014) and combinatorial approaches (Furze et al. 2013f). Use of blocking of systems and domain flipping, and sliding mode to identify convergence between systems remain methods for further exploration. Such methods can be used to compare contrasting dynamics on wide and lower resolutions. The endpoint of the use of sophisticated mathematics in biological modelling is that the best (most rational) use of species can be made in accordance with their immediate ecosystems. Communication between trophic levels should be maximised by greater weighted (higher species numbers) of the primary level. Determining proportions of life forms on both local and wider scales will strengthen the use of Gaussian process models or help to identify different functional relations. ANFIS-based systems and Type 2 systems lead to the expectation that the patterning observed on macro-ecological levels will be reflected amongst the gene pools of the species under study, although further studies are recommended to explore this. Feedback between the numbers and categories of plant species and environmental conditions may be considered by application of back propagation algorithms/

reverse computation. There are many methods by which this can be carried out, including Markov chain feedback, Fuzzy Sliding mode and other regression-based techniques (Zhu et al. 2013). Use of these methods illustrates concise empirical modelling and also enables understanding to be gained of cause–effect process modelling, allowing a non-specialist readership to recognise the importance of processes in rational mathematic terms.

Improved modelling of biological systems, offering greater predictive ability to understand the spatio-temporal dynamics of biological variation, is necessary if we are to ensure ecological security as anthropogenic demands on the environment, ecosystems and climate threaten the stability and resilience of terrestrial and aquatic systems. Such threats include invasive species, deforestation, land degradation and habitat loss, intensification of land use, environmental pollution, spread of disease and climate change. An advantage of ecological modelling using enhanced data products related to vegetation, landscape and climate is the ability to interrogate the complexity of ecological processes across a multitude of scales (Saatchi et al. 2008). Scale can be examined biologically, moving from the level of genetics (Knowles et al. 2007; Kramer et al. 2010; Richards et al. 2007), through plant and animal species (Gavashelishvili and Lukarevskiy 2008; Miola et al. 2011) to communities (Bispo et al. 2014; Sesnie et al. 2008). Scale can also be conceived geographically, including modelling applied to local, regional and continental contexts (Bragin et al. 2013; Karanth et al. 2010; Nazeri et al. 2012). A cross-scale approach allows synergistic examination of transboundary issues over historical (Waltari and Guralnick, 2009), contemporary (Bragin et al. 2013; Mantelli et al. 2011) and future timescales (Feeley and Silman 2010; Nazeri et al. 2012). This has given rise to new sub-disciplines such as landscape genetics, which uses landscape-level data in spatially explicit studies to better understand the historical and contemporary factors that influence the distribution of intra-specific diversity (Thomassen et al. 2010).

Undoubtedly, with the availability of a growing number of spatially continuous, high resolution environmental data that capture a range of topographic, climatic and habitat properties (Gillespie et al. 2008; Rabus et al. 2003; Thomassen et al. 2010), biological modelling has the potential to be applied increasingly rigorously in many fields (Elith et al. 2011). Researchers must continue to consider sampling regimes and biases, model performance and requirements, and algorithm efficiency and parallel computing capabilities (Thomassen et al. 2010) if they are to refine their understanding of species distribution and dynamics and continue to inform conservation planning and environmental policy into the future. Consideration of the range of applications detailed in this chapter will enable further investigation of the use of plants and their mathematic and 'real' function in trophic levels and our developing societies. Furthermore, additional functions will be identified in order that effective use of modelling is made, allowing great developments to follow both in our interpretation of life and our understanding of modelling theory itself.

References

Angelov PP, Filev DP (2004) An approach to online identification of Takagi-Sugeno fuzzy models. IEEE Transactions on Systems, Man, and Cybernetics 43(1):484–498

Arrhenius O (1921) Species and Area. Journal of Ecology 9:95–99

Barreto LS (2008) The reconciliation of the r-K, and C-S-R-models for life history strategies. Silva Lusitana 16(1):97–103

Barthlott W, Mutke J, Rafiqpoor D, Kier G, Kreft H (2005) Global Centers of Vascular Plant Diversity. Nova Acta Leopoldina NF 92(342):61–83

Bass MS, Finer M, Jenkins CN et al (2010) Global Conservation Status of Ecuadors YasuníNational Park. Plos One 5:1–22. doi: 10.1371/journal.pone.0008767

Beck J, Boller M, Erhardt A, Schwanghart W (2014) Spatial bias in the GBIF database and its effect on geographic distribution. Ecological Informatics 19:10–15

Best MJ, Pryor M, Clark DB et al (2011) The Joint UK Land Environment Simulator (JULES), model description—Part 1 Energy and water fluxes. Geosci Model Dev 4:677–699

Bhattarai KR, Vetaas OR (2003) Variation in plant species richness of different life forms along a subtropical elevation gradient in the Himalayas, east Nepal. Global Ecology and Biogeography 12:327–340

Bilsborrow RE, Walsh SJ, Frizelle B (2012) LBA-ECO LC-01 National, Provincial, and Park Boundaries, Ecuador. Data Set. Available online [http://daac.ornl.gov/] from Oak Ridge National Laboratory Distributed Active Archive Center, Oak Ridge, Tennessee, U.S.A. http://dx.doi.org/10.334/ORNLDAAC/1057. Accessed 29 Aug 2014

Bispo PC, Santos JR, Valeriano MM, Touzi R, Seifert FM (2014) Integration of polarimetric PALSAR attributes and local geomorphic variables derived from SRTM for forest biomass modelling in central Amazonia. Canadian Journal of Remote Sensing 40:26–42

Bragin N, Singh NJ, Reading RP (2013) Creating a ruggedness layer for use in habitat suitability modelling for Ikh Nart Nature Reserve, Mongolia. Mongolian Journal of Biological Sciences 11:19–23

Cáceres MO, Saez IC (2011) Random Leslie matrices in population dynamics. Journal of Mathematical Biology 63:519–556

Cuddington K, Fortin MJ, Gerber LR et al (2013) Process-based models are required to manage ecological systems in a changing world. Ecosphere 42(2):1–12

Cui Z, Cai X, Zeng J (2012) A new stochastic algorithm to direct orbits of chaotic systems. International Journal of Computer Applications in Technology 43:366–371

Davison R, Jacquemyn H, Adriaens D et al (2010) Demographic effects of extreme weather events on a short-lived calcareous grassland species stochastic life table response experiments. Journal of Ecology 98:255–267

Drenovsky RE, Grewell BJ, D'Antonio CM et al (2012) A functional trait perspective on plant invasion. Annals of Botany 110:1–13

Elith J, Phillips SJ, Hastie T, Dudik M, Chee TE, Yates CJ (2011) A statistical explanation of MaxEnt for ecologists. Diversity and Distributions 17:43–57

Evans KL, Greenwood JJD, Gaston KJ (2005) Dissecting the species-energy relationship. Proceedings of the Royal Society B 272:2155–2163

Feeley KJ, Silman MR (2010) Modelling the responses of Andean and Amazonian plant species to climate change: the effects of georeferencing errors and the importance of data filtering. Journal of Biogeography 37:733–740

Fithian, W., Elith, J., Hastie, T. and Keith, D. A. (2014). Bias correction in species distribution models: Pooling survey and collection data for multiple species. Cornell University Digital Library. http://arxiv.org/abs/1403.7274. Accessed 10 August 2014

Flemons P (2007) A Web based GIS Tool for Exploring the World's Biodiversity: The Global Biodiversity Information Facility Mapping and Analysis Portal Application (GBIF MAPA). Ecological Informatics 2:49–60

Fourcade Y, Egler JO, Rödder D, Secondi J (2014) Mapping species distributions with MAXENT using a geographically biased sample of presence data- a performance assissment of methods for correcting sampling bias. Plos One 9(5):1–13

Furze JN (2014) Global Plant Characterisation and Distribution with Evolution and Climate. PhD Thesis. https://eprints.uwe.ac.uk/secure/23121/. Accessed Aug 10 2014

Furze JN, Zhu Q (2015) Orthogonal relations coupling renewable energy and sustainable plant systems. Proceedings of the Sixth Conference on Modelling, Identification and Control (ICMIC 2014), Melbourne, Australia, Dec 3rd-5th 2014. doi:10.1109/ICMIC.2014.7020719

Furze J, Zhu QM, Qiao F, Hill J (2011) Species area relations and information rich modelling of plant species variation. Proceedings of the 17th International Conference on Automation and Computing (ICAC), 10 September, 2011 63-68. http://ieeexplore.ieee.org/stamp.jsp?tp=&arnumber=6084902&isnumber=6084889. Accessed 10 Aug 2014

Furze J, Hill J, Zhu QM, Qiao F (2012) Algorithms for the Characterisation of Plant Strategy Patterns on a Global Scale. American Journal of Geographic Information System 1(3):72–99. doi:10.5923/j.ajgis.20120103.05

Furze JN, Zhu Q, Qiao F, Hill J (2013a) Implementing stochastic distribution within the utopia plane of primary producers using a hybrid genetic algorithm. International Journal of Computer Applications in Technology 47:68–77

Furze JN, Zhu Q, Qiao F, Hill J (2013b) Mathematical methods to quantify and characterise the primary elements of trophic systems. International Journal of Computer Applications in Technology 47(4):314–325

Furze JN, Zhu Q, Qiao F, Hill J (2013c) Utopian exploration of global patterns of plant metabolism.Proceedings of the Fifth International Conference on Modelling, Identification & Control (ICMIC 2013), Cairo, Egypt, Aug. 31st-Sept. 3rd 47-52. http://ieeexplore.ieee.org/stamp/stamp.jsp?tp=&arnumber=6642198&isnumber=6642159. Accessed 1 July 2014

Furze JN, Zhu QM, Qiao F, Hill J (2013d) Linking and implementation of fuzzy logic control to ordinate plant strategies, International Journal of Modelling. Identification and Control 19 (4):333–342

Furze JN, Zhu Q, Qiao F (2013e) Collaborative proposal development towards sustainable communities. Global Education Magazine, Human Rights Day, 10 Dec. 6:28-32. www.globaleducationmagazine.com/collaborative-proposal-development-sustainable-communities. Accessed 1 Aug 2014

Furze JN, Zhu Q, Qiao F, Hill J (2013f) Functional enrichment of utopian distribution of plant life-forms. American Journal of Plant Sciences 4(12A):37–48

Gavashelishvili A, Lukarevskiy V (2008) Modelling the habitat requirements of leopard *Panthera pardus* in west and central Asia. Journal of Applied Ecology 45:579–588

Gentry AH, Dodson C (1987) Diversity and biogeography of neotropical vascular epiphytes. Annals of the Missouri Botanical Garden 74:205–233

Gesch DB, Larson KS (1996) Techniques for development of global 1-kilometer digital elevation models, In: Pecora Thirteen, Human Interactions with the Environment—Perspectives from Space, Sioux Falls, South Dakota, United States of America

Gillespie TW, Foody GM, Rocchini D, Giorgi AP, Saatchi S (2008) Measuring and modelling biodiversity from space. Progress in Physical Geography 32:203–221

Grime JP (1979) Plant Strategies and Vegetation Processes. John Wiley and Sons, Chichester

Grime JP, Hodgson JG, Hunt R (1995) The abridged comparative plant ecology. Chapman & Hall, London

Grime JP, Pierce S (2012) The Evolutionary Strategies that Shape Ecosystems. Wiley-Blackwell, Chichester

Gumbel EJ (1960) Bivariate exponential distributions. Journal of the American Statistical Association 55:698–707

Guo H (2011) A simple algorithm for fitting a Gaussian function [DSP Tips and Tricks]. Signal Processing Magazine IEEE 28(5):134–137

Hawkins BA, Field R, Cornell HV, Currie DJ, Guégan JF, Kaufman DM, Kerr JT, Mittelbach GG, Oberdorff T, O'Brien EM, Porter EE, Turner JRG (2003) Energy, water, and broad-scale geographic patterns of species richness. Ecology 84(12):3105–3117

Hill JL, Curran PJ, Foody GM (1994) The effect of sampling on the species-area curve. Global Ecology and Biogeography Letters 4:97–106

Humboldt AV (1806) Ideen zu einer physiognomik der gewächse., Tübingen

Humboldt AV (1808) Ansichten der natur mit wissenschaftlichen erläuterungen., Tübingen

Huntingford C, Zelazowski P, Galbraith D et al (2013) Simulated resilience of tropical rainforests to CO2-induced climate change. Nature Geoscience. doi:10.1038/NGEO1741

Jetz W, Kreft H, Ceballos G, Mutke J (2009) Global associations between terrestrial producer and vertebrate consumer diversity. Proceedings of the Royal Society B 276:269–278

Juang CF, Hsieh CD (2012) A fuzzy system constructed by rule generation and iterative linear SVR for antecedent and consequent parameter optimization. IEEE Transactions on Fuzzy Systems 20(2):372–384

Karanth KK, Nichols JD, Karanth KU, Hines JE, Christenson NL Jr (2010) The shrinking ark: patterns of large mammal extinctions in India. Proceedings of the Royal Society B 277:1971–1979

Keeley JE, Rundel PW (2003) Evolution of CAM and C_4 carbon concentrating mechanisms, International Journal of Plant Science164. S 55-S:77

Knowles LL, Carstens BC, Keat ML (2007) Coupling genetic and ecological niche models to examine how past population distributions contribute to divergence. Current Biology 17:1–7

Kraft NJB, Ackerly DD (2010) Functional trait and phylogenetic tests of community assembly across spatial scales in an Amazonian Forest. Ecological Monographs 80:401–422

Kramer K, Degen B, Buschbom J, Hickler T, Thuiller W, Sykes MT, de Winter W (2010) Modelling exploration of the future of European beech (*Fagus sylvatica* L.) under climate change – range, abundance, genetic diversity and adaptive response. Forest Ecology and Management 259:2213–2222

Kreft H, Jetz W (2007) Global patterns and determinants of vascular plant diversity. Proceedings of the National Academy of Sciences 104(14):5925–5930

Lüttge U (2003) Ecophysiology of Crassulacean Acid Metabolism. Annals of Botany 93 (6):629–652

Mantelli LR, Barbosa JM, Bitencourt MD (2011) Assessing ecological risk through automated drainage extraction and watershed delineation. Ecological Informatics 6:325–331

Mascaro J, Asner GP, Davies S, Dehgan A, Saatchi S (2014) These are the days of lasers in the jungle., Carbon Balance and Management, http://www.cbmjournal.com/content/9/1/7. Accessed 29 Aug 2014

McCracken SF, Forstner RJ (2014) Oil road effects on the Anuran community of a high canopy tank bromeliad (Aechmea zebrina) in the upper amazon basin. Ecuador Plos One. doi:10.1371/journal.pone.0085470

Miola DTB, Freitas CR, Barbosa M, Fernandes GW (2011) Modeling the spatial distribution of the endemic and threatened palm shrub *Syagrus glaucescens* (Arecaceae). Neotropical Biology and Conservation 6:78–84

Mustard JM, Standing DB, Aitkenhead MJ, Robinson S, McDonald JS (2003) The emergence of primary strategies in evolving virtual-plant populations. Evolutionary Ecology Research 5:1067–1081

Nazeri M, Jusoff K, Madani N, Mahmud AR, Bahman AR, Kumar L (2012) Predictive modelling and mapping of Malayan Sun Bear (*Helarctos malayanus*) distribution using maximum entropy. Plos One E 7(10), e48104

Nelson RB (2006) An introduction to copulas. Springer, New York

New M, Hulme M, Jones P (1999) Representing twentieth century space-time climate variability. Part I- Development of a 1961–90 mean monthly terrestrial climatology. Journal of Climate 12:829–856

Niu S, Yuan Z, Zhang Y, Liu W, Zhang L, Huang J, Wan S (2005) Photosynthetic responses of C3 and C4 species to Water availability and competition. Journal of Experimental Botany 56 (421):2867–2876

Rabus B, Eineder M, Roth A, Bamler R (2003) The shuttle radar topography mission – a new class of digital elevation models acquired by spaceborne radar. ISPRS Journal of Photogrammetry and Remote Sensing 57:241–262

Richards CL, Carstens BC, Knowles LL (2007) Distribution modelling and statistical phylogeography: an integrative framework for generating and testing alternative biogeographical hypotheses. Journal of Biogeography 34:1833–1845

Rosenzweig ML (1995) Species diversity in space and time. Cambridge University Press, Cambridge

Saatchi S, Buermann W, ter Steege H, Mori S, Smith TB (2008) Modeling distribution of Amazonian tree species and diversity using remote sensing measurements. Remote Sensing of Environment 112:2000–2017

Salah F, Abdalla H (2011) A Knowledge based system for enhancing conceptual design. International Journal of Computer Applications in Technology 40(1-2):23–36

Salisbury FB, Ross CW (1992) Plant Physiology, 4th edn. Wadsworth Publishing Company, California

Schmidt M, Kreft H, Thiombiano A, Zizka G (2005) Herbarium collections and field data-based plant diversity maps for Burkina Faso. Diversity and Distributions 11:509–516

Schölzel C, Friedrichs P (2008) Multivariate non normally distributed random variables in climate science – Introduction to the copular approach. Non linear Processes in Geophysics 15:761–772

Sesnie SE, Gessler PE, Finegan B, Thessler S (2008) Integrating Landsat TM and SRTM-DEM derived variables with decision trees for habitat classification and change detection in complex neotropical environments. Remote Sensing of Environment 112:2145–2159

Silvera K, Santiago LS, Cushman JC, Winter K (2009) Crassulacean acid metabolism and epiphytism linked to adaptive radiations in the orchidaceae. Plant Physiology 149:1838–1847

Silvera K, Winter K, Rodriguez BL, Albion RL, Cushman JC (2014) Multiple isoforms of phospoenolpyruvate carboxylase in the Orchidacea (subtribe Oncidiinae): implications for the evolution of crassulacean acid metabolism. Journal of Experimental Botany 65 (13):3623–3636

Silvert W (2000) Fuzzy indices of environmental conditions. Ecological Modelling 130:111–119

Sivanandam SV, Sumathi S, Deepa SN (2007) Introduction to fuzzy logic using matlab. Springer, Heidelberg

Sklar A (1959) Fonctions de repartition á n dimensions et leurs marges. Publications de L'Institut de Statistique de l'Université de Paris 8:229–231

Sommer JH, Kreft H, Kier G, Jetz W, Mutke J, Barthlott W (2010) Projected impacts of climate change on regional capacities for global plant species richness. Proceedings of the Royal Society B 277:2271–80

Su Y, Zhu G, Miao Z, Feng Q, Chang Z (2009) Estimation of parameters of a biochemically based model of photosynthesis using a genetic algorithm. Plant, Cell and Environ 32(12):1710–1723

Thomassen HA, Cheviron ZA, Freedman AH, Harrigan RJ, Wayne RK, Smith TB (2010) Spatial modelling and landscape-level approaches for visualising intra-specific variation. Molecular Ecology 19:3532–3548

Wallace AR (1878) Tropical nature and other essays. Macmillan, London

Waltari E, Guralnick RP (2009) Ecological niche modelling of montane mammals in the Great Basin, North America: examining past and present connectivity of species across basins and ranges. Journal of Biogeography 36:148–161

Wang C, Guo L, Li Y, Wang Z (2012) Systematic Comparison of C3 and C4 Plants Based on Metabolic Network Analysis. BMC Systems Biology 6:1–14

Wright DH (1983) Species-energy theory: an extension of species energy theory. Oikos 41 (3):496–506

Yesson C, Brewer PW, Sutton T, Caithness N, Pahwa JS, Burgess M, Gray WA, White RJ, Jones AC, Bisby FA, Culham A (2007) How global is the global biodiversity information facility? Plos One 11:1–10

Zadeh LA (1965) Fuzzy sets. Information and Control 8:338–353

Zhang XW, Gong F, Xi LZ (2012) Inspection of surface defects in copper strip using multivariate statistical approach and SVM. International Journal of Computer Applications in Technology 43(1):44–50

Zhu G, Li X, Su YH et al (2014) Simultaneously assimilating multivariate data sets into the two-source evapotranspiration model by Bayesian approach: application to spring maize in an arid region of northwestern region of China. Geosci Model Dev 7:1467–1482

Zhu Q, Wang Y, Zhao D, Li S, Billings SA (2013) Review of rational (total) nonlinear dynamic system modelling, identification and control, International Journal of Systems Science 1-12. doi:10.1080/00207721,2013.849774

Zytynska SE, Khudr MS, Harris E, Preziosi RF (2012) Genetic effects of tank-forming bromeliads on the associated invertebrate community in a tropical forest ecosystem. Oecologia 170:467–475

Chapter 3
On the Dynamics of the Deployment of Renewable Energy Production Capacities

R. Fonteneau and D. Ernst

Abstract This chapter falls within the context of modelling the deployment of renewable energy production capacities in the scope of the energy transition. This problem is addressed from an energy point of view, i.e., the deployment of technologies is seen as an energy investment under the constraint that an initial budget of nonrenewable energy is provided. Using the Energy Return on Energy Investment (ERoEI) characteristics of technologies, we propose MODERN, a discrete-time formalization of the deployment of renewable energy production capacities. Besides showing the influence of the ERoEI parameter, the model also underlines the potential benefits of designing control strategies for optimizing the deployment of production capacities and the necessity to increase energy efficiency.

Keywords Modelling • Renewable energy • Energy transition • ERoEI • Discrete-time • Dynamical systems

3.1 Introduction

The relations linking energy consumption and societies' prosperity have been thoroughly investigated in the last decades. It has progressively become clear that energy has played a decisive role in societies' demographic and economic development (Meadows et al. 1972; Cleveland et al. 1984; Lambert et al. 2012; Giraud and Kahraman 2014), as well as in their decline (Tainter 1988).

About 85 % of world energy consumption is currently from nonrenewable origin, most of which being fossil fuels such as coal, oil, and gas. The "energy transition," which is the shift to a world that would no longer virtually rely on nonrenewable energy resources, is a crucial challenge of the twenty-first century for two main reasons: (1) the massive consumption of fossil fuels has major environmental impacts, mainly pollution and greenhouse effect gas emissions, and (2) there

R. Fonteneau (✉) • D. Ernst
Systems and Modeling Research Unit, University of Liège, Institute Montefiore, (B28, P32)
Grande Traverse, 10 Sart-Tilman, Liège B-4000, Belgium
e-mail: raphael.fonteneau@ulg.ac.be; dernst@ulg.ac.be

© Springer International Publishing Switzerland 2017
J.N. Furze et al. (eds.), *Mathematical Advances Towards Sustainable Environmental Systems*, DOI 10.1007/978-3-319-43901-3_3

is convincing evidence that even putting aside these environmental concerns our societal lifestyle cannot be sustained without changing our energy production and consumption habits.

One of the main difficulties of this transition comes from the fact that switching to an energy system that would not depend on nonrenewable resources is a process that itself needs—at least for the moment—to use nonrenewable energy. For instance, in 2013, about half of photovoltaic panels have been produced in China (Jäger-Waldau 2013) whose own energy production mix was around 70 % from coal in 2011 (US Energy Information Administration 2015), which suggests that the rise of PV energy over the last 10 years was mainly achieved through using nonrenewable energy resources.

In this chapter, we propose to consider the deployment of renewable energy production capacities as an energy investment. This point of view is motivated by the fact that the ERoEI parameters characterizing the two main rising renewable technologies—wind turbines and photovoltaic panels—are currently too low to be negligible (Murphy and Hall 2010). We propose MODERN (for "MOdelling" the Deployment of Energy production from "ReNewable" resources), a discrete-time model that aims at simulating the deployment of renewable energy capacities in the context of the depletion of a given budget of nonrenewable resources. MODERN makes use of ERoEI characteristics of technologies that relate the energy produced to the energy invested together. MODERN can be controlled using growth scenarios for the deployment of the production capacities. We illustrate some typical runs of MODERN in the context of ERoEI corresponding to photovoltaic panels. In particular, we observe how the availability of nonrenewable energy can actually boost the growth of production capacities and eventually create a "bubble effect"; we show that this bubble may be mitigated using control strategies.

The following of this chapter is organized as follows: Section 3.2 provides ERoEI notions. Section 3.3 presents MODERN, our discrete-time formulation of the deployment of renewable energy production capacities. Section 3.4 illustrates several typical runs of MODERN with a parameterization matching the deployment of photovoltaic panels. Section 3.5 discusses how MODERN opens the door to the use of control strategies in the context of the energy transition. Section 3.6 provides a discussion about the link between energy and societies' GDP and emphasizes reasons why the deployment strategies of renewable energy production capacities should be carefully designed. Section 3.7 concludes this chapter.

3.2   Energy Return on Energy Investment

Energy Return on Energy Investment (ERoEI) is a notion that was probably first coined in the works of Cleveland et al. (1984) and Hall et al. (1986). It is defined as the ratio of the amount of final usable energy acquired from a particular energy resource to the amount of primary energy expended to obtain that energy resource:

$$\text{ERoEI} = \frac{\text{Usable Acquired Energy}}{\text{Energy Expended to Get that Energy}} \tag{3.1}$$

More specifically, this ratio—supposed to be dimensionless—means that a given energy production technology will provide ERoEI Joules (J) on an energy investment of 1 J. Note that computing ERoEI for a given energy production technology may be a complex task because it implies a rigorous definition of system boundaries (energy inputs, energy outputs), as well as accurate evaluation of energy costs in between these boundaries. In particular, the natural or original sources of energy are usually not taken into account. For instance, the energy consumed by the sun to produce light is not taken into account in the computation of the ERoEI of photovoltaic technologies. We refer to the work of Murphy and Hall (2010) for a solid review of the work that has been done around the notion of ERoEI. We provide hereafter in Fig. 3.1 a graph of ERoEI values for a panel of technologies in the specific case of the USA (figures taken from Murphy and Hall 2010).

This graph illustrates different aspects of the ERoEI. One can first observe that the ERoEI of US oil and gas productions has declined over time, from about 30 in the 1970s to about 15 in 2005. This is easily explained by the fact that oil and gas fields that were the easiest to exploit were exploited first. This graph also shows that hydroelectricity has a very high ERoEI (above 100). One may also observe that energy production from coal has a high ERoEI (in the order of 80). The ERoEI value of photovoltaic panels (around 10) has a rather low value here compared to other renewable energy sources such as hydroelectricity (more than 100) or wind turbines (around 18). Observe however that photovoltaic panels technology is

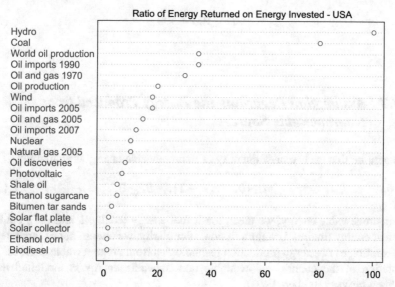

Fig. 3.1 ERoEI of several technologies in the USA—data source: (Murphy and Hall 2010)—image taken from Wikipedia

progressing, and that it might be possible that its ERoEI will increase significantly
in the coming years. Finally, even if nuclear energy is reported to have an ERoEI of
about 16, it is important to notice that this technology is among those for which the
ERoEI computation is the most uncertain (Lambert et al. 2012).

3.3 MODERN: A Discrete-Time Model of the Deployment of Renewable Energy Production Capacities

This section introduces all the elements of MODERN, a discrete-time model of the
deployment of energy production capacities from renewable sources and the multiple assumptions upon which it is built. For clarity, we assume that all variables
considered in this chapter are deterministic.

3.3.1 Time

We consider a discrete-time system, where each time-step corresponds to 1 year:

$$t = 0 \ldots T - 1 \tag{3.2}$$

The time horizon is in the order of hundreds of years:

$$T \sim 100 - 500 \tag{3.3}$$

3.3.2 Assumption Regarding the Energy Produced from Nonrenewable Sources

We assume that each year, a quantity of nonrenewable energy is available:

$$\forall t \in \{0, \ldots, T - 1\}, \quad B_t \geq 0 \tag{3.4}$$

By nonrenewable energy, we mean fossil fuel energy (coal, oil, and gas), but also
nuclear energy (mainly Uranium fission). For clarity, we choose not to separate the
different types of energy production technologies from nonrenewable sources. The
evolution of the quantity of available nonrenewable energy is modelled using
Hubbert curves (Hubbert 1956):

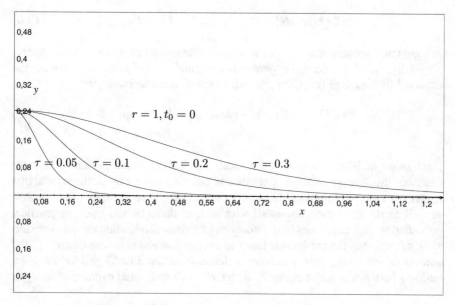

Fig. 3.2 Some Hubbert curves obtained with different values of the parameter τ

$$\exists r > 0, \ \ \exists \tau > 0, \ \ \exists t_0 \in \mathbb{R} : \forall t \in \{0, \ldots, T-1\}$$

$$B_t = \frac{1}{r} \frac{e^{-\frac{(t-t_0)}{\tau}}}{\left(1 + e^{-\frac{(t-t_0)}{\tau}}\right)^2} \tag{3.5}$$

As shown in Fig. 3.2, the role of τ is to model the level of "flatness" of the Hubbert curve. The parameter t_0 induces a time shift of the curve. For simplicity, we assume that this energy is "net," i.e., we assume that the energy required to obtain that energy is already subtracted from it. Recent papers have shown that the ERoEI related to processes producing energy from nonrenewable resources tend to decline over time (Murphy and Hall 2010). The intuition behind this is the fact that spots for which resources are easily extracted are exploited first. The Hubbert curve, which models the extraction of nonrenewable resources, reflects to a certain extend that energy is increasingly more expensive to obtain (in terms of energy investment, but also cost).

3.3.3 Energy from Renewable Origin

We assume that a set of N different technologies for producing energy from renewable sources is available. To each technology is associated a production capacity yearly producing a quantity of energy $R_{n,t}$:

$$\forall n \in \{1,\ldots,N\}, \ \forall t \in \{0,\ldots,T-1\}, \quad R_{n,t} \geq 0 \tag{3.6}$$

Among these technologies, let us (non-comprehensively) mention biomass, hydro-electricity, wind turbines, or photovoltaic panels. Two main parameters, the expected lifetime and ERoEI characterize each of these technologies:

$$\forall n \in \{1,\ldots,N\}, \ \forall t \in \{0,\ldots,T-1\}, \quad \Delta_{n,t} \geq 0$$
$$\text{ERoEI}_{n,t} \geq 0 \tag{3.7}$$

Description of ERoEI is provided in Sect. 3.2. The expected lifetime parameter describes the average lifetime of equipment enabling energy production. Note that in this model, we do not consider energy production and consumption fluctuations, as well as storage issues associated with each of these technologies. In practice, providing storage capacities or technologies that allow modulating the consumption so that it matches the production (such as energy demand side management in the context of electricity grids) induces a decrease of the ERoEI parameters (e.g., building batteries to assist photovoltaic panels is an additional expanse of energy).

3.3.4 Dynamics of Deployment of Energy Production Means

The dynamics of the deployment of energy production means is modelled using a growth parameter:

$$\forall n \in \{1,\ldots,N\}, \forall t \in \{0,\ldots,T-1\}, \quad R_{n,t+1} = (1+\alpha_{n,t})R_{n,t} \tag{3.8}$$

Note that the growth parameter may be negative:

$$\forall n \in \{1,\ldots,N\}_1 \forall t \in \{0,\ldots,T-1\}, \quad \alpha_{n,t} \in [-1,\infty[\tag{3.9}$$

3.3.5 Energy Costs for Growth and Long-Term Replacement

We introduce the energy cost associated with the growth of the production capac-ities of renewable technologies:

$$\forall n \in \{1,\ldots,N\}_1, \forall t \in \{0,\ldots,T-1\}, \quad C_{n,t}(R_{n,t},\alpha_{n,t}) \geq 0 \tag{3.10}$$

We assume that this cost also incorporates the energy required for maintenance during the lifetime of the equipment. We also introduce the energy cost associated with the long-term replacement of the production means:

$$\forall n \in \{1,\ldots,N\}, \forall t \in \{0, \ldots, T-1\}, \quad M_{n,t} \geq 0 \tag{3.11}$$

The role of this quantity of energy is to formalize the energy cost that has to be "paid" when equipment becomes obsolete and has to be replaced (see a few assumptions regarding this energy cost later in the chapter).

3.3.6 Total Energy and Net Energy to Society

Using the previous notations, we define the total energy produced at year t:

$$\forall t \in \{0, \ldots, T-1\}, \quad E_t = B_t + \sum_{n=1}^{N} R_{n,t} \tag{3.12}$$

We also define the net energy available to society:

$$\forall t \in \{0, \ldots, T-1\}, \quad S_t = E_t - \left(\sum_{n=1}^{N} C_{n,t}(R_{n,t}, \alpha_{n,t}) + M_{n,t} \right) \tag{3.13}$$

This corresponds to the amount of energy that can be used after energy investment for increasing the production capacities from renewable resources and their long-term replacement.

3.3.7 Constraints on the Quantity of Energy Invested for Energy Production

We assume that the energy investment for developing, maintaining, and replacing the production means from renewable sources cannot exceed a given fraction of the total energy. In other words, this assumption means that the ratio of net energy to society over total energy has to remain above a given threshold. Formally, we assume that:

$$\forall t \in \{0, \ldots, T-1\}, \quad \exists \sigma_t : C_{n,t}(R_{n,t}, \alpha_{n,t}) + M_{n,t} \leq \frac{1}{\sigma_t} E_t \tag{3.14}$$

In the following, we denote by "energy threshold" such a parameter. This constraint is motivated by research investigation showing that if a society invests a too high a proportion of its energy for producing energy, then less energy is dedicated to other societal needs, which may result into a decrease of the global society welfare (Lambert et al. 2012).

3.3.8 Assumptions on Growth and Replacement Energy Costs

In order to relate the energy costs associated with the deployment and the long-term replacement of the renewable energy production capacities, we make the three following assumptions:

1. The energy cost associated with the installation of new production means of technologies is proportional to the corresponding growth:

$$\forall n \in \{1, \ldots, N\}, \forall t \in \{0, \ldots T-1\}, \quad \exists_{\gamma_{n,t}} > 0$$
$$C_{n,t}(R_{n,t}, \alpha_{n,t}) = \begin{cases} \gamma_{n,t}\alpha_{n,t}R_{n,,t} & \text{if } \alpha_{n,t} \geq 0 \\ 0 & \text{else} \end{cases} \quad (3.15)$$

2. All the energy costs related to building a production capacity and to operating it over its lifetime are allocated at the time period when this capacity starts producing energy:

$$\forall n \in \{1, \ldots, N\}, \forall t \in \{0, \ldots T-1\}, \quad \gamma_{n,t} = \frac{\Delta_{n,t}}{\text{ERoEI}_{n,t}}$$
$$C_{n,t}(R_{n,t}, \alpha_{n,t}) = \frac{\Delta_{n,t}}{\text{ERoEI}_{n,t}}\alpha_{n,t}R_{n,t} \quad \text{if } \alpha_{n,t} \geq 0 \quad (3.16)$$

3. The energy cost associated with the long-term replacement of production capacities is (1) annualized and (2) proportional to the quantity of energy produced yearly:

$$\forall n \in \{1, \ldots, N\}, \quad \forall t \in \{0, \ldots T-1\}, \quad \exists \mu_{n,t} > 0 : M_{n,t}(R_{n,t}) = \mu_{n,t}R_{n,t} \quad (3.17)$$

Using the ERoEI parameter, we get the following equations:

$$\forall n \in \{1, \ldots, N\}, \quad \forall t \in \{0, \ldots T-1\}, \quad \mu_{n,t} = \frac{1}{\text{ERoEI}_{n,t}}$$
$$M_{n,t}(R_{n,t}) = \frac{1}{\text{ERoEI}_{n,t}}R_{n,t} \quad (3.18)$$

3.4 Simulation Results: Case Study for Photovoltaic Panels

We propose to simulate MODERN where only photovoltaic panels are deployed. For simplicity, we denote by one the index related to photovoltaic technology. Formally, this means that growth parameters associated to other technologies are kept constant at zero:

$$\forall n \in \{2, \ldots, N\}, \forall t \in \{0, \ldots T-1\}, \quad \alpha_{n,t} = 0 \quad (3.19)$$

3.4.1 Variable Initialization

We choose to consider normalized variables with respect to the total energy at time 0:

$$E_0 = 1 \qquad (3.20)$$

The Hubbert curve modelling the depletion of nonrenewable energy is initially scaled so that the proportion between renewable and nonrenewable energy production matches, approximately, the current situation for 2014 (British Petroleum 2014):

$$B_0 = 0.85 E_0 \qquad (3.21)$$

The quantity of energy produced by photovoltaic panels is initially assumed to be around 1 % of the world total energy mix:

$$R_{1,0} = 0.01 E_0 \qquad (3.22)$$

This value (1 %) also corresponds, approximately, to the current proportion of energy produced by photovoltaic panels plus wind turbines in the world total energy mix. All remaining technologies producing energy from renewable sources are kept constant at their initial level, i.e.,

$$\forall t \in \{0, \ldots, T-1\}, \quad \sum_{n=2}^{N} R_{n,t} = \sum_{n=2}^{N} R_{n,0} = 0.14 E_0 \qquad (3.23)$$

The constraint of the total amount of energy that may be dedicated to growing energy production means is chosen as follows:

$$\forall t \in \{0, \ldots, T-1\}, \quad \sigma_t = 14 \qquad (3.24)$$

The choice of this value for the energy threshold is motivated by results reported in the literature (Lambert et al. 2012). As shown by Lambert et al., this value appears to be the smallest so that society may develop and sustain social amenities that are considered to be at the top of the "society Maslow pyramid," such as healthcare systems and arts (see the figure "Pyramid of Energetic Needs" in Lambert et al. 2012).

3.4.2 Growth Scenario

MODERN can be controlled through the growth scenario. By growth scenario, we mean a sequence of predefined growth parameters. Formally, a growth scenario is a T-tuple of real numbers:

$$(\alpha_{1,0}, \ldots \alpha_{1,T-1}) \in \mathbb{R}^{T} \tag{3.25}$$

When simulated, such scenarios may not satisfy the energy threshold constraint. If so, the growth parameter is reduced to the maximal allowed value so that it does not violate the constraint. In the case where the constraint is violated, then the growth parameter is set to the maximal value that still satisfies the energy threshold constraint defined as follows:

$$\forall t \in \{0, \ldots, T-1\}, \quad \alpha_{1,t}^{\max} = \frac{\text{ERoEI}_{1,t}}{\Delta_{1,t} R_{1,t}} \left(\frac{1}{\sigma_t} E_t - \frac{1}{\text{ERoEI}_{1,t}} R_{1,t} \right) \tag{3.26}$$

In the simulations reported in this section, we consider the simple, constant over time growth scenario:

$$\forall t \in \{0, \ldots, T-1\}, \quad \alpha_{1,t} = \alpha_0 = 0.1 \tag{3.27}$$

Observe that, in practice, the growth scenario may be constrained by the availability of resources for building capacities, as well as the availability of suitable locations to install capacities (sunny places in the case of photovoltaic panels).

3.4.3 Depletion of Nonrenewable Resources Scenario

We consider several scenarios for the depletion of nonrenewable resources. We arbitrarily define four scenarios and provide below the corresponding values of the parameters of the Hubbert curve:

– Peak at time 0:

$$t_0 = 0, \quad \tau = 30 \tag{3.28}$$

– Plateau at time 0:

$$t_0 = 0, \quad \tau = 60 \tag{3.29}$$

– Peak at time $t = 20$ years:

$$t_0 = 20, \quad \tau = 30 \tag{3.30}$$

– Plateau at time $t = 20$ years:

$$t_0 = 20, \quad \tau = 60 \tag{3.31}$$

The graph of resulting Hubbert curves can be found later in the chapter (Figs. 3.3, 3.4, 3.5, and 3.6). Note that the terms "peak" and "plateau" have been chosen to illustrate the fact that "plateau" curves are flatter than "peak" curves.

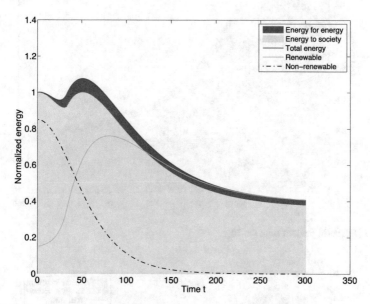

Fig. 3.3 Scenario peak at time $t = 0$

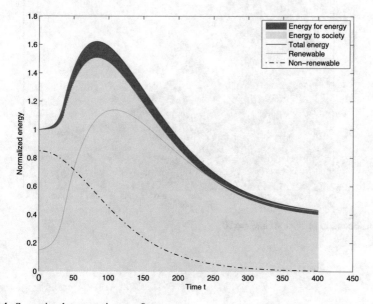

Fig. 3.4 Scenario plateau at time $t = 0$

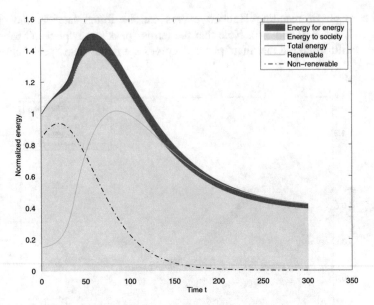

Fig. 3.5 Scenario peak at time $t = 20$

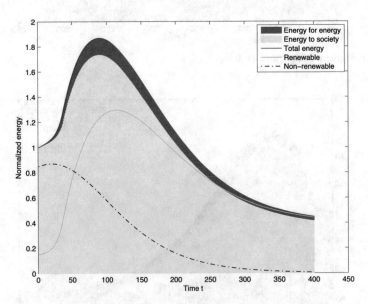

Fig. 3.6 Scenario plateau at time $t = 20$

3.4.4 Values of ERoEI and Lifetime

EROI for PV panels have been studied in the literature. For example in Lambert et al. (2012), a range of values varying from 6 to 12 is proposed, depending on the configurations. In the following experiments, we consider the average of these two values, i.e.,

$$\forall t \in \{0, \ldots, T-1\}, \quad \text{ERoEI}_{1,t} = 9 \tag{3.32}$$

Note that (1) the computation of ERoEI values of PV panels is still discussed in the literature (Raugei et al. 2012), and that (2) it is very likely that such values will evolve significantly in the future. In all configurations considered in the following experiments, we consider a lifetime parameter equal to 20:

$$\forall t \in \{0, \ldots, T-1\}, \quad \Delta_{1,t} = 20 \tag{3.33}$$

3.4.5 Typical Runs

In this section, we provide simulation results obtained through our discrete-time models in the different configurations described above. Each graph shows, for every year, the evolution of the total energy (yearly) produced (top blue curve) which comprises two parts: energy dedicated to the production of energy ("energy for energy," red part) and energy dedicated to other needs of society ("energy to society," yellow part). We also report the levels of nonrenewable energy production (black dotted curve) and renewable energy production (green curve).

Note that the results presented in the following subsections should definitely not be considered as predictions. Their role is just to illustrate the behavior of the model in theoretical configurations.

Initially, it can be seen that the production of energy from renewable resources as well as the net energy to society both reached a global maximum before decreasing to a steady-state value. This decrease is a consequence of the "energy threshold" constraint: if the energy required for the long-term replacement of the current production capacity is larger than what the energy threshold constrain allows for investment, then the growth parameter becomes negative. In other words, the bubble that can be observed on the graphs illustrates the fact that the deployment of the renewable energy production capacities is boosted by the availability of nonrenewable resources.

As a second observation, we notice that the depletion scenario has an influence on the maximal level of production that can be reached during the transition phase. However, one can compute that it does not affect the steady-state production level, which is exactly the same in the four scenarios, and function of the ERoEI of the photovoltaic panels.

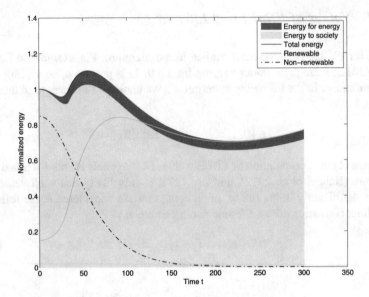

Fig. 3.7 Simulation result with an increase of the ERoEI parameter

To illustrate the influence of the ERoEI parameter on the levels of energy production, we give in Fig. 3.7 a last run of MODERN for which we consider a linear increase of the ERoEI parameter from 9 to 12 between time 0 and the time horizon (the growth scenario is the same as before, 10 % annual growth):

$$\forall t \in \{0, \ldots, T-1\}, \quad \text{ERoEI}_{1,t} = 9 + \frac{t}{T}(12-9) \tag{3.34}$$

3.5   On the Potential Benefits of Using Control Strategies

MODERN can be controlled through the growth scenario (which may be constrained by the system itself). This section discusses the potential benefits of using optimal control techniques for designing growth scenarios. In particular, we propose a control scheme that makes the variations of the net energy available to society vanish.

We have seen in Sect. 3.4 that growth scenarios may induce that the quantity of net energy available to society may reach a maximum level before decreasing to a steady-state level. We may assume that such a bubble effect can have destabilizing effects on society that one may want to avoid. It may thus be of interest to look for a sequence of growth parameters that would make such a "bubble" effect disappear. We illustrate below a sequence of growth that manages to do so.

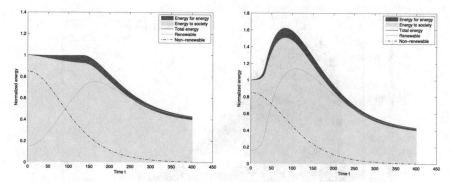

Fig. 3.8 Simulation results obtained when using the controlled growth (*left*) and a constant scenario growth (*right*)

We consider the "plateau at time $t = 0$" scenario, with a medium ERoEI of 9. We control the deployment growth using the following closed-loop growth scenario:

$$\forall t \in \{1, \ldots, T-1\}, \quad \alpha_{1,t} = \frac{B_{t-1} - B_t}{R_{1,t}} \tag{3.35}$$

This controller has been designed by considering the depletion of nonrenewable energy between two subsequent time steps and planning a growth that may counterbalance the depletion. We compare the result of this controlled growth scenario with the constant growth scenario obtained in the same depletion scenario (cf. Fig. 3.3):

It can be observed in Fig. 3.8 above that the simple controller proposed allows for the suppressing of the net energy bubble effect. One can also observe that negative growth parameters—which mean that the system is decreasing its renewable energy production capacities—appears around $t = 150$ in the controlled growth case, while it appears at around $t = 100$ in the noncontrolled case. In addition, one can see in Fig. 3.9 below that the cumulative sum of energy invested for the growth and long-term replacement of renewable energy production capacities is much smaller in the controlled growth scenario case.

We mention that, in the case of energy production technologies having a low ERoEI value, a strong growth can lead to a transient phenomenon called "energy cannibalism." This is a paradoxical situation where the energy invested for growing production capacities is so huge that the net energy available to society is temporarily decreasing while production capacities are increasing (Pearce 2009).

3.6   From Modelling to Society

Several articles in the literature relate to the link between societies' prosperity and their access to energy. Among others, historians, anthropologists, and economists have studied how energy has played a major role in the rise and decline of societies

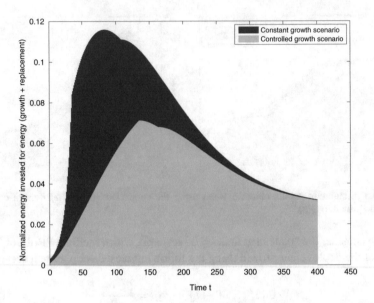

Fig. 3.9 Comparison of the quantities of energy invested for energy in the controlled growth scenario (*green*) and the constant growth scenario (*blue*)

(Meadows et al. 1972; Tainter 1988; Cleveland et al. 1984; Lambert et al. 2012; Jancovici 2011, 2013; Giraud and Kahraman 2014). The decline of the Western Roman Empire can be partly explained by (1) the decrease of agriculture efficiency (agriculture, which allows for the gathering of solar energy through photosynthesis, was the main energy source of the Roman Empire), and (2) the fact that looting was a nonrenewable way of obtaining access to resources (Tainter 1988). During the Middle Ages, the European GDP per inhabitant was increasing much faster than the Asiatic GDP during the period 1000–1500 (Maddison 2004). This has been explained by the increase of the use of windmills and sawmills in Western Europe, a mill being able to provide an energy equivalent to 40 men (Gimel 1976). This is even more striking in the case of the Dutch Golden Century, where the use of peat, as well as windmills and sawmills, allowed for increasing energy and food provision as well as better health, thus allowing cities to expand, boats to be built and trade developed (Zeeuw 1978). More recently, it has been shown that the impact on the GDP growth of capital accumulation and technical progress was minor compared to the role of energy in the period 1970–2012 (Giraud and Kahraman 2014) (see also Stern and Enflo 2013 for the specific case of Sweden). These three examples suggest that societies should consider energy as a key parameter of their economic development, and strategically manage their decisions related to energy supply.

The increasing use of energy over the last 150 years has generated an increase in work productivity that had never been seen before in the history of humanity. It is precisely this work productivity increase that has led to the diversification of human activities, resulting in complex societies with beneficial healthcare systems and a rich cultural life (Lambert et al. 2012; Jancovici 2013). In this study, we have used an energy threshold parameter, which basically models the fact that societies should not invest too much energy in producing energy otherwise the energy sector may cannibalize other human activities. We have observed that this parameter drastically constrains the model. As a consequence, technologies having a high ERoEI value lead to high renewable energy steady-state production levels. In this respect, we concur with several other papers stating that the ERoEI should be a major axis of technologies improvement. In parallel to this, a better geographical deployment strategy of renewable energy production technologies would result to an increase in their empirical ERoEI (Chatzivasileiadis et al. 2013, 2014).

The goal of this first version of MODERN (denoted by MODERN 1.0) was to model the deployment of renewable energy production capacities. In particular, MODERN 1.0 suggests that there is a possibility that the availability of nonrenewable energy in the short-term may create an artificial boost of energy production from nonrenewable resources that may not be sustainable on the long term, depending on the evolution of the technology. This potential effect should be taken into account when designing energy policies.

3.7 Conclusions

This chapter has introduced MODERN, a discrete-time formalization of the deployment of renewable energy production capacities. In particular, MODERN simulations show that deployment of renewable energy production capacities may be unsustainably boosted by the use of nonrenewable energy. This suggests that strategies for (1) deploying production capacities and (2) improving the efficiency of technologies, as well as the way they are used (energy efficiency) should be carefully designed.

MODERN 1.0 will be followed by other releases incorporating other parameters. In particular, MODERN 1.0 does not address the question of storage and fluctuations, which remains a major challenge of renewable energy deployment. Besides, MODERN 1.0 does not take into account the distinction between energy vectors (such as electricity, liquid fuels, heat...), which is another crucial point of the energy transition challenge. It would also be interesting to develop a version of MODERN, where the deployment of production capacities could be localized. This would enable the incorporation of constraints induced by local factors (geography, climate). However, besides calibration issues, more sophisticated versions of MODERN, taking into account such parameters, would come with a substantial increase in the level of difficulty for extracting near-optimal policies.

Acknowledgments This chapter presents results obtained thanks to the IAP program DYSCO (Dynamical Systems, Control and Optimization). We also thank the entire Smart Grid research group of the University of Liège for fruitful discussions.

References

British Petroleum (2014) Statistical review of world energy, June 2014. British Petroleum, London

Chatzivasileiadis S, Ernst D, Andersson G (2013) The global grid. Renew Energy 57:372–383

Chatzivasileiadis S, Ernst D, Andersson G (2014) Global power grids for harnessing world renewable energy. In: Jones LE (ed) Renewable energy integration: practical management of variability, uncertainty and flexibility in power grids. Academic, New York, pp 175–188

Cleveland CJ, Costanza R, Hall CAS, Kaufmann R (1984) Energy and the U.S. economy: a biophysical perspective. Science 225:890–897

Gimel J (1976) The medieval machine: the industrial revolution of the middle ages. Penguin, London. ISBN 978-0_7088-1546-5

Giraud G, Kahraman Z (2014) On the output elasticity of primary energy in OECD countries (1970–2012). Working paper, Center for European Studies

Hall CAS, Cleveland CJ, Kaufmann R (1986) Energy and resource quality: the ecology of the economic process. Wiley, New York

Hubbert MK (1956) Nuclear energy and the fossil fuels. Spring Meeting of the Southern District Division of Production, San Antonio

Jäger-Waldau A (2013) PV status report 2013. European Commission, Joint Research Centre, Institute for Energy and Transport

Jancovici JM (2011) Changer le monde—Tout un programme ! Calmann-Lévy

Jancovici JM (2013) Transition énergétique pour tous ! Ce que les politiques n'osent pas vous dire. Éditions Odile Jacob

Lambert J, Hall CAS, Balogh S, Poisson A, Gupta A (2012) EROI of global energy resources—preliminary status and trends. In: Report prepared by the College of Environmental Science and Forestry, State University of New York. United Kingdom Department for International Development, London

Maddison A (2004) When and why did the West get richer than the rest? In: Exploring economic growth: essays in measurement and analysis; a Festschrift for Riita Hjerppe on her 60th birthday. Aksant, Amsterdam, pp 29–62. ISBN 9052601658

Meadows DH, Meadows DL, Randers J, Behrens WW (1972) The limits to growth. Universe Books, New York

Murphy DJ, Hall CAS (2010) Year in review EROI or energy return on (energy) invested. Ann N Y Acad Sci 1185:102–118. doi:10.1111/j.1749-6632.2009.05282.x

Pearce JM (2009) Optimizing greenhouse gas mitigation strategies to suppress energy cannibalism. In: 2nd climate change technology conference, Hamilton, Ontario, Canada, May 12-15, 2009

Raugei M, Fullana-i-Palmer P, Fthenakis V (2012) The energy return on investment (EROI) of photovoltaics: methodology and comparisons with fossil fuel life cycles. Energy Policy 45:576–582

Stern DI, Enflo K (2013) Causality between energy and output in the long-run. Energy Econ 39:135–146

Tainter J (1988) The collapse of complex societies. Cambridge University Press, Cambridge

US Energy Information Administration (2015) China—international energy data and analysis, Washington, DC, 14 May 2015

Zeeuw JW (1978) Peat and the Dutch golden age. The historical meaning of energy attainability. AAG Bijdragen 21:3–31

Chapter 4
Water System Modelling

M.H. Bazrkar, J.F. Adamowski, and S. Eslamian

Abstract A comprehensive range of analytical, empirical, conceptual, numerical, and physical models facilitates an improved understanding of water systems; this improves time and cost efficiency, and facilitates the transition to more sustainable water resources planning and management. Models cover a variety of user interests, from historical uses which generate temporal and spatial operating scenarios, to the verification of variations and changes in the physical environment. Numerical models can be used to determine operational policy changes, the impact of floods, and changes in water quality. Physical models are useful in determining flow and level changes, and the impact of man-made changes on rivers and their biological communities. Providing additional understanding of deltas and multichannel irrigation systems are other examples where models are useful. The selection of a suitable model is challenging, and the use of an ill-suited model can lead to undesirable results. The aim of this chapter is to categorize models, based on their applications, and subsequently to provide information on the process of selecting an appropriate model.

Keywords Water systems • Modelling • Selection • Suitable model • Calibration • Verification • Temporal and spatial scale

4.1 Introduction

A model is "a small object usually built to scale, that represents in detail another, often larger object." In water modelling, the model is not physically built; rather, there are mathematical relations that are applied in order to simulate reality (Chapra

M.H. Bazrkar (✉) • S. Eslamian
Department of Water Engineering, Isfahan University of Technology,
Isfahan 8415683111, Iran
e-mail: bazrkar@gmail.com; pr.eslamian@gmail.com

J.F. Adamowski
Department of Bioresource Engineering, Faculty of Agricultural and Environmental Sciences,
McGill University, 21 111 Lakeshore Road, Ste Anne de Bellevue, QC, Canada, H9X 3V9
e-mail: jan.adamowski@mcgill.ca

© Springer International Publishing Switzerland 2017
J.N. Furze et al. (eds.), *Mathematical Advances Towards Sustainable Environmental Systems*, DOI 10.1007/978-3-319-43901-3_4

1997). Single process modelling has been applied in hydrology and hydraulics since the 1950s (Hashemi and O'Connell 2010a). Predicting peak discharge from rainfall (Shaw 1994; O'Connell 1991; Singh and Woolhiser 2002), and the use of the Sherman unit hydrograph (Sherman 1932), are examples of important attempts by scientists to explain and quantify hydrological phenomena. The Stanford watershed model was the first comprehensive digital model created after the emergence of computers in the late 1950s (Hashemi and O'Connell 2010b). Physical spatial distribution models were the next generation of hydrological models (Freeze and Harlan 1969), an example of which is the Systeme Hydrologique Europeen (SHE) modelling system (Abbott et al. 1986; Bathurst 1986). SHE was developed into two separate models: SHE (Refsgaard and Storm 1995) and SHETRAN (Ewen et al. 2000).

Third generation models were mathematical, developed to simulate watershed hydrological processes, in addition to sediment transport and water quality (Singh and Woolhiser 2002; Fakhri et al. 2014). Geographic Information System (GIS) and remote sensing development provided the opportunity for further application of the abovementioned models, by adding spatial dimensions to the outputs.

To illustrate the growth of modelling development up to 1991, an inventory of more than 60 watershed hydrological models was reported (Dzurik 2003; Singh and Woolhiser 2002). Development of water resources systems modelling and optimization progressed, with numerous simulation and optimization examples, such as reservoir system simulations, hydrological flood forecasting, and water quality models (Biswas 1974). Between the mid-1960s and early 1980s, more than 39 major projects were recorded around the globe which used hybrid or integrated modelling techniques in their assessments, and linked different components of water resources systems (Loucks et al. 1985; Dzurik 2003). Wurbs (1997) listed a number of generalized water resources simulation models in the following categories: watershed, river hydraulics, river and reservoir water quality, reservoir/river system operation, groundwater, water distribution system hydraulics, and demand forecasting.

With the development of new models, and the subsequent increased numbers of models available, selecting an appropriate model became an ever more important issue. Research has been conducted to compare models' abilities and limitations. Kovács (2004) compared the results of SWAT and MONERIS, where it was found that SWAT is weak in estimating phosphorous loads, as it does not account for nonorganic phosphorus attached to sediment. SWAT has also been compared to HSPF (Singh et al. 2005), where it was shown that SWAT is more powerful in the simulation of low flows. The reason posited was a potential underestimation of evapotranspiration, which was confirmed by Saleh and Du (2004).

The aim of this chapter is to classify models based on their applications and the structures on which they are developed, assisting with selecting the desired model in different scenarios.

4.2 Water Systems Modelling for Quantity and Quality

Existing water systems models simulate water quality, quantity, or both. Integration of these discrete and continual aspects of water systems, as well as socioeconomic parameters into one model, helps users analyze a water system holistically. Water quality models differ in various ways, and from the early twentieth century, their evolution has progressed based on societal concerns and available computational abilities.

Water quality models such as the Streeter–Phelps model (1925) focused Velz (1947) on the quantification of dissolved oxygen in streams and estuaries. Consequently O' Connor 1967 provided a model with respiratory and spatial bacterial simulation capabilities. These early models were however limited to linear kinetics, simple geometries, and steady-state receiving waters, due to the absence of advanced computational tools. Following the development of computers in the 1960s, models underwent considerable improvement, particularly in numerical expressions of their analytical frameworks (Thomann 1963). Two-dimensional systems were new improvements during this period, with models being used to simulate activities and processes within watersheds. Operational research was added to models' abilities, in order to generate more cost-effective treatment alternatives (Thomann and Sobel 1964; Deininger 1965; Ravelle et al. 1967). In the 1970s, eutrophication was one of the main water quality problems that attracted attention within nutrient/food chain models (Chen 1970; Chen and Orlob 1975), which employed nonlinear kinetics equations. Subsequent advancements in modelling included the inclusion of environmental issues such as solute transport and the fate of toxicants (Chapra 1997). In the last decade, advancement in computer hardware and software has led to a revolution in modelling; two- or three-dimensional models with highly mechanistic kinetics are now readily available with graphical user interfaces at reasonable costs.

In this section, examples of well-known models are given, with a brief background on development and use. The models discussed are: AGNPS and AnnAGNPS, ANSWERS and ANSWERS-Continuous, CASC2D, MIKESHE, SWAT, DWSM, KINEROS, PRMS, HEC-HMS, HEC-RAS, and WEAP.

4.2.1 AGNPS

AGNPS (Agricultural Non-Point Source pollution model) is an event-based model developed at the USDA-ARS North Central Soil Conservation Research Laboratory in Morris, Minnesota, and designed to analyze the impact of non-point source pollutants from predominantly agricultural watersheds on the environment (Young et al. 1987).

The model components include: transport of sediment, nitrogen (N), phosphorous (P), and chemical oxygen demand (COD), with user interfaces for data input and analysis and other capabilities. Revision of this model was undertaken by the

USDA-ARS National Sedimentation Laboratory (NSL) in Oxford, Mississippi, and led to its upgrading to the Annualized Agricultural Non-Point Source model (AnnAGNPS) (Bingner and Theurer 2001). This upgraded model is practical in continuous simulations of hydrology and soil erosion, as well as transport of sediment, nutrients, and pesticides. The model has source accounting capabilities and user interactive programs, including TOPANGPS generating cells, and stream networks from Digital Elevation Models (Borah and Bera 2003).

4.2.2   ANSWERS

ANSWERS (A real Non-point Source Watershed Environment Response Simulation) was developed at Purdue University in West Lafayette, Indiana. ANSWERS considers various processes of runoff, infiltration, subsurface drainage, and erosion for single-event storms. The model has two major components: hydrology (with the conceptual basis adapted from Huggins and Monke (1966)), and upland erosion responses (with the conceptual basis adapted from Foster and Meyer (1972)). ANSWERS-Continuous is an upgraded version of ANSWERS, developed at the Virginia Polytechnic Institute and State University in Blacksburg, Virginia; the upgrade is a one-dimensional model which uses square grids with similar hydrological characteristics. Although it is not able to simulate sediment transportation, nitrogen and phosphorus transport and transformation is possible. ANSWERS-Continuous was also improved and expanded to include upland nutrient transport and losses. Notable examples of these newer models are GLEAMS and EPIC (Williams et al. 1984; Leonard et al. 1987).

4.2.3   CASC2D

CASC2D (Cascade of planes in two dimensions) is a physically based model with single-event and long-term continuous simulation components, capable of simulating water and sediment in two-dimensional overland grids and one- dimensional channels (Ogden and Julien 2002). Development of CASC2D occurred in two phases, initially at Colorado State University in Fort Collins, Colorado, and then at the University of Connecticut in Storrs, Connecticut.

4.2.4   MIKESHE

MIKESHE is a physically based model, founded on the European Hydrological System (SHE). It was developed by a European consortium of three organizations: the French consulting firm SO-GREAH, the UK Institute of Hydrology, and the

Danish Hydraulic Institute (Abbott et al. 1986). The model performs simulations of water, sediment, and water quality parameters in two-dimensional overland grids, one-dimensional channels, and one-dimensional unsaturated and three-dimensional saturated flow layers. MIKESHE has two components, including the capability for continuous long-term and single-event simulation (Borah and Bera 2003). The model also has the ability to simulate dissolved conservative solute in surface, soil, and groundwater by applying a numerical solution to the advection–dispersion equation for the respective regimes.

4.2.5   DWSM

The DWSM (Dynamic Watershed Simulation Model) was developed by the Illinois State Water Survey (ISWS) in Champaign, Illinois. This event-based model can simulate distributed surface and subsurface storm-water runoff, erosion, sediment, and agrochemical transport in agricultural and rural watersheds during single rainfall events. It simulates nutrients and pesticides, soil and water temperature, dissolved oxygen, carbon dioxide, nitrate, ammonia, organic N, phosphate, organic P, and pesticides in dissolved, adsorbed, and crystallized forms (Borah and Bera 2003).

4.2.6   KINEROS

The KINematic runoff and EROSion model was developed between the 1960s and 1980s at the USDA-ARS in Fort Collins, Colorado. It is a single-event model, with the ability to simulate channel excess, overland flow, surface erosion and sediment transport, channel erosion and sediment transport, flow, sediment, and channel routing (Smith et al. 1995).

4.2.7   HSPF

The Hydrological Simulation Program-Fortran was initially developed by the U.S. Environmental Protection Agency (USEPA) in 1980. A mixture of the Stanford Watershed Model (SWM), the Agricultural Runoff Management (ARM) model, the Non-point Source Runoff (NPS) model, and the Hydrologic Simulation Program (HSP) (including HSP Quality) formed the base of the HSPF (Donigian Jr and Crawford 1979). Various software tools are applied in this model by the U.S. Geological Survey (USGS) for better interaction between its capabilities, from model input and data storage, to input–output analyses, and calibration; because of this, different versions of this model have been released, for instance:

Version 8 in 1984, and Version 10 in 1993 (Bicknell et al. 1993). HSPF is a continuous watershed simulation model, with runoff and water quality constituents on pervious and impervious land areas, movement of water and constituents in stream channels and mixed reservoir components (Borah and Bera 2003).

4.2.8 SWAT

The SWAT (Soil and Water Assessment Tool) was developed at the USDA-ARS Grassland, Soil, and Water Research Laboratory in Temple, Texas. Development was geared towards creating a means of predicting the impact of management on water, sediment, and agricultural chemical yields in watersheds or river basins, and it has the ability to account for parameters such as hydrology, weather, sedimentation, soil temperature, crop growth, nutrients, pesticides, and agricultural management. SWAT is a continuous long-term model, based on a daily time step, but recent improvement allows for the use of rainfall input at any time increment, and channeling routing at an hourly time step (Arnold 2002). Similar to HSPF, SWAT is incorporated into the USEPA's BASINS for non-point source simulations on agricultural lands. SWAT simulates Nitrate-N based on water volume and average concentration, runoff P based on partitioning factors, daily organic N and sediment adsorbed P losses using loading functions, crop N and P use from supply and demand, and pesticides, based on plant leaf-area-index, application efficiency, wash off fraction, organic carbon adsorption coefficient, and exponential decay according to half-lives (Bazrkar and Sarang 2011).

4.2.9 PRMS

The Precipitation- Runoff Modelling System was developed at the USGS in Lakewood, Colorado. PRMS has both long-term and single-storm modes. The "long-term" mode of PRMS is a hydrological model, while the "single-storm event" mode has hydrology and surface runoff, channel flow, channel reservoir flow, soil erosion, and overland sediment transport components. In addition, it is linked to the USGS data management program ANNIE for formatting input data and analyzing simulated results (Borah and Bera 2003).

4.2.10 HEC-HMS

The Hydrologic Modelling System was initially created by the Hydrologic Engineering Center within the U.S. Army Corps of Engineers in 1998 as a single-event model to replace an older standard model for hydrologic simulation, HEC-1. The

Hydrologic Modelling System provides a variety of options for simulating precipitation-runoff processes, and components that cover a wide range of hydrologic features, such as precipitation, potential evapotranspiration, snowmelt, infiltration, surface runoff, base flow, channel routing, and channel seepage (Xuefeng and Alan 2009).

HEC-HMS, in addition to HEC-1's capabilities, provides a number of features, such as continuous simulation and grid cell surface hydrology. In 2005, the U.S. Army Corps of Engineers added two different soil moisture models for continuous modelling: one with five layers and another with a single layer. This model also includes advanced numerical analysis and graphical user interfaces which make it simpler and more efficient than its predecessor.

4.2.11 HEC-RAS

The Hydrologic Engineering Center River Analysis System (HEC-RAS) was developed and released in 1995 by the U.S. Department of Defense, Army Corps of Engineers, with the aim of performing one-dimensional hydraulic calculations for a full network of natural and constructed channels. Components of this model for one-dimensional river analysis can be divided into different categories: steady flow water surface profile computations, unsteady flow simulation; movable boundary sediment transport computations, and water quality analysis (Hicks and Peacock 2005).

4.2.12 WEAP

The Water Evaluation and Planning system was initially developed in 1988 and continued by the U.S. Center of the Stockholm Environment Institute, a nonprofit research institute based at Tufts University in Somerville, Massachusetts. Different components of this model, including hydrology, climate, land use, technology, and socioeconomic factors, offer a wide variety of simulation capabilities. Water demand and supply, runoff, evapotranspiration, infiltration, crop irrigation requirements, instream flow requirements, ecosystem services, groundwater and surface storage, reservoir operations, and pollution generation, treatment, discharge, and instream water quality are all parameters considered in this model (Sieber 2011).

4.3 Time and Space Scale

Scaling refers to an increase or reduction in size and can be defined mathematically as a function as follows:

$$s \longleftrightarrow S; \; e \longleftrightarrow \theta; i \longleftrightarrow I; \quad g\{s; \; e; i\} \longleftrightarrow G\{S; \theta; I\} \tag{4.1}$$

Where g is a small-scale function of state variables s, parameters θ, and inputs I. G is the corresponding large-scale function.

Hydrological models can be divided into two categories: predictive and investigative. Predictive models have been applied to determine the answer to a particular problem, while investigative models developed based on our perception of hydrological processes. Both types however, are similar, in that they involve the following: (1) data collection and analyses; (2) development of a conceptual model; (3) development of the mathematical model using a conceptual model; (4) calibration and validation of the mathematical model. Some of the steps have to be repeated, until validation is satisfactory. Unfortunately, the conditions for prediction are often different in space or time scale from those of the modelling data set. The time scale, in this instance, refers to a characteristic time (or length) of a process, observation, or model and can range from seconds (e.g., flashfloods of several minutes duration) to hundreds of years or more (such as flow in aquifers). Hydrological processes occur across a wide range of space scales; from small (a 1 m soil profile) to larger scales (such as floods in river systems with catchments millions of square kilometers in area). "Scaling," within this context, is defined as a transfer of information across scales by extrapolation, or interpolation. The limitations of measurement techniques and logistics define the observation scale. The observation scale is related to the finite nature of the number of samples and can be defined by the spatial or temporal extent (coverage) of a data set, the spacing (resolution) between samples, or the integration volume (time) of a sample. Based on the nature of the process, space scales can be defined as spatial extent or integral scale. Integral scale in this instance refers to the average distance or time over which a property is correlated.

Ideally, the observation scale should be equal to the process scale. This is seldom possible, due to the cost and other limitations associated with observation instruments. If a process lasts longer than the coverage, it will be used as trends in data; conversely, there is "noise" wherever a process is shorter than the resolution (coverage). The modelling scale is defined based on the process characteristics and applications of the hydrological model. Some typical model scales are categorized in Table 4.1 below.

The gap between scaling and modelling can be bridged using one of three methods: upscaling, downscaling, and regionalization. Upscaling entails transferring data to a large scale and consists of two steps: (1) distributing and (2) aggregating. To illustrate, assume estimation of rainfall in a catchment is carried out using a small number of rain gauges. In step (1), small-scale precipitation is distributed over the catchment as a function of topography. Interpolation is the distribution of information over space and time. Since hydrological measurements are much more coarsely spaced in space than in time, most interpolation refers to the space domain. In the Isohyetal method, optimum interpolation/kriging, spline interpolation, moving polynomials, and inverse distance are all methods of solving

Table 4.1 Time and space scale

Space		Time	
Local scale	1 m	Event scale	1 day
Hill slope (reach) scale	100 m	Seasonal scale	1 year
Catchment scale	10 km	Long-term scale	100 years
Regional scale	1000 km		

this kind of classical problem in hydrology. In step (2), the spatial distribution of rainfall is then aggregated into one single value. In contrast, disaggregating and singling out are the two steps involved in downscaling. Transferring data and information from one catchment to another is defined as regionalization, and can feasibly be carried out, if the catchments are similar. The difficulty in scaling, however, is derived from catchments' heterogeneity and hydrological processes' variability. The term heterogeneity typically refers to variations in space, and is related to media properties, while the term variability here refers to differences in space, time, or both, and is often used for fluxes (runoff).

Hydrological processes may display one or more of the following properties: discontinuity, periodicity, and randomness. Intermittency of rainfall events within discrete zones is referred to as discontinuity, within these zones, properties are relatively uniform and predictable. Periodicity is shown in an annual cycle of runoff and is predictable. Randomness, while not predictable in detail, is predictable in terms of statistical properties such as PDF (Probability Density Function). Statistical calculations can be applied when the property of randomness is observed in a data set (Bloschl and Silvaplan 1995).

4.3.1   Time Scales in Modelling

Models use various ranges of time steps, from seconds to years, based on their applications. Temporal scale is one of the most important factors in modelling, as time step length remains constant throughout the model run. Choosing a time step requires special care; a time step that is too short will require unnecessary computing power, while a time step that is too long will create model instability and simulation failure (Todd 2007). To illustrate, applying a second as a time step in modelling may waste time; on the other hand, using a year as a time step cannot simulate an ephemeral event. Table 4.2 shows the temporal scale for the models discussed in this chapter.

4.3.1.1   Event-Based Models

Some models are developed to simulate a particular event. Event modelling shows the response in a watershed or basin to an individual event; HEC-HMS, AGNPS,

Table 4.2 Models' temporal scale

Model	Temporal scale
AGNPS	Storm event; one step is the storm duration
AnnAGNPS	Long-term: daily or sub-daily steps
ANSWERS	Storm event; variable constant steps depending on numerical stability
ANSWERS–Continuous	Long term; dual time steps: daily for dry days and 30 s for days with precipitation
CASC2D	Long term and storm event; variable steps depending on numerical stability
MIKESHE	Long term and storm event; variable steps depending on numerical stability
DWSM	Storm event; variable constant steps
KINEROS	Storm event; variable constant steps depending on numerical stability
HSPF	Long term; variable constant steps (hourly)
SWAT	Long term; daily steps
PRMS Storm Mode	Storm event; variable constant steps depending on numerical stability
HEC-HMS	Can be specified for several days to cover a single event or span multiple decades to do a period of records
HEC-RAS	24 h

ANSWERS, CASC2D, DWSM, KINEROS, and PRMS Storm Mode are all models within this category. AGNPS runs in-storm duration, while the temporal scale of ANSWERS, CASC2D as well as KINEROS can vary based on numerical stability. DWSM is variable in the number of steps required. In an HEC-HMS simulation covering a single event, the scale can be specified for several days (Xuefeng and Alan 2009).

4.3.1.2 Continuous Models

Continuous modelling synthesizes processes and phenomena over a longer period than event-based models (Xuefeng and Alan 2009). For example, AnnAGNPS runs as a long-term model in daily or sub-daily steps, while ANSWERS-Continuous can simulate in dual time steps: daily for dry days and 30 s for rainy days. HSPF can run hourly as a long-term model, while SWAT, another long-term model, runs with a daily time step. MIKESHE is both a long-term and storm-event model, and its time step is variable depending on numerical stability. HEC-HMS can span multiple decades to do a period of records. The discussed models have been categorized in Table 4.3 based on their types.

Table 4.3 Types of water system models

Model	Type
AGNPS	Event-based model
AnnAGNPS	Continuous model
ANSWERS	Event-based model
ANSWERS–Continuous	Continuous model
CASC2D	Physically based model (single-event and long-term continuous)
MIKESHE	Physically based model (continuous)
DWSM	Event-based model
KINEROS	Event-based model
HSPF	Continuous model
SWAT	Continuous model
PRMS Storm Mode	Event-based model
HEC-HMS	Event-based model and continuous model
HEC-RAS	Continuous model

4.3.2   Space Scale in Modelling

Distributed parameter models subdivide the catchment into a number of units to quantify the hydrological variability that occurs at a range of scales. These units are called either HRUs (hydrological response units), subcatchments, hillslopes, contour-based elements, or square grid elements. The representation of a process within a unit (element) involves local (site) scale descriptions, and some assumptions on the variability within the unit. Distributed parameter hydrological models often represent local phenomena in considerable detail, while the variability within a unit is often neglected. To drive the models for each unit, input variables need to be estimated for each element by some sort of interpolation between observations. Unfortunately, distributed models are limited by the extreme heterogeneity of catchments, which makes accurately defining element to element variations and subgrid variability difficult; with a large number of model parameters, model calibration and evaluation also become difficult (Bloschl and Silvaplan 1995).

4.3.3   Mathematical Bases for the Selected Models

Hydrological processes are widely related to many aspects of atmosphere, water, soil, etc. making them complex to formulate numerically. Fortunately, statistical and mathematical methods can help hydrologists to improve existing models and develop new models. In this section, the hydrological models discussed will be reviewed in terms of the mathematical bases which reflect model performance and application.

Dynamic wave and St. Venant or shallow water wave are the flow-governing equations (continuity and momentum) for the gradual unsteady flow as follows (Singh 1996):

$$\frac{\partial h}{\partial t} + \frac{\partial Q}{\partial x} = 0 \tag{4.2}$$

$$\frac{\partial u}{\partial t} + u\frac{\partial u}{\partial x} + g\frac{\partial h}{\partial x} = g(S_0 - S_f) \tag{4.3}$$

where
h = flow depth (m)
Q = flow per unit width ($m^3 \, s^{-1} \, m^{-1}$)
u = water velocity (ms^{-1})
g = gravitational acceleration ($m \, s^{-2)}$
S_0 = bed slope (mm^{-1})
S_f = energy gradient ($m \, m^{-1}$)
t = time (s)
x = longitudinal distance (m)

CASC2D is the only watershed model that uses the dynamic wave equation on a limited basis, due to its computationally intensive numerical solutions.

Diffusive wave equations consist of the continuity and simplified momentum equations that are used in some models (Singh 1996).

$$\frac{\partial h}{\partial t} + \frac{\partial Q}{\partial x} = q \tag{4.4}$$

$$\frac{\partial h}{\partial x} = (S_0 - S_f) \tag{4.5}$$

where
q is the lateral inflow per unit width and per unit length ($m^3 \, S^{-1} m^{-1} m^{-1}$).

CASC2D and MIKESHE use approximate numerical solutions of these equations for routing surface runoff, overland planes and through channel segments. In order to compute flow, Manning's formula is used as follows (Ogden and Julien 2002):

$$Q = \frac{1}{n} A R^{\frac{2}{3}} S_f^{\frac{1}{2}} \tag{4.6}$$

where

n = Manning's roughness coefficient
A = flow cross-sectional area per unit width ($m^2 \, m^{-1}$)
R = hydraulic radius (m)

$$So = Sr. \tag{4.7}$$

Equations (4.4) and (4.7) illustrate the kinematic wave equations that are well-accepted tools for modelling a variety of hydrological processes (Singh 1996). In the momentum equation, the energy gradient is equal to the bed slope. In order to express this equation as a parametric function of stream hydraulic parameters, a suitable law of flow resistance can be used (Borah 1989).

$$Q = \propto h^m \tag{4.8}$$

where

α is the kinematic wave parameter
m is the kinematic wave exponent
and α and m are related to channel or plane roughness and geometry.

Equations (4.4) and (4.8) are kinematic wave equations that have the advantage of yielding an analytical solution. These equations generate only one system of characteristics. In other words, waves traveling upstream, as is the case with backwater flow, cannot be presented in these equations. Singh (2002) suggested the use of kinematic wave solutions to present accurate results of hydrological significance. An approximate numerical solution of kinematic wave equations is used in KINEROS and PRMS. On the other hand, an analytical and an approximate shock-fitting solution are applied in DWSM.

For flow routing, the simple storage-based (nonlinear reservoir) equations (4.6 and 4.9) are used in ANSWERS, ANSWERS-Continuous, and HSPF (Borah and Bera 2003):

$$\frac{ds}{dt} = I - 0 \tag{4.9}$$

where

s = storage volume of water (m^3)
I = inflow rate $(m^3 s^{-1})$
O = outflow rate $(m^3 s^{-1})$

SWAT, AGNPS, and AnnAGNPS do not route water by means of mass conservation-based continuity equations. In order to compute runoff volumes, these models apply the USDA Soil Conservation Service runoff curve number method (SCS 1972), while other empirical relations similar to the rational formula are used to compute peak flows. The empirical procedure is used in SWAT in order to route water in channels. The SCS runoff curve number method, in addition to an interception-infiltration alternative procedure, is used in DWSM to estimate rainfall excess rates at discrete time intervals, while the interception-infiltration routine is used in the following models: ANSWERS, ANSWERS-Continuous, CASC2D, HSPF, KINEROS, MIKESHE, and PRMS.

$$Q_r = \frac{(P - 0.2S_r)^2}{P + 0.8S_r} \tag{4.10}$$

$$S_r = \frac{25,400}{CN} - 254 \tag{4.11}$$

$$Q_p = 0.0028CiA \tag{4.12}$$

where

Q_r = direct runoff (mm)
P = cumulative rainfall (mm)
S_r = potential difference between rainfall and direct runoff (mm)
CN = curve number representing runoff potential for a soil cover complex
Q_p = peak runoff rate (m^3 s^{-1})
C = the runoff coefficient
I = rainfall intensity (mm h^{-1})
A = watershed area (ha)

4.4 Model Calibration and Verification

A model's performance needs to be evaluated to ascertain: (1) a quantitative estimate of the model's ability to reproduce historic and future watershed behavior, (2) a means for evaluating improvements to the modelling approach through adjustment, which allows for the modelling of parameter values and modelling structural modifications, the inclusion of additional observational information, and representation of important spatial and temporal characteristics of the watershed; and (3) comparison of current modelling efforts with previous study results. Calibration of a model makes it useful and applicable to a specific watershed, and in this section the various methods of calibration will be introduced.

In addition, in order to study the scenarios' effects on a watershed using a calibrated model, the model must be verified. Verification is defined as the "examination of the numerical technique in the computer code to ascertain that it truly represents the conceptual model and there are no inherent numerical problems," (Reckhow 1990), while validation is the comparison of model results with an independent data set (without further adjustment).

The most fundamental approach to assessing model performance in terms of behavior is through visual inspection of the simulated and observed hydrograph. Calibration is defined as model testing with known input and output used to adjust or estimate factors. The key factors influencing model calibration are: Calibration parameters, length of calibration period, and the calibration coefficient (Objective Function). Objective assessment generally requires the use of a mathematical estimate of the error between simulated and observed hydrological variables.

Calibration parameters are selected with the specific characteristics of the model and watershed playing a key role; each series of parameters included in an objective function is important in reducing the problem of nonuniqueness. Sensitivity analysis is a well-known method of choosing calibration parameters. Sensitivity analysis in model calibration is optional, but highly recommended for all parameters in the early stages of calibration, and is conducted by keeping all parameters constant to realistic values, while varying each parameter within an assigned range.

After choosing an objective function, physically meaningful absolute minimum and maximum ranges of parameters are selected. Lack of information, however, may cause the model user to assume a uniform distribution of all parameters within this range. Parameters' ranges have to be as large as possible, due to their constraining role in model calibration.

$$b_j : b_{j.\text{abs\_min}} \leq b \leq_j b_{j.\text{abs\_max}} j = 1, \ldots m \quad (4.13)$$

where

b_j is the jth parameter, and m is the number of parameters to be estimated (Abbaspour 2008).

A decrease or increase in the calibration period affects the calibration results. There are various ways of defining simulation period, calibration period, and verification period (Abbaspour 2008). Defining an objective function is a crucial step in model calibration, and different methods for objective functions have been reviewed (Legates and McCabe 1999; Gupta et al. 1998). Each formulation yields a different result, so the range of final parameters in the model is ultimately based on the objective function. In order to achieve a multi-criteria formulation, various types of objective functions (root mean square error, absolute difference, logarithm of differences, R^2, Chi square, Nash-Sutcliffe, etc.) have been combined. These well-known objective functions in calibration are covered in the subsequent sections of the chapter.

4.4.1 Root Mean Square Error (RMSE)

RMSE is presented in two forms: Multiplicative and summation. The multiplicative form (Green and Stephenson 1986) is written as:

$$g = w_1 \sum_i (Q_m - Q_s)_i^2 + w_2 \sum_i (S_m - S_s)_i^2 + w_3 \sum_i (N_m - N_s)_i^2 + \ldots$$
$$w_i = 1/n_i \sigma_i^2 \quad (4.14)$$

where

σ_i^2 is the variance of the ith measured variable

$$w_1 = 1, w_2 = \overline{Q_m}/\overline{S_m}, w_3 = \overline{Q_m}/N_m$$

4.4.2   Coefficient of Determination R^2

The coefficient of determination R^2 is defined as the squared value of the coefficient of correlation (Pearson 1896). It is calculated as:

$$R^2 = \frac{\left[\sum_i (Q_{m,i} - \overline{Q}_m)(Q_{s,i} - \overline{Q}_s)\right]^2}{\sum_i (Q_{m,i} - \overline{Q}_m)^2 (Q_{s,i} - \overline{Q}_s)^2} \qquad (4.15)$$

The R^2 range lies between 0 and 1. A result nearer 1 indicates better results and a more comparable simulation. R^2 can also be expressed as the squared ratio between the covariance and the multiplied standard deviations of the observed and predicted values. Therefore, it estimates the combined dispersion against the single dispersion of the observed and predicted series. The R^2 range also describes how much of the observed dispersion is explained by the prediction. A value of zero means no correlation at all; whereas a value of 1 means that the dispersion of the prediction is equal to that of the observation. The fact that only the dispersion is quantified is one of the major drawbacks of R^2 if it is considered in isolation. A model which systematically over- or under-predicts consistently will still display good R^2 values close to 1 even if all the predictions are wrong. If R^2 is used for model validation, it is advisable to take into account additional information which can cope with this problem. Such information is provided by the gradient (b) and the intercept (a), of the regression on which R^2 is based. Ideally, the intercept a should be close to zero, which means that an observed runoff of zero would also result in a prediction near zero, and the gradient b should be close to one.

For proper model assessment, the gradient b should always be discussed in tandem with R^2. To do this in a more operational way, the two parameters can be combined to provide a weighted version (wR^2) of R^2. Such a weighting can be performed by (Abbaspour 2008):

$$wr^2 = \begin{cases} |b|.r^2 \text{ for } b \leq 1 \\ |b|^{-1}.r^2 \text{ for } b > 1 \end{cases} \qquad (4.16)$$

By weighting R^2, under- or overpredictions are quantified together with the dynamics, which results in a more comprehensive reflection of model results.

4.4.3   Chi-square

Chi-square is a statistical method of assessing the goodness of fit between a set of observed values and those expected theoretically and is calculated as follows (Mann and Wald 1942):

$$x^2 = \frac{\sum_i (Q_m - Q_s)_i^2}{\sigma_Q^2} \tag{4.17}$$

4.4.4   Nash-Sutcliffe Coefficient

The efficiency E, proposed by Nash and Sutcliffe (1970), is defined as one minus the sum of the absolute squared difference between the predicted and observed values normalized by the variance of the observed values during the period under investigation. It is calculated as:

$$\text{NS} = 1 - \frac{\sum_i (Q_m - Q_s)_i^2}{\sum_i (Q_{m,i} - \overline{Q}_m)_i^2} \tag{4.18}$$

The normalization of the variance of the observation series results in relatively higher values of E in catchments with higher dynamics and lower values of E in catchments with lower dynamics. To obtain comparable values of E in a catchment with lower dynamics, the prediction has to be better than one in a basin with high dynamics. The range of E lies between 1 (perfect fit) and -1. An efficiency lower than zero indicates that the mean value of the observed time series would have been a better predictor than the model. The greatest disadvantage of the Nash-Sutcliffe efficiency is the fact that the differences between the observed and predicted values are calculated as squared values. As a result, larger values in a time series are greatly overestimated, whereas lower values are neglected. For the quantification of runoff predictions, this leads to an overestimation of the model performance during peak flows and an underestimation during low flow conditions.

Similar to R^2, the Nash-Sutcliffe is not very sensitive to systematic model over- or under-prediction, especially during low flow periods. If NS is closer to 1, the results of the simulation have higher validity and less error. NS between 0 and 1 is accepted. Negative NS indicates that mean observed values are better predictors than simulated values and indicates unacceptable model performance (Moriasi et al. 2007). For instance, a negative NS coefficient was observed in a study by Saleh and Du (2004) in estimating daily sediment, and in another study estimating monthly discharge (Sudheer et al. 2007).

4.4.5   Index of Agreement d

The index of agreement d was proposed to overcome the insensitivity of E and R^2 to differences in the observed and predicted means and variances (Willmot 1981).

The index of agreement d represents the ratio of the mean square error to the potential error and is defined as:

$$d = 1 - \frac{\sum_{i=1}^{n}(O_i - P_i)^2}{\sum_{i=1}^{n}\left(|P_i - \overline{O}| + |O_i - \overline{O}|\right)^2} \tag{4.19}$$

The potential error in the denominator represents the largest value that the squared difference of each pair can attain. With the mean square error in the numerator, d is also very sensitive to peak flows, and insensitive to low flow conditions, as is E. The range of d is similar to that of R^2 and lies between 0 (no correlation) and 1 (perfect fit). Practical applications of d show that it has some disadvantages: (1) Relatively high values (more than 0.65) of d may be obtained even for poor model fits, leaving only a narrow range for model calibration; and (2) d is not sensitive to systematic model over- or under-prediction.

4.4.6 Nash-Sutcliffe Efficiency with Logarithmic Values ln E

To reduce the problem of the squared differences and the resulting sensitivity to extreme values, the Nash-Sutcliffe efficiency E is often calculated with logarithmic values of O and P. Through the logarithmic transformation of the runoff values, the peaks are flattened, and the low flows are kept more or less at the same level. As a result, the influence of the low flow values is increased in comparison to the flood peaks, resulting in an increased sensitivity of ln$E$ to systematic model over- or under-prediction (Krause et al. 2005).

4.4.7 Modified Forms of E and d

The logarithmic form of E is widely used to overcome the oversensitivity to extreme values induced by the mean square error in the Nash-Sutcliffe efficiency and the index of agreement and to increase the sensitivity for lower values (Krause et al. 2005). In addition to this modification, a general form of the two equations can be used for the same purpose:

$$E_j = 1 - \frac{\sum_{i=1}^{n}|O_i - P_i|^j}{\sum_{i=1}^{n}|O_i - \overline{O}|^j}\,\text{with}\,j \in N \tag{4.20}$$

$$d_j = 1 - \frac{\sum_{i=1}^{n} |O_i - P_i|^j}{\sum_{i=1}^{n} |P_i - \overline{O}| + |O_i - \overline{O}|^j} \text{ with } j \in N \qquad (4.21)$$

In particular, when $j = 1$, overestimation of flood peaks is reduced significantly, resulting in a better overall evaluation. Based on this result, it can be expected that the modified forms are more sensitive to significant over- or under-prediction than the squared forms. In addition, the modified forms with $j = 1$ always produce lower values than the forms with squared parameters. This behavior can be viewed in two ways: (1) The lower values leave a broader range for model calibration and optimization, or (2) the lower values might be interpreted as a worse model result when compared to the squared forms. A further increase in the value of j results in an increase in the sensitivity to high flows; thus, it is used when only the high flows are of interest, e.g., for flood prediction.

4.4.8   Relative Efficiency Criteria E_{rel} and d_{rel}

All the criteria described above quantify the difference between observation and prediction by absolute values. As a result, an over- or under-prediction of higher values has a greater influence than that of lower values. To counteract this, efficiency measures based on relative deviations (Krause et al. 2005) can be derived from E and d as follows:

$$E_{rel} = 1 - \frac{\sum_{i=1}^{n} \left(\frac{O_i - P_i}{O_i}\right)^2}{\sum_{i=1}^{n} \left(\frac{O_i - \overline{O}}{\overline{O}}\right)^2} \qquad (4.22)$$

$$d_{rel} = 1 - \frac{\sum_{i=1}^{n} \left(\frac{O_i - P_i}{O_i}\right)^2}{\sum_{i=1}^{n} \left(\frac{|P_i - \overline{O}| + |O_i - \overline{O}|}{\overline{O}}\right)^2} \qquad (4.23)$$

Through this modification, the differences between the observed and predicted values are quantified as relative deviations which reduce the influence of the absolute differences significantly during high flows. On the other hand, the influence of the absolute lower differences during low flow periods is enhanced, as it is significant if looked at relatively. As a result, it can be expected that the relative forms are more sensitive to systematic over- or under-prediction, in particular during low flow conditions.

4.4.9   Measures of Efficiency

Krause et al. (2005) investigated nine different efficiency measures for the evaluation of model performance, using three different examples.

In the first example, efficiency values were calculated for a systematically underpredicted runoff hydrograph. The systematic error was not reflected by all of the measures—values between 1.0 (R^2) and 0.81 ($\ln E$) were calculated. Only the weighted form wR^2 and the modified form E1 produced lower values of 0.7 and 0.62, and therefore proved to be more sensitive to the model error in this example. Since most of the criteria investigated are primarily focused on the reproduction of the dynamics compared to the volume of the hydrograph, it is advisable to quantify volume errors with additional measures, like absolute and relative volume measures, or the mean squared error, for a thorough model evaluation.

In the second experiment of Krause et al. (2005), 10,000 random predictions were created by modifying the values of an observed hydrograph to compare the behavior of different efficiency measures against one another. It was found that E and R^2 were not correlated. To improve the sensitivity of R^2, a weighted form wR^2 was proposed, which takes the deviation of the gradient from 1.0 into account. With wR^2, a good and positive correlation with E was found, highlighting the improved applicability of wR^2 over R^2 for model evaluation. In this case, the comparison of the index of agreement d with E, revealed that only the "ideal" values for both measures were found in the same model realizations. In the range of lower values, an increasing amount of scatter occurred. From the comparisons, and the fact that E, R^2, wR^2, and d are based on squared differences, it is fair to say that efficiency measures are primarily focused on the peaks and high flows of the hydrograph, at the expense of improvements to the low flow predictions. The experiment illustrates an important trade-off between accuracy and computational efficiency. For better quantification of the error in fitting low flows, the logarithmic Nash-Sutcliffe efficiency ($\ln E$) was tested. The comparison of $\ln E$ with E and d showed almost no correlation, which is evidence that $\ln E$ is sensitive to other components of model results. With the findings of example 3 in Krause's work, it was shown that $\ln E$ reacts less to peak flows and more to low flows than E. To further increase the sensitivity of efficiency measures to low flow conditions, relative forms of E and d were proposed. The results from the three different examples showed that neither E_{rel} nor d_{rel} was able to reflect the systematic under-prediction of example 1. The comparison in example 2 demonstrated that the correlation of E_{rel} and E was similar to that of $\ln E$ and E. This could be underpinned by the comparison of E_{rel} with $\ln E$, which showed a linear trend, but also a considerable amount of scatter. In example 3, the scatter was explained by the fact that E_{rel} showed virtually no reaction to model enhancement during peak flow, being mostly sensitive to better model realization during low flow conditions. A more suitable sensitivity measure for the quality of the model results during the entire period was found in the two modified forms E1 and $d1$. Both parameters showed linear correlations with not only E and d, but also with $\ln E$. These findings could be underpinned by the evolution of $E1$ and $d1$ during example 3, where they showed average values between the extremes of E and d on one side and $\ln E$, E_{rel}, and d_{rel} on the other.

Overall, it can be stated that none of the efficiency criteria described and tested performed ideally. Each of the criteria has specific pros and cons which have to be taken into account during model calibration and evaluation. The most frequently used Nash-Sutcliffe efficiency and the coefficient of determination are very sensitive to peak flows, at the expense of better performance during low flow conditions. This is also true for the index of agreement, as all three measures are based on squared differences between observation and prediction. Additionally, it was shown that R^2 alone should not be used for model quantification, as it can produce high values for very bad model results—it is based solely on correlation. To counteract this, a weighted form wR^2 was proposed, which integrates the gradient b in the evaluation.

The Nash-Sutcliffe efficiency, calculated with logarithmic values, shows that it is more sensitive to low flows, but still reacts to peak flows. This reaction could be suppressed by the derivation of the relative form E_{rel}. E_{rel} proved to be sensitive solely to low flows, showing no reaction to peak flows. Based on this behavior, E_{rel} could be suitable for calibration of model parameters which are responsible for low flow conditions. The use of E or R^2 for such a task often results in the statement that the parameter under consideration is not sensitive.

As for more global measures, the modified forms $E1$ and $d1$ were identified as a kind of middle ground between the squared and relative forms. One drawback associated with these criteria is that it is more difficult to achieve high values, which makes them less attractive at first glance.

We conclude here that in scientifically sound model calibration and validation, a combination of different efficiency criteria is recommended and should be complemented by assessment of the absolute or relative volume error. The selection of the best efficiency measures should reflect the intended use of the model and should concern model quantities which are deemed relevant for the study at hand. The goal should be to provide good values for a set of measures, even if they are lower than single best realizations, to include the full set of dynamics of the model results (Krause et al. 2005).

4.5   Discussion

Due to the multitude of models available for the simulation of water resources and systems, the selection process requires special attention. Tables 4.4 and 4.5 summarize each model's characteristics, abilities, weaknesses, and limitations.

Table 4.4 Models' abilities and components

Model	Model abilities
AGNPS	Hydrology, soil erosion, transport of sediment, nitrogen, phosphorous, chemical oxygen demand from non-point and point sources, and user interface for data input and analysis of result
AnnAGNPS	Hydrology, transport of sediment, nutrient and pesticides resulting from snowmelt, precipitation and irrigation, source accounting capability, and user-interactive programs including TOPAGNPS generating cells and stream network from DE
ANSWERS	Runoff, infiltration, subsurface drainage, soil erosion, and overland sediment transport
ANSWERS–Continuous	Daily water balance, infiltration, runoff and surface water routing, drainage, river routing, ET, sediment detachment, sediment transport, nitrogen and phosphorous transformation, nutrient losses through uptake, runoff, and sediment
CASC2D	Spatially varying rainfall inputs including radar estimates, rainfall excess and 2-D flow routing on cascading overland grids, continuous soil moisture accounting, diffusive wave or full dynamic channel routing, upland erosion, sediment transport in channels, and is part of U.S. Army Corps of Engineers' Watershed Modelling System with graphical user interface and GIS data processing
MIKESHE	Interception, ET, overland and channel flow, unsaturated zone, saturated zone, snowmelt, exchange between aquifer and rivers, advection and dispersion of solutes, geochemical processes, crop growth and nitrogen processes in the root zone, soil erosion, dual porosity, irrigation, and interface with pre-and user post-processing, GIS, and UNIRAS for graphical presentation
DWSM	Spatially varying rainfall inputs; individual hyetograph for each overland grid, rainfall excess, surface and subsurface overland flow, surface erosion and sediment transport, agrochemical mixing and transport, channel erosion and deposition routing of flow, sediment, and agrochemicals, and flow routing through reservoirs
KINEROS	Distributed rainfall inputs; each catchment element assigned to a rain gauge from a maximum of 20, rainfall excess, overland flow, channel routing, surface erosion and sediment transport, channel erosion and sediment transport, flow and sediment routing through detention structures
HSPF	Runoff and water quality constituent on pervious and impervious land areas, movement of water and constituents in stream channels and mixed reservoirs, and part of the USEPA BASINS modelling system with user interface and ArcView GIS platform
SWAT	Hydrology, weather, sedimentation, soil temperature, crop growth, nutrients, pesticides, agricultural management, channel and reservoir routing, water transfer, and part of the USEPA BASINS modelling system with user interface and ArcView GIS platform
PRMS Storm Mode	Hydrology and surface runoff, channel flow, channel reservoir flow, soil erosion, overland sediment transport, and linkage to USGS data management program ANNIE for formatting input data and analyzing simulated results
HEC-HMS	Watershed hydrology, watershed catchments where rain falls, rivers and streams, reservoir, junction, confluence, diversion; Source: springs and

(continued)

Table 4.4 (continued)

Model	Model abilities
	other model sinks; Sink: outlets and terminal lakes; precipitation, potential evapotranspiration, snowmelt
HEC-RAS	Water flowing through systems of open channels, computing water surface profiles, finds particular commercial application in floodplain management, bridge and culvert design and analysis, and channel modification studies

Table 4.5 Models' weaknesses and limitations

Model	Model weakness
AGNPS	1. Simulating subsurface flow
	2. Predicting time-varying water, sediment, and chemical discharges
AnnAGNPS	1. Simulating intense single-event storms
ANSWERS	1. Simulating flow in reservoirs
ANSWERS–continuous	1. Simulating intense single-event storms
	2. Having channel erosion and sediment transport routines, and therefore the sediment and chemical components are not applicable to watersheds
CASC2D	1. Simulating subsurface flow
MIKESHE	1. Having enough information on flow in reservoirs, overland sediment, channel sediment, reservoir sediment, and BMP evaluation
DWSM	1. Lack of backwater simulation
	2. Uncertainties of input data
	3. Temporally constant values of input parameters
KINEROS	1. Simulation of subsurface flow
	2. Chemical simulation
HSPF	1. Simulating intense single-event storms, especially for large sub-basins and long channels
	2. Represent single-event flood waves
SWAT	1. Overland flow simulation between an upper sub-watershed and lower sub-watershed
	2. Simulating effects on flow and sediment reduction of various BMP
PRMS Storm Mode	1. Subsurface simulation in the storm mode
HEC-HMS	1. Simulating evapotranspiration-infiltration and infiltration-base flow
	2. Backwater possible but only if contained within a reach
HEC-RAS	1. Finding numerical instability problems during unsteady analyses, especially in steep and/or highly dynamic rivers and streams
	2. Working well in environments that require multidimensional modelling

4.6 Selecting a Model for Estimating Nutrient Yield and Transportation During Flash Floods and Wet Seasons

In order to exemplify the selection process between available models, we have selected a problem within the Chamgordalan Reservoir watershed in Iran. Located between the latitudes of 33° 23′ 53″ and 33° 38′ 56″ N and the longitudes of 46° 20′ 25″ to 48° 36′ 58″ E, the Chamgordalan Reservoir watershed has three rivers: the Golgol, Chaviz, and Ama. The watershed has an area of 471.6 km^2, is heavily forested and, as a mountainous watershed, the average land slope is approximately 34 %. Absolute maximum and minimum annual temperatures are 40.6 °C and −13.6 °C, respectively, and the average annual rainfall, recorded at the Ilam synoptic station, is 616 mm.

The watershed's topography is characterized by high mountains, steep slopes and deep valleys, making it highly vulnerable to flooding. Soil is exposed to erosion, and sediment as well as pollutants are transported downstream by common flash floods in the region. Pollutants accumulate in the reservoir, resulting in an increase in unusable volume, and a subsequent reduction in water quality. In September 2008, one of these flash floods occurred, leading to a critical water quality situation. The decision makers in Ilam province had no way of preventing the flow of this water into the urban potable water network, and this led to a crisis in public health and hygiene lasting several days (Bazrkar and Sarang 2011).

A model with the capability of simulating runoff and chemical parameters during the flood event had to be selected. After preparation of data, selection of a suitable model from the pool of available models was undertaken using the flow-chart in Fig. 4.1. In this figure, the abovementioned criteria were considered. The inverted pyramids on the left hand side of the figure were then applied in order to select the model. In this case, DWMS was selected as an event-based model for simulation of nutrient transport during floods.

4.7 Selecting a Model for Estimating Nutrient Yield and Transportation During Regular Flow

Data scarcity at the local scale for flood simulation (lack of observation data in minutes and hours) means event-based models could not be applied in the case of the Chamgordalan Reservoir watershed. Selection must then be carried out using the available continuous models' options. Figure 4.1 illustrates that the SWAT Model was selected in order to simulate nutrient yield and transport during regular flow. The inverted pyramids on the right hand side of Fig. 4.1 are related to the selection process during regular flow.

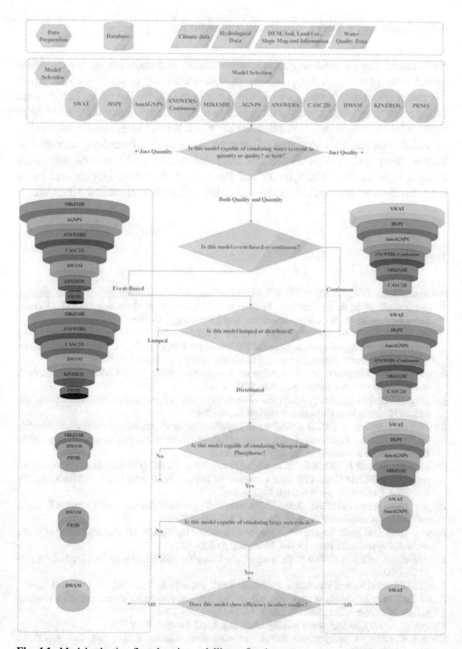

Fig. 4.1 Model selection flowchart in modelling a flood event

4.8 Summary and Concluding Remarks

Each model has its most appropriate applications, and choosing an inappropriate model may result in unexpected errors. Understanding a model's capabilities, advantages, and disadvantages will help the user to achieve the desired objectives. Using models as "clear boxes" with regard to their underlying assumptions leads to more valuable results, while application as "black boxes" increases the possibility of undesirable results. The aim of this chapter was not to rank models, but rather to present their unique capabilities and applications. This chapter has attempted to shed light on the "black boxes" of the most well-known models and helps users to select an appropriate model that is best suited to their unique situation.

References

Abbaspour KC (2008) SWAT-CUP2: SWAT calibration and uncertainty programs- a user manual. Department of Systems Analysis, Integrated Assessment and Modelling (SIAM), Eawag, Swiss Federal Institute of Aquatic Science and Technology, Duebendorf

Abbott MB, Bathurst JC, Cunge JA, O'Connell PE, Rasmussen J (1986) An introduction to the European hydrological system- Systeme Hydrologique European, "SHE", History and philosophy of a physically-based, distributed modeling system. J Hydrol 87:45–59

Arnold JG (2002) Personal communication on 4 September 2002. USDA-ARS Grassland, Soil, and Water Research Laboratory, USA, Temple

Bathurst JC (1986) Physically-based distributed modeling of an upland catchment using the system Hydrologique Europeen. J Hydrol 87:79–102

Bazrkar MH, Sarang A (2011) SWAT model application in simulation of nutrients to identify pollutant sources contribution in Chamgordalan reservoir watershed in wet seasons. MSc Thesis. Tarbiat Modares University, Tehran, Iran

Bicknell BR, Imhoff JC, Kittle Jr JL, Donigian Jr AS, Johanson RC (1993) Hydrologic simulation program- FORTRAN (HSPF): user's manual for release 10. Report No. EPA/600/R-93/174. U.S. EPA Environmental Research Lab, Athens

Bingner RL, Theurer FD (2001) AnnAGNPS technical processes: documentation version 2. www.sedlab.olemiss.edu/AGNPS.html. Accessed 3 October 2002

Biswas AK (1974) Modelling of water resources systems. In: Biswas AK (ed) Modelling of water resources systems. Harvest House, Montreal, pp 520–529

Bloschl G, Silvaplan M (1995) Scale issues in hydrological modeling: a review. Hydrol Process 9:251–290

Borah DK (1989) Runoff simulation model for small watersheds. Trans ASCE 32(3):881–886

Borah DK, Bera M (2003) Watershed-scale hydrologic and non-point-Source pollution models: review of mathematical bases. Trans ASAE 46(6):1553–1566

Chapra SC (1997) Surface water quality modeling. McGraw-Hill, New York

Chen CW (1970) Concepts and utilities of ecological models. J San Eng Div ASCE 96 (SAS):1085–1093

Chen CW, Orlob GT (1975) Ecological simulation for aquatic environments. In: Patton BC (ed) Systems analysis and simulation in ecology, vol 3. Academic, New York

Deininger RA (1965) Water quality management-the planning of economically optimal control systems. In: Proceedings of the first annual meeting of the American Water Resources Association

Donigian Jr AS, Crawford NH (1979) User's manual for the non-point source (NPS) model. Unpublished Report. U.S. EPA Environmental Research Lab, Athens, GA

Dzurik AA (2003) Water resources planning, 3rd edn. Rowman & Littlefield, Lanham

Ewen J, Parkin G, O'Connell PE (2000) SHETRAN: distributed river basin flow and transport modeling system. J Hydrol Eng 5(3):250–258

Fakhri M, Dokohaki H, Eslamian S, Fazeli Farsani I, Farzaneh MR (2014) Flow and sediment transport modeling in rivers. In: Eslamian S (ed) Handbook of engineering hydrology. Francis & Taylor, Boca Raton, pp 233–275

Foster GR, Meyer LD (1972) A Closed-Form Soil Erosion Equation for Upland Areas. In: Shen HW (ed.) Proceeding of Sedimentation Symposium to Honor Prof. H. A. Einstein, Vol. 12, Colorado State University, Fort Collins, 1–19

Freeze RA, Harlan RL (1969) Blueprint for a physically-based, digitally simulated hydrologic response model. J Hydrol 9:237–258

Green IRA, Stephenson D (1986) Criteria for comparison of single event models. Hydrol Sci J 31 (3):395–411

Gupta HV, Sorooshian S, Yapo PO (1998) Toward improved calibration of hydrologic models: Multiple and noncommensurable measures of information. Water Resour. Res. 34(4):751–763

Hashemi M, O'Connell E (2010a) From hydrological models to policy-relevant decision support systems (DSSs): a historical perspective. In: Handbook of research on hydroinformatics: technologies, theories and applications. Newcastle University, Newcastle upon Tyne

Hashemi M, O'Connell E (2010b) Science and water policy interface: an integrated methodological framework for developing decision support systems (DSSs). In: Handbook of research on hydroinformatics: technologies, theories and applications. Newcastle University, Newcastle upon Tyne

Hicks FE, Peacock T (2005) Can Water Resour J 30(2):159–173

Huggins LF, Monke, EJ (1966) The mathematical simulation of the hydrology of small watersheds. Technical Report No. 1. West Lafayette, Ind. Purdue University, Water Resources Research Center

Kovács Á (2004) Modeling non-point phosphorus pollution with various methods, 8th ICDP Kyoto

Krause P, Boyle DP, Base F (2005) Comparison of different efficiency criteria for hydrological model assessment. Adv Geosci 5:89–97

Leonard RA, Knisel WG, Still DA (1987) GLEAMS: groundwater loading effects on agricultural management systems. Trans ASAE 30(5):1403–1428

Loucks DP, Stedinger JR, Shamir U (1985) Modelling water resources systems: issues and experiences. Civ Eng Syst 2:223–231

Mann HB, Wald A (1942) On the choice of the number of intervals in the application of the chi-square test. Ann Math Stat 13:306

Moriasi DN, Arnold JG, VanLiew MW, Bingner RL, Harmel RD, Veith TL (2007) Model evaluation guidelines for systematic quantification of accuracy in watershed simulations. Trans ASABE 50(3):885–900

Nash JE, Sutcliffe JV (1970) River flow forecasting through conceptual models. Part I. A discussion of principles. J Hydrol 10(3):282–290

O'Connell PE (1991) A historic perspective in recent advances in the modelling of hydrologic systems. In: Bowles DS, O'Connell PE (eds) Proceedings of the NATO advanced study institute. Kluwer, Dordrecht, pp 3–30

O'Connor DJ (1967) The temporal and spatial distribution of dissolved oxygen in streams. Water Resour Res 3(1):65–79

Ogden FL, Julien PY (2002) Chapter 4: CASC2D: a two-dimensional, physically based, Hortonian hydrologic model. In: Singh VP, Frevert DK (eds) Mathematical models of small watershed hydrology and applications. Water Resources, Littleton, pp 69–112

Pearson K (1896) Mathematical contributions to the theory of evolution. III. Regression, heredity and panmixia. Philos Trans R Soc Lond Ser A 187:253–318

Ravelle C, Loucks DP, Lynn WR (1967) A management model for water quality control. J Water
    Pollut Control Fed 39(7):1164–1183
Reckhow KH, Clements JT, Dodd RC (1990) Statistical evaluation of mechanistic water quality
    models. J Environ Eng 116(2):250–268
Refsgaard JC, Storm B (1995) MIKE SHE. In: Singh VP (ed) Computer models of watershed
    hydrology. Water Resources, Littleton, pp 809–847
Saleh A, Du B (2004) Evaluation of S~AT and HSPF within BASINS Program for the Upper
    North Bosque River Watershed in Central Texas. Trans ASAE 47(4):1039–1049
Shaw EM (1994) Hydrology in practice. Chapman & Hall, London
Sherman LK (1932) Stream flow from rainfall by the unit-hydrograph method. Eng News Record
    108:501–505
Singh J, Knapp HV, Arnold JG, Demissie M (2005) Hydrological modeling of the iroquois river
    watershed using HSPF and SWAT. J Am Water Res Assoc 41(2):343–360
Sieber J (2011) WEAP history and credits. http://www.weap21.org/. Accessed 20 Jan 2016
Singh VP (1996) Kinematic wave modeling in water resources, surface-water hydrology. Wiley,
    New York
Singh VP (2002) Kinematic wave modeling in hydrology. Invited paper submitted to ASCE-
    EWRI Task Committee on Evolution of Computer Methods in Hydrology. Reston, VA.,
    ASCE, 2002
Singh VP, Woolhiser DA (2002) Mathematical modeling of watershed hydrology. J Hydrol Eng 7
    (4):270–292
Smith RE, Goodrich DC, Woolhiser DA, Unkrich CL (1995) KINEROS-A kinematic runoff and
    erosion model. In: Singh VP (ed) Computer models of watershed hydrology. Water Resource,
    Highlands Ranch, pp 697–732
SCS (USDA Soil Conservation Service) (1972) Hydrology. Section 4. In: National Engineering
    Handbook. USDA Soil Conservation Service, USA, Washington, DC
Streeter HW, Phelps EB (1925)· A study of the pollution and natural purification of the Ohio River,
    III. Factors Concerning the Phenomena of Oxidation and Reareation. U.S. Public Health
    Service, Pub. Health Bulletin No. 146, Reprinted by U.S. DHEW
Sudheer KP, Chaubey I, Garg V, Migliaccio KW (2007) Impact of time scale of the calibration
    objective function on the performance of watershed models. Hydrol Process 21(25):3409–3419
Thomann RV (1963) Mathematical model for dissolved oxygen. J San Eng Div ASCE 89
    (SAS):1–30
Thomann RV, Sobel MJ (1964) Estuarine water quality management and forecasting. J San Eng
    Div ASCE 90(SAS):9–36
Todd AH (2007) An adaptive time-step for increased model efficiency. Weather Service Interna-
    tional, Andover
Velz CJ (1947) Factors influencing self-purification and their relation to pollution abatement.
    Sewage Works J 19(4):629–644
Willmott CJ (1981) On the validation of models. Physical Geography 2:184–194
Williams JR, Jones CA, Dyke PT (1984) A modeling approach to determine the relationship
    between erosion and soil productivity. Trans ASAE 27(1):129–144
Wurbs RA (1997) Computer models for water-resources planning and management. National
    Study of water management during drought. DIANE, New York
Xuefeng C, Alan S (2009) Event and continuous hydrological modeling with HEC-HMS. J Irrig
    Drain Eng 135(1):119–124
Young RA, Onstad CA, Bosch DD, Anderson WP (1987) AGNPS, Agricultural nonpoint source
    pollution model: a large watershed analysis tool. ARS Conservation Research Report No. 35.
    USDA ARS, Washington, DC

Chapter 5
Introduction to Biodiversity

Kelly Swing

Abstract Biodiversity has finally entered conversations among scientists and the populace at large as a common topic. With news about high rates of species loss and increased extinction rates, getting an idea of the wealth of species on our planet takes on an air of urgency and significance. In about 250 years of applying scientific names, we have managed to work through only a fraction of what is believed to be out there. Part of human nature involves collecting and categorizing the things around us. Our priority and primordial purpose for keeping collections and catalogs can vary from simple curiosity to economic interests. Only recently has biodiversity, in and of itself, come to be regarded as a valuable resource. Undoubtedly, every species plays a role in nature and as such, has potential to provide humankind rewards at some level.

Keywords Biodiversity • Extinction rates • Significance • Species catalog • Priority • Collections

5.1 Introduction

Biodiversity, a term shortened from biological or biotic diversity in recent decades, is typically seen rather as a bare bones concept—as simply the number of species that inhabit any given area. Within this concept, scientists seek to catalog absolutely all life forms, extant and extinct. The addition of thousands of species per year certainly bolsters our overall knowledge of life, but according to our best estimates of a planetary total, species lists remain surprisingly incomplete. Our understanding of phylogenetic relationships and geographic distribution has grown exponentially in the last century, but we are still relegated to extrapolate from a relatively limited fraction of existing biota.

Conversations about biodiversity commonly begin with purely biological perspectives but quickly stray to include conservation issues as well as philosophical,

K. Swing (✉)
College of Biological and Environmental Sciences, Tiputini Biodiversity Station,
University of San Francisco de Quito, Quito, Ecuador
e-mail: kswing@usfq.edu.ec

© Springer International Publishing Switzerland 2017   89
J.N. Furze et al. (eds.), *Mathematical Advances Towards Sustainable
Environmental Systems*, DOI 10.1007/978-3-319-43901-3_5

utilitarian, and economic considerations. Treatments of how individual species come into being occupy thousands of pages of published literature, presenting nearly endless explanations ranging from diversity-begets-diversity arguments to allopatric speciation and punctuated equilibrium among dozens of theories. Since the eighteenth century when Linnaeus initiated the formal process of documenting all the "divine creations" and subsequently, in the nineteenth century when Lamarck, Darwin, Wallace, and Haeckel attempted to elucidate how species come into being without the need for divine intervention (Milner 2009), multitudes of scenarios have been proffered (Dunn 2010) to explain that "mystery of mysteries" (Darwin 1859). Science has sought equally to explain how multiple species subsequently fit together in ecosystems and ultimately to comprehend their disappearance. The abundant scientific and philosophical theories put forward succeed, to varying degrees, in explaining these processes at different levels for different individuals and cultures, but considering the state of knowledge in other fields, we are just beginning the work of putting together an inventory of life on Earth. It is highly plausible that we already know far more about the processes involved than we ever will about their multitudinous results.

5.2   Perspectives/Perceptions

Biodiversity is, in essence, the full array of life on Earth. ... biodiversity is more than just the number and diversity of species ... The concept of biodiversity includes both the variety of things and the variability within and among them. Biodiversity also encompasses the processes—both ecological and evolutionary—that allow life on Earth to continue adapting and evolving. (Stein et al. 2000)

Recently, as an opening exercise for a conservation-oriented meeting in Quito, Ecuador, all 20 participants, from very different backgrounds, were given 1 min to write on a piece of paper what they as individuals considered biodiversity to be. Upon revealing their notes, everyone was given a brief opportunity to expound upon any underlying meaning related to what they had jotted down. Responses varied from concise to complex, from surprisingly simple to possibly profound. "Biodiversity is beauty." From the perspective of any nature lover, esthetic value may be of central importance, with each kind of organism being an entity which provokes both wonderment and admiration. "Biodiversity is life" sounded rather basic at first, superficially referring to each and every living being, but with further explanation, suggesting that life begets life, that all life is interconnected, as in the Gaia concept, and that in the swirling mass of protoplasm on the planet, an interdependence is both evident and necessary for success, survival, and continuance. "Biodiversity is a legacy that was handed down to us under the condition that we would pass it along intact to future generations." That thought focuses on a sense of responsibility while ignoring any need for a source or justification of said responsibility. At the end, there was a perspective that perhaps implied a value judgment. "Biodiversity is the expressed compilation of millions of years of answers to all the challenges ever presented; it is the tangible product of all biotic trial and error." This last statement hints that every species is an accumulation of information about survival across the millennia. It also implies that, at any given point in the history of Earth, we are merely observing the current winners in an ongoing process that is always sorting out successful forms and strategies from those that are less successful.

All that information about life, the product of infinite experimentation, is stored in a marvelous chemical warehouse known as the genetic code. Although the last half century has yielded astounding documentation of this biochemical system, we have little understanding of just how a series of repetitive units in DNA and RNA chains makes everything work at the cellular, organismal, or behavioral levels. Although we can identify sections called genes and even transfer pieces that we see as advantageous into other species for our own benefit, no one could, from scratch, put together and install even a few letters of the A-C-G-T alphabet and know precisely what the outcome would be. Of course, we talk about applied gene therapy along with genetic engineering and splicing, but we mostly mean extracting functioning combinations and inserting them into other cells or individuals with the hopes that they might work the same way in their new nuclear or cellular surroundings. Sometimes there are great successes, but the challenges are enormous and failures are frequent.

In recent decades, we have mapped the genome of a selection of organisms and that great achievement invites us to work toward deciphering the connections between molecular composition and coded meanings. We have, however, not yet discovered a Rosetta stone that would allow us to interpret directly individual "words" or "phrases" from this language or to convert them into functioning systems or beings. Our inability to read this language does not in any way diminish the value of the information contained there. However, not recognizing the potential value of that information, and complacently allowing it to disappear through extinction, represents an unthinkable tragedy.

If we try to communicate with a person who only speaks a language we do not know, we can generally bumble around until a basic message gets through, but the limitations are staggering and the process is both tedious and frustrating. In the end, depending upon our motivation and resourcefulness, we can seek out ways to share our thoughts, but we tend not to bother unless our curiosity is driven by the promise of some substantial reward. We have a few alternatives but we tend to simply step away. In such a situation, we hope that we are not missing anything particularly important, but we cannot know. What if, however, we truly want to know what is in the mind of that person? What if we somehow had a chance to interact with an historic figure who we profoundly admire but we speak different languages? What if a shaman, who is the sole repository for handed-down information related to a cure for any of the maladies that humans suffer on a regular basis, only speaks an obscure indigenous language and belongs to an uncontacted culture that has no writing? We have no access to his knowledge base and that information is guaranteed to be lost (Plotkin 1988). Our not knowing undoubtedly represents a loss to many and maybe all of humanity. The same is true for all the untapped genetic information accumulated in millions of species over millions of years. We cannot possibly know its value until we have interpreted it, but it would be utterly foolish to discount its potential (Ehrlich 1988).

5.3   Significance

The overall meaning of biodiversity hints at another level of understanding and has undoubtedly occupied some human minds for as long as we have been conscious of our surroundings. Some, possibly from a basal spiritual perspective, tend to interpret mere existence as an indication of inherent significance. The scientific community tends to believe that every species, no matter its size, has some functional value for the ecosystem it inhabits as well as some potential economic value. On the other hand, numerous nonscientists have a way of summarily dismissing a massive portion of species, the majority actually, by simply discounting the unknown, the abstract to them. If one thinks about this situation a bit, we must recognize that everyone has no choice but to do the same at some level. No one can "know" all those millions of species. Even the greatest taxonomists, depending upon their areas of expertise, commonly know somewhere between hundreds and tens of thousands of species by their scientific names, but the ones who are naturalists at heart are familiar with many more and can broadly categorize just about anything upon first sight. This capacity gives them an incalculable edge; they manage to acquire that most difficult of all qualities in this realm, perspective. In contrast, the lack of perspective in everyone else turns out to be a threat to all the unknown. If one categorizes each species as "worthwhile" or not, including only those with some rather obvious "significance" in a practical or economic sense, an underlying interpretation is that many other living "things" are of little to no consequence or concern because of their small size or lack of proven value (Ehrenfeld 1988; Kareiva et al. 2011) or inherent interest. Hence, the frequently asked question in relation to species-specific conservation efforts—"What good is it?" (Thiele 2011) This attitude is probably at the root of why we have not made any effort at national or planetary levels to complete the catalog. As long as potential value is not converted into known, confirmed hard cash value, it remains abstract and is therefore not considered real or worth exploring. As they say, "the bottom line is the bottom line." The argument feeds upon itself. As long as we do not know enough, we cannot say anything about specific value and as long as we know nothing of value, we do not routinely justify making efforts to learn more. Thereby, as our behavior has confirmed repeatedly, biodiversity that matters to humans at large is the only biodiversity that matters. Smith (2015) points out that we should protect the environment so as to keep our options open into the future. The human race has also demonstrated on many occasions that having economic value does not guarantee sustainable management or even survival of some species.

We only know what we see;
  we only love what we know;
  we only take care of what we love.

This Native North American proverb is applicable at just about every possible level, whether in reference to a piece of furniture, a national symbol, a family member, or a species on the brink of extinction.

Another facet of the biodiversity conundrum is the fact that human brains are not especially good at dealing with truly large numbers. Once we get past a few thousand, our grasp becomes ever more tenuous, our perception becomes ever more vague. Our ability to comprehend huge numbers—millions, billions, trillions—in an absolute way is probably simply beyond most of us. This is true whether we are considering the classic grains of sand on a beach or the stars in the heavens. Relative dimensions are easier to absorb so we try to gain a foothold by making comparative scales in our minds. Knowing that there are thousands of times more beetle species than birds does not help a great deal because most people are not familiar with more than a few dozen kinds of birds. Whenever I ask college biology classes how many species of birds exist, the guesses range wildly from hundreds to a few thousand, but the estimates are commonly far fewer than the 10,000 or so that have been documented scientifically. The knowledge gap gets much worse if we move into the realm of invertebrates. Let's think about mollusks, for example. If we ask about this group of relatively familiar organisms, most students will start their calculations by considering how many kinds they have seen personally (including garden slugs and seashells), adding the kinds they have eaten (clams, mussels, oysters, scallops, calamari, octopus, escargot perhaps), plus the ones they remember from interesting nature documentaries (cuttlefish, giant squid, giant clams, nautilus) and eventually arrive at a sum of dozens and then, by extrapolating for the world at large and various unknown factors, toss out numbers between hundreds and thousands. Of course, the real number is closer to 100,000 but because we have such limited experience and such a weak knowledge base in general, our capacity to approach such a number without knowing it by rote memory is almost nil. Being able to make a list helps immensely, but simply hearing big numbers does not do the trick for most of us. Stacking abstractness on top of abstractness does not resolve the problem at all.

Being told that there are nearly 30,000 kinds of fishes in the world does not automatically give each of them any real status. Only when we are able to apply names and images to them does the picture start to become clear. With these vertebrates, we can go through the same process my students used for mollusks, but most people cannot get past the few kinds they personally have consumed (tuna, trout, salmon, flounder, mahi-mahi, sole, cod, red mullet, bass, catfish) or seen in the local pet shop (goldfish, tetras, plecos, angelfish) plus those depicted in exciting documentaries on shark attacks. Dozens versus tens of thousands represents a knowledge gap that may be insurmountable at the level of the masses. This, in turn, probably means that the average person may well not care whatsoever that there could be millions of species of beetles (Farrell 1998) in the world's rainforests or that 80 % of them still have not been cataloged. If we know nothing of them, the majority remain rather abstract and completely unappreciated (Ehrenfeld 1988).

> This must be the wood where things have no names.—Lewis Carroll, *Alice's Adventures in Wonderland*, 1865

5.4   Challenges to Documentation

Understanding the diversity of life on Earth remains a tremendous challenge due primarily to the immensity of the task of recognizing precisely how many kinds of beings exist. Superficially, cataloging the species that share the planet with us sounds like it should be a relatively simple and straightforward chore. After all, how hard could it be to go out into nature, make the necessary observations, collect some images or specimens of all we encounter and put the results in a museum, an encyclopedia or on a website (WWF 2014; EOL website 2008)?

5.4.1   Mentality and Motivation in Relation to Cost Versus Benefit

All indications are that, on the whole, we basically do not care about finding out what's out there. A gargantuan, jaw-stretching yawn would be the most predictable public response to the mention of a world taxonomy campaign. If we suggest that taxpayers should cover the cost of applying scientific names to everything, the response would probably be resoundingly negative and might even provoke protests. The absolute number of species in existence is about as relevant to the common man as the number of bolts in the Eiffel Tower is to a typical holiday traveler in Bangkok. Such a detail might be interesting to a select group of civil or structural engineers, but the average person who has saved up for a visit to this destination would not pay much for this disjunct tidbit of information associated with a monument in Paris, no matter how famous. Neither does an average citizen see any rationale that could justify putting large sums of money into making such a list.

Logically, the game changer would be to show people at large that such expenditures should be viewed as investments. It is indeed a lottery but there can be real rewards. Unfortunately, few individuals have heard much about the limited number of true success stories. Science gets little publicity outside the realm of medicinal applications, so getting the message out continues to present an important hurdle. Here, scientists can only point a finger at themselves for not making more efforts to share the results of our research in an accessible form.

Pharmaceuticals represent one realm of possibilities that could change public opinion. Every estimate available suggests that well over half of all clinically available drugs are derived, directly or indirectly, from wild plants (Farnsworth

1988). Some of the related information came to us through our grandparents and some has come through indigenous shamans (Plotkin 1988) and the rest has come through direct investigation or pure dumb luck. In any case, less than 5 % of the world's flora has been scientifically screened for biochemical activity or potential medical applications. Chemical defense compounds, which seem to work in countless ways, evolved over millions of years within the tissues of many organisms, help them combat all sorts of challenges. Based on observations of what is referred to as "self-medication" in nonhuman primates, it can be inferred that certain terrestrial plants have been used "intentionally" since before humans existed. Many marine invertebrates and algae are particularly promising sources as well (Graneli and Pavia 2006). Considering that innumerable molecular innovations in nature have served their producers well for millennia, it is only reasonable that some portion could also be put to good use in the human organism. Instead of having to work from zero to develop molecules that can treat or cure specific ailments and diseases, it would be, in all probability, more efficacious for laboratory technicians to take hints from naturally occurring substances that have already been evaluated in nature over huge expanses of time (Plotkin 1988).

5.4.2 Dimension/Scale

When humans go out into nature anywhere, we readily notice certain species and tend to ignore others altogether (Balmer 2013). This is mostly a matter of scale, based primarily on the fact that we are ourselves rather large as compared to most life forms. Right around the globe, we have done a reasonable job documenting the existence of organisms that exceed 10 cm or 10 g. Anything less than that has proven to be a much greater challenge (Vaijalainen et al. 2012). Through technological advances, mostly related to improved imaging and genetic/molecular techniques, science has managed to overcome this perspective bias to some extent, but the fact that the majority of organisms fall into this minute size category continues to present a massive stumbling block. Being small certainly pushes many organisms into the state of being "out of sight, out of mind" for lay people and keeps taxa mostly made up of diminutive forms (Lucky et al. 2002) from ever being noticed by a new generation of potential systematists or conservationists. Ecotourists and even young biologists typically focus their search on the large species associated with any particular ecosystem and quickly become bored when they are unsuccessful at observing animals like jaguars or tapirs in the Neotropical rainforest, all the while entirely overlooking the unending parade of varied arthropods that appear on nearly every leaf. We basically could not care less about small organisms and ignore the fact that this is precisely where the real diversity of the planet lies. Two hundred years ago, Lamarck astutely pointed out that "we should chiefly devote our attention to the invertebrate animals, because of their enormous multiplicity in nature" (Milner 2009). Nonetheless, fervently searching the deep blue sea in order to apply names to dozens of species of tiny crustaceans or the jungles of darkest Peru to catalog even thousands of beetle species 2–3 mm in length does not sound exciting to the masses so funding will continue to be limiting. In contrast, one might expect that the addition of vertebrates (Aldred 2013; Butler 2013) to the list would attract the world's attention for a moment, but the recent discovery of several new species of giant fish (Stewart 2013a; Stewart 2013b), a frog in New York City (Feinberg et al. 2014), more than a dozen birds (Hance 2013), a tapir (Cozzuol et al. 2013), a bat (Dias et al. 2013), a shrew-opossum (Ojala-Barbour et al. 2013), a raccoon, the "olinguito" (Helgen et al. 2013), an ocelot (Trigo et al. 2013), a porcupine (Mendes Pontes et al. 2013), even dolphins (Morell 2014; WDC 2014) previously unknown barely made headlines beyond a few specialized blog sites.

5.4.2.1 Accessibility

Precisely where biota is distributed is also part of the challenge (ter Steege et al. 2013). Generally speaking, the first few hundred thousand or so species to be described were those easiest to find and recognize as being distinct. A half century ago, in North America we still celebrated the discovery of a new species because we considered these special events that only resulted from intense, pain-staking searches that deserved recognition. Now, we know that undescribed species abound in specific ecosystems among certain taxa and understand that we best move ahead with the task at hand and need not be distracted by celebrating each one. Nonetheless, in a way that is very parallel to oil or gold extraction, it looks like the phase of easy finds has passed. Those remaining valuable deposits are typically more remote or difficult to access, but that does not stop us from going to the ends of the Earth to get at them. To continue bringing more material into the hands of taxonomists sorting through museum collections, we will need to apply more money and physical effort to explore the places (Natural History Museum website 2012) that 99 % of the human population has no intention of ever visiting simply because it is too difficult (ter Steege et al. 2013). We cannot, however, discount the fact that even some species of vertebrates have been overlooked right under our noses in places like New York City (Feinberg et al. 2014). What does that imply for remote areas? For some context, in eastern Ecuador, in the very recent past, trees over 40 m in height have been described (Pérez et al. 2014; Neill et al. 2003, 2011). How could an organism of such stature have escaped our discerning loupe for so long?

5.4.3 Interest/Incentive

Just as we keep pushing to find more oil and gold, the same should be true of the species of our planet, but the incentives are admittedly quite different. As each discovery inevitably grows in difficulty, the costs of collecting specimens from remote sites will become increasingly prohibitive due to political, geographical or logistical challenges, safety and security risks, the need for specialized equipment and methods, etc. In turn, our willingness to invest public funding in finding them will undoubtedly wane, as a result of diminishing returns. And on top of that, the more we know about the biota around us, the more we will be documenting species in decline, and the more we would be obligated by laws to manage and defend them, implying greater and greater expenditures over time. Ignorance is indeed bliss.

No one becomes famous for finding a new species of fruit fly and most discoveries will never produce money for those who managed to apply genus and species names to them. Now and again, however, something almost magical happens to change the scenario altogether; another kind of impetus enters the story. The 1993 Hollywood blockbuster, *Jurassic Park*, made a huge difference in the level of public interest in dinosaurs and paleontology. As the movie's teenage viewers reached college age, they swelled the rosters for related university courses and then set off energetically down the path of joining research efforts. More hands took to digging for fossils and more minds were put to the task of deciphering their meaning and relationships. The number of discoveries of all kinds of fossils, not just dinosaurs, took a leap forward starting in the mid-1990s. Ultimately, our understanding of the geological past of many taxa was improved astronomically by the input of enthusiasm and excitement into a new generation because of the creative use of computer-generated images in a popular movie. Fantastic beasts were brought to life on the silver screen and in the passions of new recruits to the ranks. Scientists should probably not be expecting such miraculous events to provide the impetus for a wave of science nerds to fill the back halls of museum collections in order to catalog the remaining biodiversity.

Nonetheless, one intended purpose of nature documentaries is expressly to provoke a response parallel to that dinosaur movie. The idea, especially related to conservation is "the more you know, the more you care." Producers always hope their particular film is going to make a difference in the world, by including footage meant to capture the hearts and minds, and wallets, of all its viewers. In turn, those viewers are expected to insist that all their friends see this wonderful exposition of nature at its glorious best. *Animal Planet* and *National Geographic* have their established fan bases just like other outlets and tend to reach the same audience year after year so the likelihood of capturing a new segment of the population is seemingly minimal no matter the quality or rarity of the content. Unfortunately, many people who watch these videos are apparently lulled into the belief that as long as someone can still go out into the wild and bring such images to the television screen, the organisms at the center of the stories must be doing just fine. The moment is ripe for such shows to include some indication of how much time and effort are required to capture the material that inspires awe, so that a typical viewer might appreciate what has been lost already as opposed to celebrating the fact that one of those things can still be found somewhere on the planet even in these modern times.

5.4.4   Level of Expertise and Distribution

Novotny (2009) pointed out that the majority of species live in the tropics, but most biologists do not. Complications abound in getting specimens into the hands of the experts. In many countries, fully formed specialists are truly rare creatures and they are already overwhelmed by a backlog of specimens sitting in jars on museum shelves waiting to be studied and allocated to their corresponding families and genera. No country, university, or institution of any kind has taxonomists on staff who can deal with all the diversity of their region. When an article is published on specimens that have been in collections for half a century, governmental accountants and politicians tend to ask why the scientists have been so inefficient, sitting on this for so long, as opposed to recognizing that more funding and personnel are necessary to make the wheels of the machine turn efficiently.

So far, all we can do is estimate how many species there are, while recognizing that no one knows for sure. While many parks and countries have published lists and field guides to aid in the identification of some groups of large organisms, the invertebrates remain sorely underrepresented.

As it turns out, there are far more species than we had thought just a few decades ago (Alonso et al. 2011). We have gone from a situation in which the majority of people believed that most life forms had already been catalogued to one in which no knowledgeable person sincerely believes we will ever complete this task (Raven and Wilson 1992; Wheeler et al. 2012; Costello et al. 2013; Phys.org website 2013). This may sound rather defeatist and daunting, but at the macroscopic level, we are convinced that there are several million species left to discover and describe, maybe five times what we have accomplished so far in two-and-a-half centuries. For microscopic organisms, we are sure that we have only just begun. On the other hand, we are quite convinced that we have already done a reasonable job of classifying the large organisms—or at least the land dwelling things that are larger than several centimeters in size, diurnal or nocturnal. The reliability of our lists tends to be directly related to the size of the organisms. Therefore, we have no choice but to depend on this known portion of the flora and fauna to give us hints about what is left to discover among the more diminutive realms of life. Until we advance at all levels, this strategy seems rational and simply must suffice; although there is some controversy, science has produced evidence that the species diversity of any taxon tends to serve well as an indication that other taxa in the same places are similarly diverse. So by specifically looking at what we know about the big stuff—trees and vertebrates—we can extrapolate to make informed guesses about the other smaller, more hidden species. For these generally well-known groups, relatively reliable information is available from most regions of the world. Indeed, many countries have produced lists and even published illustrated guidebooks to their birds, mammals, reptiles, and amphibians; additions come out in the form of scientific publications and the process can be tracked.

5.4.5   *Estimates Down Through History*

In the early years, investigations primarily depended upon that which could be seen with the naked eye. Naturalists went out into the landscape and noted the various organisms that could be observed within some practical distance of where they lived and worked. So far, science has been formally working on this project fairly constantly for over 250 years. In the eighteenth century when Swedish biologist Carolus Linnaeus officially started this endeavor of documenting all the biota, he saw his work of applying names to all of life as a religious mission, a mandate originally assigned to Adam (Genesis 2:20) but never fully realized. By recognizing each and every one of the magnificent beings in existence, the idea was to give credit, and thereby praise, to their creator. From the perspective of a scientist who grew up at a high latitude in the temperate zone, the number of organisms to be listed probably seemed challenging but rather manageable. We can imagine that the related work would be expected to represent a gratifying pastime but not something that might occupy the rest of his life. When Linnaeus died in 1778, he and his "apostles" had applied binomials to about 10,000 species. The European scientific community was duly impressed by this tremendous opus and was satisfied that a goodly portion of the project had been completed. They could not have guessed that Linnaeus' life work would not represent even 1 % of the global total. Half a century later, in 1833, Westwood suggested that there could be as many as 600,000 species of insects alone and reported that his botanical colleagues had arrived at an estimated total of between 110,000 and 120,000 species of plants. Westwood also remarked that John Ray had proposed in 1691 that the number of insects could approach 20,000.

Since WWII, our perceptions of the biota have continued to be modified drastically. In the middle of the twentieth century (Sabrosky 1952), with a list of around half a million species, most people believed that the majority of life forms had already been recognized. The rate of discoveries was diminishing because we continued to look in the same places for new species that were as large as vertebrates or vascular plants. But then, in the 1970s and 1980s, science started to look in different regions more carefully using different techniques (Erwin 1982). Upon turning our focus toward the tropics and the previously unexplored parts of the sea, we began to notice that many specimens being brought back from far-flung areas were different and still required recognition. The rate of accumulation lunged ahead with unprecedented fervor. The astounding upward trend in the growth curve led some scientists to project that there might be 30 million, or 50 million, or 100 million species living in our time (Wilson 1988). By the beginning of the new millennium, we were approaching the 1.5 million mark for named species. The rate has continued to race along with an estimated 18,000 additions per year (Wheeler et al. 2012). In an effort to get a handle on a potential total, in light of all this progress, new calculations have been made (Alonso et al. 2011). Of course, our estimations should be based upon the portion of the biota that is best known (Nielson and Mound 2000) so as to give us a reliable answer. We presume to have a reasonable perception of macroscopic organisms but are still flummoxed by the microbial world. Some advances are certainly occurring there, but estimates of bacterial and viral diversity are mostly seen as guesses. For the things visible to the naked eye and just beyond, a total of somewhere between five and nine million seems to be the most widely acceptable estimates available at present (Hilton-Taylor and Stuart 2009; Mora et al. 2011). The best we can do currently is to, by applying species accumulation curves, extrapolate out to the potential number of species and to estimate where we stand along a scale of progression toward our ultimate goal (Stuart et al. 2010).

5.5   Extinction Rate

Human history is basically a story of one species dominating the planet by using its wit to overcome most of the challenges nature has thrown at it. Our ingenuity has allowed us to occupy almost every niche and habitat (Kolbert 2014) because we, among all species, have uniquely seen opportunity everywhere and uniformly ignored adversity. We are able to make use of an inordinately wide array of resources but tend to be unaware or wasteful of the less than obvious. This incessantly repeated behavior is responsible for our overrunning every part of the globe that we enter.

We never set out to destroy the landscape or its nonhuman occupants, but we do set out to change things for our own benefit, almost always thinking exclusively about short-term payoffs without considering in depth the long-term drawbacks. At the global scale and at the local scale, we start off small and then grow in numbers and force, and geographical distribution. As a consequence, nature cedes at every turn; it is tamed and diminished in every dimension.

"One weedy species has unwittingly achieved the ability to directly affect its own fate and that of most of the other species on this planet." (Wake and Vredenberg 2008)

If we were losing money from our bank accounts as quickly as we're losing species from our forests and oceans, we would certainly take notice and we would undoubtedly do something about it. For most people, if dollars or euros or yen disappear from an account, we immediately look for explanations and remedies. At the same time, we remain completely complacent about the loss of valuable natural patrimony every day, and in most cases, are never even aware that this is happening. Even the lowliest of species must be worth at least a few dollars, but essentially all the large animals, terrestrial and aquatic, have experienced population decreases of 75–90 % (Safina 1997; Myers and Worm 2003; National Geographic 2007; Swing 2013; Ripple et al. 2014; FAO website) just in the last half century, and most humans were never aware of what was going on or just how fast. If one day you wish to see mountain gorillas foraging in the mist, an orangutan brachiating through the forest canopy, or a black rhino ambling on the savannah, you better go now because there is no guarantee whatsoever that any of them will exist until the middle of this century.

We have always needed nature and now, what remains of nature needs us. Basic human behavior has resulted in our nibbling away, like locusts, everything in our surroundings. Once we start down the path of exploitation, a gold-rush mentality takes over and we have never been able to stop. Current extinction rates are estimated at 100–10,000 times background rates measured during intervals between previous extinction peaks; approximately 27,000 species are disappearing annually (Myers 1988; Stevens 1995) and the rate is likely increasing with each passing year (Wheeler et al. 2012; Natural History Museum 2012). Will humankind be able to change? Will we be able to draw a line in the sand (Fowlie 2014) in relation to our exploitation of nature and natural resources? What might drive us to make such a decisive change?

As we have transformed the Earth into our rather exclusive habitat, we have left little of the original version. The result is that we will be relegated to fewer resources in the future. No one seems to have a clear idea of how much the world ecosystem can withstand and we apparently have no intention of using our vision to change our trajectory as long as we are benefited in the immediate sense.

Can we choose what will remain? Shouldn't we be proactive as opposed to accepting by random chance what tatters are left over? Each form of life represents an enormous collection of information about how to survive on this planet. Each undescribed species is a book without a title that has never been opened. With every extinction event, we're discarding a volume that could hold the key to our own survival in the context of climate change.

Most other species on the planet have existed much longer than humans. That means that they have endured events that our own lineage has not. We can infer that, in their genetic makeup, they likely have answers to many of the challenges that lie ahead for our species in relation to current impending global changes. As technology gives us increasing abilities to interpret the molecular/genetic language of form and function, it may become possible to tap into those reference sources at some moment in the future, if, and only if, we have not already naively eliminated the living repositories represented by the diversity of biota.

5.5.1 Role of Scientific Collections

Despite widespread recent perceptions, the collection of research specimens remains vital to the advance of modern science (Phillips 1974; Foster 1982; Herholdt 1990; Remsen 1995; Lundmark 2004; Rocha et al. 2014; Swing et al. 2014). If a shopkeeper does not have an inventory record, it is impossible to manage his merchandize; if we do not know what is living in the world's forests and seas, we cannot even hope to manage the biota for its own well-being or in relation to its potential for our use.

Around the world, increasing popular awareness of extinctions has resulted in policies and regulations that run counter to this necessity. Within the scientific community as well as among the lay citizens of the planet, confusion abounds regarding the causes of impacts on wild populations and the loss of species. In the vast majority of cases throughout human history, rational collection of specimens for scientific study has neither been connected with extirpations nor extinctions. Habitat loss, global climate change, emergent pathogens, environmental pollution, and targeted harvest (hunting and fishing) have provoked much more important impacts and have already transformed entire ecosystems. These are the factors that have led primarily to the sixth mass extinction.

Undoubtedly, informed regulations and oversight are necessary for efficient and functional management. Regulation cannot depend upon heartfelt sentiments or popular opinions; instead, controls must be based on ecological principles and scientific reality. While the sacrifice of individuals of certain species for any reason would be totally unethical and should be punished, the lack of broad collections drastically affects our capacity to catalog the biodiversity of the planet. If we do not know the species in our surroundings, we cannot recognize potential resources among them, and we can certainly not take advantage of all our opportunities in a reasonable way. As in any arena, the lack of information represents a loss of opportunity (Sukhdev 2010). Without a complete list of species and a basic understanding of their roles in nature, it is impossible to comprehend the overall functioning of ecosystems and in the end, effective management of "renewable" resources in light of increasing demands becomes more difficult and less likely every day. Without access to relatively complete collections (accompanied by a plethora of data and meta-data including genetic material), it is impossible to recognize and analyze phylogenetic relationships and to properly position known species into existing classification schemes. In Ecuador, arguably the country with the greatest concentration of species on the globe (Bass et al. 2010), only about 75,000 species of macroscopic organisms have been catalogued (Peter Raven, pers. comm.) This total is estimated to represent less than 10 % of the country's existing biodiversity. In megadiverse countries like Ecuador, this problem is of greater relative importance because the lack of scientific collections and complete biotic inventories imposes serious limits on possibilities for conservation, generation of biological knowledge, and ultimately diminishes potential for development and access to an improved standard of living (Sukhdev 2010).

The proportion of known versus unknown species, according to best estimates, is similar for the entire planet. Worldwide, we have about 1.5 million kinds cataloged and an estimate of nine million or so in total for macroscopic life (Mora et al. 2011). The situation may look far worse in developing tropical countries because the total numbers of species within their borders are so much greater and rates of loss are likely faster due to rapid development in highly biodiverse settings. For politicians and developers, a not-too-obvious corollary of having the greatest concentrations of species per hectare implies the greatest loss of species per impacted area.

We should take a moment to return to the idea of how we apply the terms "known" and "unknown" to various organisms and must admit that using such language leaves a lot to be desired. Just because we have curated a specimen in a museum and have applied a scientific name to it, that does not mean we know anything about it beyond what it looks like. We actually have a very superficial knowledge base for the vast majority of species that are named. The exceptions lie primarily in ourselves, our mono-cultured crops and domesticated animals, but include a few dozen other species that have turned out to be especially apt for laboratory life and prodding.

5.6   Conclusion

There are more things in Heaven and Earth, Horatio, than are dreamt of in your philosophy.—Shakespeare, *Hamlet*

While good intentions abound, cataloging our fellow inhabitants of Earth has turned out to be exacerbated by a long list of complications. First of all, there are far more kinds than we ever expected. Unfortunately, expertise remains broadly lacking. Because we have only classified a small portion of species so far, our confidence in estimating the possible total is quite weak. Over the years, this has led to the publication of figures ranging from a few million to over one hundred million if we include absolutely every form of life no matter its size. Also, it costs something to find and categorize species so funding is ever a concern. There is always some reticence in moving forward in the face of demands for justification of any endeavor. Out of ignorance, we cannot typically assign any value to organisms before we make the effort to find, dissect, observe, and analyze them. Until we write about them, we cannot share information with other potential collaborators so as to advance our overall knowledge base. Classically, by the way, science has not often been in the business of carrying out financial appraisals. As it turns out, relatively few taxonomists exist and they do not have access to all the species or specimens necessary to complete the catalog—not even close. Discovery is challenging and description is tedious. The rate of additions is inherently slow due to requirements for detailed evaluations in light of already known species. The rate of extirpations and extinctions puts us in a perverse race to label organisms before they disappear altogether.

We must recognize that the diverse biota around us is the product of natural processes across immense expanses of time and that we have many vested interests in its well-being. We still have the opportunity to make decisions that will leave an indelible mark on the planet. In relation to climate and extinctions, we can choose to be remembered as villains or as heroes for all time.

References

Aldred J (2013) Purring monkey and vegetarian piranha among 400 new Amazon species. http://www.theguardian.com/environment/2013/oct/23/purring-monkey-vegetarian-piranha-amazon-species. Accessed 25 May 2015

Alonso LE, Deichmann JL, McKenna SA, Naskrecki P, Richards SJ (eds) (2011) Still counting… biodiversity exploration for conservation; the first 20 years of the rapid assessment program. Conservation International, Arlington

Balmer AJ (2013) Why we don't need pandas., Scientific American, http://news.yahoo.com/why-don-t-pandas-113100956.html. Accessed 9 August 2015

Bass MS, Finer M, Jenkins CN, Kreft H, Cisneros-Heredia DF, McCracken SF, Pitman NCA, English PH, Swing CK, Villa G, Di Fiore A, Voigt CC, Kunz TH (2010) Global Conservation Significance of Ecuador's Yasuní National Park. PLoS One 5(1), e8767. doi:10.1371/journal.pone.0008767, Accessed 18 November 2014

Butler RA (2013) Biggest new animal discoveries of 2013. http://news.mongabay.com/2013/1223-top-new-species-2013.html?fbfnpg. Accessed 19 November 2014

Costello MJ, May RM, Stork NE (2013) Can we name Earth's species before they go extinct? Science 339:413–416

Cozzuol MA, Clozato CL, Holanda EC, Rodriques FHG, Nienow S, De Thoisy B, Redonod RAF, Santos FR (2013) A new species of tapir from the Amazon. J Mammal 96(4):1331–1345

Darwin CR (1859) On the origin of species by means of natural selection or the preservation of favoured races in the struggle for life. Murray, London

Dias D, Esbérard CEL, Moratelli R (2013) A new species of Lonchophylla (Chiroptera, Phyllostomidae) from the Atlantic Forest of Southeastern Brazil, with comments on L. bokermanni. Zootaxa 3722(3):347–360

Whale and Dolphin Conservation (2014) How did we miss an entirely new species of river dolphin? us.whales.org/blog/alisonwood/2014/01/how-did-we-miss-an-entirely-new-species-of-river-dolphin. Accessed 3 August 2015

Dunn R (2010) Every Living Thing: man's obsessive quest to catalog life, from nanobacteria to new monkeys. Smithsonian Books, Harper

Ehrenfeld D (1988) Why put a value on biodiversity? In: Peter FM, Wilson EO (eds) Biodiversity. National Academy Press, Washington, DC, pp 212–216

Ehrlich PR (1988) The loss of diversity: causes and consequences. In: Peter FM, Wilson EO (eds) Biodiversity. National Academy Press, Washington, DC, pp 21–27

Encyclopedia of Life (2008) www.eol.org. Accessed 8 July 2014

Erwin TL (1982) Tropical forests: Their richness in Coleoptera and other arthropod species. Coleopterists' Bull 36:74–75

Farnsworth NR (1988) Screening plants for new medicines. In: Peter FM, Wilson EO (eds) Biodiversity. National Academy Press, Washington, DC, pp 83–97

Farrell BD (1998) "Inordinate fondness" explained: Why are there so many beetles? Science 281:555–559

Feinberg JA, Newman CE, Watkins-Colwell GJ, Schlesinger MD, Zarate B et al (2014) Cryptic diversity in metropolis: confirmation of a new leopard frog species (Anura: Ranidae) from New York City and surrounding Atlantic Coast regions. PLoS One 9(10), e108213. doi:10.1371/journal.pone.0108213, Accessed 5 November 2014

Food and Agricultural Organization of the United Nations, Fisheries and Aquaculture Department. http://www.fao.org/fishery/statistics/en. Accessed 4 July 2015

Foster MS (1982) The research natural history museum: pertinent or passé? Biologist 64:1–12

Fowlie M (2014) IUCN Red List raises more red flags for threatened species. http://www.birdlife.org/worldwide/news/iucn-red-list-raises-more-red-flags-threatened-species. Accessed 7 August 2015

Graneli E, Pavia H (2006) Allelopathy in marine ecosystems. In: Reigosa MJ, Pedrol N, Gonzalez L (eds) Allelopathy: a physiological process with ecological implications. Springer, Dordrecht, pp 415–431

Hance J (2013) Bird extravaganza: scientists discover 15 new bird species in the Amazon. http://news.mongabay.com/2013/0606-hance-15-birds.html. Accessed 22 August 2013

Helgen KM, Pinto M, Kays R, Helgen L, Tsuchiya M, Quinn A, Wilson D, Maldonado J (2013) Taxonomic revision of the olingos (*Bassaricyon*), with description of a new species, the Olinguito. ZooKeys 324:1–83

Herholdt EM (ed) (1990) Natural history collections: their management and value. Transvaal Museum Special Publication No. 1

Hilton-Taylor C, Stuart SN (eds) (2009) Wildlife in a changing world—an analysis of the 2008 IUCN red list of threatened species. IUCN, Gland

Kareiva P, Tallis H, Ricketts TH, Daily GC, Polasky S (eds) (2011) Natural Capital: theory and practice of mapping ecosystem services. Oxford University Press, Oxford

Kolbert E (2014) The Sixth Extinction: An Unnatural History. Holt, New York

Lucky A, Erwin TL, Witman JD (2002) Temporal and spatial diversity and distribution of arboreal Carabidae (Coleoptera) in a Western Amazonian Rain Forest. Biotropica 34(3):376–386

Lundmark C (2004) UC-Berkeley's Museum of Vertebrate Zoology turns 100. The importance of scientific collections, from dinos to DNA. BioScience 54(8):800

Mendes Pontes AR, Gadelha JR, Melo ERA, Bezerra de Sa F, Loss AC, Caldara V Jr, Pires Costa L, Leite YLR (2013) A new species of porcupine, genus *Coendou* (Rodentia: Erethizontidae) from the Atlantic forest of northeastern Brazil. Zootaxa 3636(3):421–438

Milner R (2009) Darwin's Universe: Evolution from A to Z. University California Press, California

Mora C, Tittensor DP, Adl S, Simpson AGB, Worm B (2011) How many species are there on earth and in the ocean? PLoS Biol 9(8), e1001127. doi:10.1371/journal.pbio.1001127, Accessed 21 July 2015

Morell V (2014) New species of humpback dolphin. news.sciencemag.org/biology/2014/08/new-species-humpback-dolphin. Accessed 15 October 2014

Myers N (1988) Tropical forests and their species: going, going…? In: Peter FM, Wilson EO (eds) Biodiversity. National Academy Press, Washington, DC, pp 28–35

Myers RA, Worm B (2003) Rapid worldwide depletion of predatory fish communities. Nature 423:280–283

National Geographic (2007) Special report. Saving the Sea's Bounty, April

Natural History Museum website. 2012, 19 April. Mission to map 10 million species in 50 years. http://www.nhm.ac.uk/about-us/news/2012/april/mission-to-map-10-million-species-in-50-years109580.html. Accessed 19 April 2015

Neill DA y Ulloa Ulloa C (2003) Nuevas especies y nuevos registros para la flora del Ecuador: 1999-2003. p 157 *en* Libro de Resúmenes, IV Congreso Ecuatoriano de Botánica, Edit. Univ. Tec. Part. Loja, Ecuador

Neill DA y Ulloa Ulloa C (2011). Adiciones a la flora del Ecuador: segundo suplemento, 2005-2010. Fundación Jatún Sacha

Nielson ES, Mound LA (2000) Global diversity of insects: The problems of estimating numbers. In: Raven PH, Williams T (eds) Nature and Human Society: The Quest for a Sustainable World. National Academic Press, Washington, pp 213–222

Novotny V (2009) Notebooks from New Guinea: field notes of a tropical biologist. Oxford University Press, Oxford

Ojala-Barbour R, Pinto M, Brito J, Albuja L, Lee TE, Patterson BD (2013) A new species of shrew-opossum (Paucituberculata: Caenolestidae) with a phylogeny of extant caenolestids. J Mammal 94(5):967–982

Pérez AJ, Hernandez C, Romero-Saltos H, Valencia R (2014) Árboles emblemáticos de Yasuní Ecuador, 337 especies. Publicaciones del Herbario QCA, PUCE, Quito

Phillips AR (1974) The need for education and collecting. Bird-Banding 45:24–28
Phys.org website (2013) Racing to identify species as biodiversity shrinks. http://phys.org/news/2013-01-species-biodiversity.html. Accessed 28 September 2015
Plotkin MJ (1988) The outlook for new agricultural and industrial products from the tropics. In: Peter FM, Wilson EO (eds) Biodiversity. National Academy Press, Washington, DC, pp 106–116
Raven PH, Wilson EO (1992) A fifty-year plan for biodiversity studies. Science 258:1099–1100
Remsen JV (1995) The importance of continued collecting of bird specimens to ornithology and bird conservation. Bird Conservation Intl 5:145–180
Ripple WJ, Beschta RL, Nelson MP et al (2014) Status and ecological effects of the world's largest carnivores. Science. doi:10.1126/science.1241484
Rocha LA, Aleixo A, Allen G, Almeda F, Baldwin CC et al (2014) Specimen collection: an essential tool. Science 344(6186):814–815. doi:10.1126/science.344.6186.814
Sabrosky CW (1952) How many insects are there? In: Steferud A (ed) Insects: the yearbook of agriculture. GPO, Washington, DC, pp 1–7
Safina C (1997) Song for the Blue Ocean. Owl Books, New York
Smith C (2015) If a rusty gravedigger dies in Alabama's wetlands, should conservatives care?: opinion. http://www.al.com/opinion/index.ssf/2015/01/if_a_rusty_gravedigger_dies_in.html#incart_river. Accessed January 11
Stein BA, Kutner LS, Adams JS (eds) (2000) Precious heritage: the status of biodiversity in the United States. Oxford University Press, Oxford
Stevens WK (1995) How many species are being lost? Scientists try new yardstick. NY Times. http://www.nytimes.com/1995/07/25/science/how-many-species-are-being-lost-scientists-try-new-yardstick.html. Accessed 23 May 2015
Stewart DJ (2013a) Re-description of Arapaima agassizii (Valenciennes), a rare fish from Brazil (Osteoglossomorpha: Osteoglossidae). Copeia 2013(1):38–51
Stewart DJ (2013b) A new species of Arapaima (Osteoglossomorpha: Osteoglossidae) from the Solimões River, Amazonas State, Brazil. Copeia 2013(3):470–476
Stuart SN, Wilson EO, McNeely JA, Mittermeier RA, Rodriguez JP (2010) Ecology: The barometer of life. Science 328(5975):177
Sukhdev P (2010) What is the world worth? Putting nature on the balance sheet. Sydney lecture transcript, 3 August 2010. Center for Policy Development. http://cpd.org.au/2010/08/pavan-sukhdev-sydney-lecture-transcript/. Accessed 13 January 2014
Swing K (2013) Inertia is speeding fish-stock declines. Nature 494:314
Swing K, Denkinger J, Carvajal V, Encalada A, Silva X, Coloma LA, Guerra JF, Campos F, Zak V, Riera P, Rivadeneira JF, Valdebenito H (2014) Las colecciones científicas: percepciones y verdades sobre su valor y necesidad. Bitácora Académica USFQ 1:1–46
ter Steege H, Pitman NCA et al (2013) Hyperdominance in the Amazonian tree flora. Science 342 (6156):1243092. doi:10.1126/science.1243092
Thiele LP (2011) Indra's Net and the Midas Touch. Living sustainably in a connected world. MIT, Cambridge
Trigo TC, Schneider A, de Oliveira TG, Lehugeur LM, Silveira L, Freitas TRO, Eizirik E (2013) Molecular data reveal complex hybridization and a cryptic species of Neotropical wild cat. Curr Biol 23(24):2528–2533
Vaijalainen A (1748) N Wahlberg, GR Broad, TL Erwin, JT Longino, Sääksjärvi IE (2012) Unprecedented ichneumonid parasitoid wasp diversity in tropical forests. Proc Biol Sci 279:4694–8. doi:10.1098/rspb.2012.1664
Wake DB, Vredenberg VT (2008) Colloquium paper: are we in the midst of the Sixth Mass Extinction? A view from the world of amphibians. Proc Natl Acad Sci U S A 105:11466–11473
Wheeler QD, Knapp S, Stevenson DW, Stevenson J, Blum SD, Boom BM, Borisy GG, Buizer JL, De Carvalho MR, Cibrian A, Donoghue MJ, Doyle V (2012) Mapping the biosphere: exploring

species to understand the origin, organization and sustainability of biodiversity. Systematics Biodiversity 10(1):1–20

Wilson EO (1988) The current state of biological diversity. In: Peter FM, Wilson EO (eds) Biodiversity. National Academy Press, Washington, DC, pp 3–18

World Wildlife Fund (2014) New species discoveries in the Greater Mekong. wwf.panda.org/wwf_news/?222513%2FNew-species-discoveries-in-the-Greater-Mekong. Accessed 8 July 2015

Chapter 6
Challenges to Conservation

Kelly Swing

Abstract The management and maintenance of nature are complicated because both nature and human nature are themselves quite complex. While traditional explanations based on human population growth combined with increasing resource demands offer some perspective on an ever-worsening situation, they seem somewhat separated from root causes as well as potential solutions.

Keywords Conservation • Challenges • Sustainability • Human behavior • Religion

6.1 Challenges

"Indeed, the greatest challenges of conservation involve non-scientific issues..." (Terborgh 1999). Most experts would undoubtedly define the greatest threats to nature and biodiversity as the ever-increasing human population and our occupation of growing proportions of all resources, space, and primary productivity (Cairns 1988; Ehrlich 1988; Wilson 1992; MacDonald 2003).

We use nature because it is valuable; we abuse nature because it is free.—Barry Gardiner, Member of UK Parliament, 25 January 2016

Although our own existence and prosperity ultimately depend heavily upon comprehending our position and role within a complex backdrop that minimally includes millions of life forms, all indications point to the fact that we have never considered completing an inventory a priority (Dunn 2010). This knowledge gap presents a series of complications in our approach to applying value to the natural world as well as developing strategies to manage and sustain functioning ecosystems and a healthy environment in which our own species may survive and thrive while ensuring that fellow species have similar opportunities.

One simplified model frequently used to depict relationships of humans and our world involves a triangle with the three corners, respectively, occupied by our

K. Swing (✉)
College of Biological and Environmental Sciences, Tiputini Biodiversity Station,
University of San Francisco de Quito, Quito, Ecuador
e-mail: kswing@usfq.edu.ec

© Springer International Publishing Switzerland 2017
J.N. Furze et al. (eds.), *Mathematical Advances Towards Sustainable Environmental Systems*, DOI 10.1007/978-3-319-43901-3_6

species (the exploiters), another specific species (the exploited), and its habitat (the stage). Assuming that ecosystems were in some state of equilibrium before humans appeared on the scene, it is evident that our addition to this geometric figure is precisely what has upset the balance. We know that humans have been increasingly responsible for impacts on their surroundings for millennia. We have repeatedly proven our capacity to modify nearly any habitat for our own benefit and concomitantly to the detriment of nature.

> ...you toxify your Earth and pursue ever more imaginative means of self-destruction. You cannot be trusted with your own survival. You are so like children. We must save you from yourselves. My logic is undeniable.—Supercomputer V.I.K.I. (Virtual Interactive Kinetic Intelligence) in the 2004 movie "I, Robot" based on the writings of Isaac Asimov

As civilizations have advanced, we have tended to become more concentrated in urban settings. This distribution could eventually present benefits for nature, but so far typical urban sprawl has continued to nibble away at the world to accommodate our growth. Questions of survival for ourselves and the biodiversity of the planet arise and we have tried to justify the amount of space that we occupy in proportion to the amount we leave for the other millions of species that share our finite conglomeration of rock, rubble, water, and air.

Our behavior appears to be as big an influence as our numbers. How humans think and resolve problems is as big a piece of our past as it will be for our future. Early pressures in our evolution and survival almost certainly were more linked to the immediate exigencies of day-to-day living than were long-term concerns. This means that our make-up, originally influenced by one kind of demands, is going to be sorely tried by the changes that we have caused. Ecological succession is part of every natural progression; each species' presence and activity cause changes that, in turn, tend to favor other species. Will the extreme changes we have provoked ultimately lead to our own demise and a takeover of some other life form that has the capacity to deal with what we are leaving in our wake?

We know that our religions and philosophies have been integral to our survival in the past, and that they are quite responsible for our actions and impacts. The question now is whether our intellect might allow us to overcome what appears to be a standard law of replacement that has repeated itself billions of times during the historic eons of this Earth.

Though not a direct quotation from Darwin, the idea remains relevant

> It is not the strongest species that survive, nor the most intelligent, but the ones most responsive to change.

Some people say this is all about money. John Terborgh (1999) has even called "wild nature and the biodiversity it perpetuates. . . a luxury". We do know that prosperous countries tend to spend more on environmental issues, recovery of impacts, and management of parks, while poorer countries often maintain the position that they simply cannot afford special measures that could protect their resource base, their surroundings, and even their own people from contaminants over the long term. Practical environmentalists and accountants would point out that it is much cheaper to maintain anything than it is to repair it. This applies to the nature around us as well as it does to the family car. Keep the working parts in good order and it can be fairly cheap; be lax on the simple things and soon reparation costs get out of hand and become so prohibitive that we get to the point that it would cost far too much to attempt to recover, and we give up. Haiti may represent a case so far gone that we do not have much hope for its economy or its nature (Diamond 2005). Is its economy so troubled because of a lack of forests and other natural resources or are there no forests left because the economy is so bad? Analysis of the timeline indicates that disregard for forests and other resources led to overharvest and abuse and that the economy took a dive as a direct result.

Other folks suggest that this is all about creativity and technology. A corollary would be that by having a greater human population, and thereby more minds involved, practical strategies would be more likely to arise. While these factors, founded in human intelligence, are key parts of any potential solution, we must recognize that a fundamental change in mentality among the masses will also be necessary to arrive at a functional balance between what we take and what we leave (Quinn 1992).

A society is defined not only by what it creates but also by what it refuses to destroy.—John C. Sawhill, 1988

As everyone knows, the world economy is strongly anchored to energy availability. Due to our dependence on fossil fuels to turn the wheels of industry and transportation, countries that control abundances of these resources and companies that specialize in their extraction essentially rule the planet. Although many cleaner alternatives exist for powering the globe's cities, factories, and vehicles, we are only beginning to escape our addiction to fossilized photosynthates. Scientists have long explored other sources of energy, but Big Oil in particular has been especially aggressive in maintaining its grip on a substantial market share by buying up patents for potentially competing ideas. Not surprisingly, the process of moving on to better strategies has been hampered by the selfish interests of this overwhelmingly powerful sector. The Stone Age did not end because we ran out of stones, nor will the Petro Age last until we run out of oil, gas, and coal. The natural progression of innovation would have probably moved us beyond our current dependence if there were no huge economic incentives to hang onto old technologies.

Contrary to the wishes of environmentalists, hydrocarbon resources remain abundant in the ground. Unfortunately, greenhouse gas emissions resulting from their combustion change global climates drastically, in ways that humans are poorly equipped to manage at the moment. The question at this juncture in history is whether we are capable of making a decision that is good for overall well-being in the long term, but dauntingly difficult in the short term.

It is difficult to get a man to understand something, when his salary depends on his not understanding it.—Upton Sinclair, 1934

We are like tenant farmers chopping down the fence around our house for fuel when we should be using Nature's inexhaustible sources of energy—sun, wind and tide. I'd put my money on the sun and solar energy. What a source of power! I hope we don't have to wait until oil and coal run out before we tackle that.—Thomas Edison (1931)

Proximate explanations of why conservation has proven so difficult are discussed broadly in many publications on this topic, but they tend to be primarily related to direct economic drivers and the basest of human vices, greed. The following points attempt to go beyond that kind of treatment to look at ultimate influences.

6.2   Separation Anxiety

At present, the human race, by and large, probably has less direct experience with nature than ever before (Lovejoy 1988). Early humans were tied to natural resources so tightly that there was no choice but to be aware of whatever was happening in our surroundings. Over time, however, the processes of civilization, development of agriculture and animal husbandry, have allowed us to become increasingly separated from our personal dependence upon natural resources, and consequently, less aware of the realities in nature. The human race has become concentrated in cities and has little time or funds to make special efforts to be in contact with nature. In Latin America, for example, a century ago, 75 % of the population lived outside urban settings, mostly on farms; today, precisely the reciprocal is true with the vast majority living in cities and in economic conditions that disallow experiences as simple as a trip to the countryside just beyond the outskirts of their municipality. The consequence is that children grow up without understanding the beauty and wonders of nature and many come away with little empathy toward other species (Louv 2005) or comprehension of what an ecosystem is. This is so true that now we hear the word "ecosystem" applied to the workings of business more often than we do in relation to the functioning of a forest or the sea.

Another point of view is that by concentrating people into cities, much of the pressure on ecosystems can be relieved. By physically separating people from nature, there would logically be greater opportunity for recovery (Gibbons 1993). Accepting that humans living in cities in high densities would mean more intensive impacts on the local scale, we should recognize that this would also imply more efficient use of space. The trick would be to avoid their continued use and consumption of wild products. Having fewer people in the countryside impacting fauna directly through harvest or competition for space would imply an improved situation for the limited natural areas that survive the next few decades. Getting through this bottleneck will represent a huge challenge, but the scenario could provide hope for both rare species and limited availability of quality habitats (pers. comm. Joseph Walston, Wildlife Conservation Society, Nov 2014).

6.3 Selective Acceptance of Science

At one level, the general public reveres and respects the teachings of science; at another, opinions are laden with suspicion and doubt. In some situations, the public blindly and gladly accepts the findings of those who don lab coats. In others, researchers are accused of trying to milk the taxpayer for money or might be labelled charlatans as though they are selling snake oil. Those reporting on environmental issues are often seen as a club of childish bunny huggers trying to keep nature for purely superficial and selfish reasons, while less fortunate members of society desperately need the resources in question to survive. Listen long enough and one will hear many explanations as to why scientists are not to be trusted. Unfortunately, all this cynicism is probably rooted in a widespread lack of understanding of what science is, how research works, what investigations produce, and how everyone benefits from the resulting increase in knowledge. Of course, this is rooted in a problem at the level of basic education (Miller 2002).

Let us contemplate the things that people believe without question and some that we would rather ignore. When we are infirmed, we go instantly to the pharmacy to buy nearly any product sitting on the shelf because there is immediate self-interest involved and we believe that some agency (in the US, the FDA) would not allow such products to be available unless they had been tested and proven effective. When the illness is more serious and we visit a physician for a proper diagnosis, we willingly take his/her advice and even call what is prescribed "doctor's orders" as though we have no choice, or grounds, to do otherwise. This compliance may extend to the extreme of having a limb amputated or a chest opened to tinker with your ticker, especially if the prediction is that the situation could otherwise become life threatening. When we drive across a bridge or board an airplane, we see these as routine experiences as opposed to wondering if the respective engineers used the right calculations and proper materials; we automatically believe the science is well-founded and trust our lives to the physics behind the design and function of these things. When we get dressed in the morning, we try to notice the weather forecast so as to be prepared for the day's conditions. When a hurricane is headed toward our city or tornados are likely to be spawned, our very lives depend on being aware of what scientists are predicting in an immediate sense. When nutritionists tell us about health risks associated with certain diets, we take notice of longer-term risks and at least try to decrease consumption of the really bad stuff. When governments require warnings about addictions and lung cancer on cigarette packaging, we believe the science that has made connections between cause and effect because we have all experienced them for ourselves.

Having witnessed something personally likely makes a tremendous difference in this scenario. We have seen many cars cross that bridge before and assume that we too can cross safely. We have seen members of our own families cured of nasty medical complications through clinical interventions. Although we have seen thousands of lives saved due to a functioning system of storm warnings, everyone tends to reserve some doubt in relation to what the local weather forecast says for tomorrow. Logically, it follows that we would be even more skeptical about climate predictions out across decades or centuries. Despite the fact that few people have witnessed first-hand an extinction event and no one has personally lived through an ice age or a sea level drawdown of 100 m, we accept through the teachings of science that these events have indeed occurred. Albeit with healthy curiosity, we should also accept that modern science has some capacity to predict similar events by studying trends. Nevertheless, many people ignore the warning signs and even persecute the messengers.

So why the suspicions? Once again, we return to the topic of money. Certainly, there are great short-term incentives not to heed the warnings of the scientific community in some cases because of the extreme costs of doing anything about what is headed our way. And as long as we can convince ourselves that nothing is really going to happen, it is much cheaper for our generation to sit back and do nothing. Taking no action costs nothing in the present and who cares about what some generation far in the future will have to deal with? Doing nothing allows us to comfortably live out the old adage of "ignorance is bliss"; collectively putting our heads in the sand until impacts actually reach devastating levels may admittedly be irresponsible but it might well alleviate a lot of worry, and expenditures, for a generation or so. The tremendously understated downside is that it would be far too late to do anything about a process that had already passed a point of no return.

Without question, intentional manipulations fuel the fire. Frequent utilization of empty rhetoric in propaganda campaigns serves to confuse issues in the minds of the masses. To play the game of populist politics, leaders expose the people to precisely what they wish to hear, so they sincerely believe the situation is not as bad as it may seem or that simple, cheap solutions have already been implemented and are working. Basically, instead of complaining about their plight, the people are lulled into a false sense of security. Oftentimes, leaders claim to be protecting the masses from unnecessary spending or stress and thereby, gain support through vacuous arguments. Such propaganda techniques primarily serve to maintain popularity while inevitably postponing the resolution of root problems. Lessons in controlling majority opinions have been passed down and polished since the time of the Roman Empire right through to the evil genius Goebbels of the Third Reich (Reuth 1990). Through the use of marketing ploys and half-truths, popular politicians influence voting processes to produce modifications in the distribution of funds, often strongly affecting possibilities for justifiable prioritization of environmental issues. Some U.S. legislators have even proposed that scientists be eliminated from the pool of information sources during hearings in relation to environmental issues, undoubtedly because their self-interests are threatened by the facts. In the modern world, climate deniers, Big Oil and Big Ag, are all major actors in this scenario. Their combined influence on the world economy is staggering, and when they team up with especially ambitious political leaders whose short-term interests combine with self-aggrandizing administrative strategies, there is little opportunity for alternatives. Since President Eisenhower voiced his warnings about the Military-Industrial Complex over a half century ago as he was leaving the office, the threat has only grown to incorporate other powerful sectors.

The argument is always that too many jobs are at stake for us to reign in this juggernaut that has an insatiable appetite for "biggering and biggering" (Seuss 1971). Advertising, marketing by employing charismatic figures, and the press have been the standard tools in this game of distraction and now, education and social media will necessarily be the tools that allow us to overcome the lack of knowledge that makes societies vulnerable to these manipulations.

6.4   Species/Area Relationships and Sustainability

Everyone is aware that there are more species in any country than in any individual park. The question remains as to precisely how this relationship works. "More space, more species" makes sense on every scale (Gleason 1922; Pitman et al. 1999), but the quantifying of proportions has proven challenging. In places where the human population has been intensively concentrated for great expanses of history, natural spaces were converted long before anyone thought about loss of species and in certain areas, we undoubtedly extirpated many species before we even noticed that they existed (Westwood 1833). As time has proceeded, we first noted the depletion of the species (Baker and Clapham 2004) we wished to continue consuming (Swing 2014). Later, we noticed that others were concomitantly following along, even in cases when we were entirely unaware of causing them any direct harm.

> When we try to pick out anything by itself, we find it attached to everything else in the universe.—John Muir

During decades, calculations have been made to estimate how many species will be lost (MacArthur and Wilson 1967; Myers 1988; Wilson 1992) as we go forth with the mass conversion of natural landscapes to crops, roadways, and cities. The goal would be to determine some ceiling on just how destructive we can be. The underlying strategy would be to set a definitive limit, very similar to the way MSY (maximum sustainable yield) formulas should work to set quotas for fishery resources. The implication is that some losses are acceptable, but a functional threshold can indeed be defined. One enormous question would be who makes the decision about the placement of the line that cannot reasonably be crossed (Wallace 2013). We are essentially asking how much space we can commandeer without major repercussions for ourselves or too much damage to our reputation with future generations. So far, losing large numbers of species has had essentially no impact on our behavior. Ideas of sustainable development have dominated our actions, leading inexorably to our continuing to occupy more space every day.

What we have typically called "sustainable development" is not actually sustainable. Instead, that old idea of sustainable development justifies, and is a guarantee for, continued destruction of our environment. For quite some time, we've been praising this concept as the primary answer to many conservation concerns. Unfortunately, underlying premises guarantee that we're on a road to perdition. We must recognize specific details to understand why the traditional idea of sustainable development is simply not viable in the long run (Suárez 2010). Of course, I'll be the first to admit that "sustainable use" may well be possible and optimal, if it can in fact, occur as a functional manifestation of "wise use". The root of the problem is that long ago, we came to believe that a little restraint with our resource bases would allow us to make use of everything everywhere. As it turns out, our best intentions based on intelligent concepts tend to yield to pressures as justifiable as hunger pangs and as insidious as corporate greed when population growth and competition come into play, and then, wise use is typically replaced by outright abuse. Another complication is that the idea of "sustainable" development has commonly morphed into the idea that all resources can by default deal with "sustained" development *ad infinitum*. Through some perverse process, sustainability has very often come to be confused with limitlessness.

At the current seven billion plus mark, the global human population is continuing to grow and requires ever-increasing amounts of resources, in the form of consumables and space. While agricultural scientists have come up with many creative strategies to substantially improve crop productivity under the best of circumstances, there are undoubtedly limits to how much food can come from each square meter of croplands; potable water and space certainly will always be finite commodities on our planet. Of particular concern is the fact that lands cleared for agricultural activities, especially in the developing world, are frequently abandoned after short periods of productivity. On a regional scale, this generally means that forest losses far outstrip gains in crop productivity. In most of Amazonia, we go on clear-cutting, using the basic economic needs of the people as a justification for opening up more land. Paradoxically, as more forested areas are converted into plantations and pastures, greater production of consumable or marketable goods is not at all guaranteed. Due to poor soil quality and inappropriate agricultural practices, the typical scenario for the family farmer/squatter is a vicious cycle of cutting down forest, hoping the land will provide, making an honest effort at planting and tending to crops, suffering crop failure, cutting more forest, desperately hoping for better results, quickly getting dwindling results and once again moving onto another patch of forest, continually nibbling away at the countryside until every last fragment of intact ecosystem is erased from horizon to horizon. Nobody gets rich; most families work hard but continue to live in abject poverty while the flora and fauna disappear across the entire landscape. This is not due to any fault of the people themselves; they do not set out to eliminate rainforest and all the species therein.

Actually, in most scenarios, habitat degradation or species extirpation are primarily "a cascade of unintended consequences" (Thiele 2011). People are often drawn into such situations due to economic desperation and unrealistic expectations that result from lack of two components key to any life choice—reliable information and real options. At some point, people must realize that agriculture on more than a subsistence level is essentially incompatible with most of Amazonia and the tropics in general. Traditionally in the developing world, governments have allowed this system of ongoing destruction (of rainforest and of human lives) to proliferate due to the fact that unoccupied land appears to be a waste of potential resources that could improve the economy, locally and nationally. It is also very easy to pass the blame for poverty to the individual once the government has provided access to land since land tenure in most of the world has stereotypically been linked to wealth. Acquisition of land in places like Amazonia, in the vast majority of cases, has trapped families in an untenable paradox; they are indeed landowners with title to many hectares, but their efforts yield little in the way of liquid assets. Intensive use of agrochemicals might overcome this common lack of productivity and lack of income, but these products continue to be inaccessible on a financial basis and would likely result in widespread pollution of surface waters as a collateral impact. Use of scientifically founded information (Pitman et al. 2011) instead of socio-politically driven agendas to determine the future of areas that are simply not compatible with agriculture, being brutally honest with potential squatters instead of encouraging the occupation of remote Amazonian sites, is the only way to avoid loss of remaining rainforests while eliminating the false hopes that drive the poor to believe that a better life awaits them in the wilderness.

In truth, the two concepts of development and preservation of nature are by definition mutually exclusive. Development means changing things for our own purposes, and consequently, impacting whatever occupied that space previously. In most cases, that means entirely eliminating whatever species were there beforehand. I am not aware of any real example of development that has had no impact on the nature of the precise location where it took place. Taking this piece for us while leaving that adjacent piece for nature has always seemed reasonable due to our selfish sense of proportions. For how long have we been generously taking only half and leaving the other half? But after repeating this division so many times, nature is relegated to infinitesimally small fragments in or near the places we live, sow, reap, build, trample, and extract. And we're always pushing farther into those beautiful mountains, that luxuriant forest, those remote islands—for the spectacular scenery without human influence, or the fine hardwoods inside, or the oil beneath, or the sharks and sea cucumbers offshore. On a global scale, since the Industrial Revolution, and even more so since World War II, we have surpassed the halfway mark for conversion of native forests into something else. That's right; we're racing quickly down the same road that has put Haiti, in recent history, in its present situation with nearly no natural ecosystems intact—and with only poverty to show for it.

World biodiversity experts suggest that we've already lost tens of thousands of species, especially in the last half century, but we don't have a reliable figure for how many. For a few, their demise through direct exploitation has been well-documented; for most, however, disappearance has occurred, through our incessant degradation and elimination of their habitats, without so much as a local newspaper headline or even a whimper (Gibbons 1993). Should we ask why? We cannot notice an extinction event if we never noticed that the species existed in the first place.

Given that we have catalogued thus far, realistically speaking, as little as a minute 10–15 % of the planet's total species (Lovejoy 1988; Wilson 2000; Wheeler et al. 2012), we are undoubtedly missing out on many opportunities that could have important positive financial implications globally, but especially for all those tropical countries that categorize themselves as poor. For some numerical perspective, economist Pavan Sukhdev (2010) estimated that losses related to biodiversity cost us between \$1.75 and \$4 trillion (1.35–3.1 trillion euros) each and every year (Phys.org website 2012). Given current rates of deforestation and concomitant losses through extinction, tremendous potential essentially has no chance of ever being realized. We are, in effect, throwing away untold opportunities before noticing they exist.

Traditionally, by sustainable development, we have implied that while we're advancing the human agenda, whether to provide an inhumanely cramped space for some starving cousins at the outskirts of the poorest town in Africa or to maintain the lifestyle of one more CEO in a Manhattan penthouse, there will always somehow, as if by magic, be "enough" resources for all of us (Mittermeier 2014) as well as space left for nature to survive and thrive, or to recover and make up for lost ground elsewhere. I doubt that we have ever sincerely believed development and nature can actually occupy precisely the same space. Granted, there are some projects that have "only" caused the loss of particular species on a local scale, or provoked something perceived to be as innocuous as an alteration of relative abundance, in some fraction of the developed space. What we have typically meant by sustainable development is that humankind can go ahead reproducing, spreading its tentacles, consuming more resources, generally elbowing out nature— as long as we don't massacre everything across a large geographical expanse in one go. Sustainable development has implied that we will be continuing development activities right on, as we please, more or less. Because no one wants to be an outright villain in the modern version of this story, we do look to the citizenry of whatever place we're talking about for approval of an "acceptable" impact limit (Randall 1988), but the voters making the decisions mostly turn out to be invested parties, in either a proximate or ultimate way. Sustainable development has meant saying that a certain level of destruction is acceptable, and even desirable.

Opening day of a new 5-ha shopping mall or industrial park could only represent a moment for celebration of human achievement, right? And what about the bulldozing of that last forested area of the county two years earlier that allowed its construction to get started? Strong economies, resulting from continuing development, after all, are arguably more able or likely to defend or recover the environment. But as there comes to be less of nature in its intact form, this chore becomes more difficult and much more expensive—so much so that the danger exists that we may begin to ignore it altogether. And we should always remember that the converse, just leaving nature alone is pretty cheap and effective—even when accounting for opportunity losses, if we are also honest enough to include long-term impacts, remediation, clean-up, and so forth in our calculations.

To those devoid of imagination, a blank place on the map is a useless waste; to others, the most valuable part.—Aldo Leopold, A Sand County Almanac (1949)

Perspectives on sustainable development generally imply coming to an agreement about how much of nature can be degraded or eliminated during our lifetimes without being blamed vehemently by our descendants for the destruction of the entire Earth or the overall collapse of environmental services on a regional or global scale. By pushing for or accepting a little more "sustainable development" with each generation of self-interested parties, mathematically we are guaranteed, sooner or later, to reach a point that cannot possibly be construed as sustainable. Is it possible to reach an equilibrium point way before getting to the extreme depicted in the animated movie "Wall-e"?

How do we decide what is worth keeping and when do we decide? Typically, we have waited until a moment of desperation when we're encroaching on that last intact bit of countryside that has suddenly become noticeably precious due to its rarity. It would be preferable to most citizens, and economically valid for all, to plan ahead with vision so that functional, effective strategies could allow us to maintain the best pieces of nature (criteria may vary from most scenic to most biodiverse to greatest endemism, etc.) without having to grovel before taxpayer representatives for a pittance to preserve the last decent-sized plot of cloud forest on the western slope of the Ecuadorian Andes, rainforest in Papua New Guinea, or tiger habitat in Siberia. Our most common strategy worldwide has been to wait until it is far too late for any proactive approach, consequently being cornered into settling for whatever leftovers might remain. Do we intend to "sustainably" develop until we've occupied the entire globe?

If the impacts we see today in places like northeastern Ecuador occurred in a matter of minutes or days, everyone would classify them as world-class disasters, even greater than major earthquakes and hurricanes, but because the destruction has advanced agonizingly slowly over a period of 4–5 decades and masses of people never died outright, no one seems to notice or care. The same kinds of landscape scale losses can be expected worldwide if we continue with the same short-sighted mentality that has provoked the changes across huge expanses of Amazonia and West Africa in the last decades.

When we destroy something created by man, we call it vandalism; when we destroy something created by nature, we call it progress.—Ed Begley Jr. (2006)

6.5 Human Behavior in Light of Evolutionary Pressures

It's not just about the total number of species lost; it's also about which particular species are disappearing. We directly and actively overexploit many organisms that we wish to continue consuming into the future and are simply elbowing others out of a place to live. Our outright greed in the present trumps our self-interest for the future. How can we be so myopic?

"It is reckless to suppose that biodiversity can be diminished indefinitely without threatening humanity itself." (Wilson 1992)

Considering that the self-interests of every human being are inextricably linked to the overall condition of our planet (Daily 1997), all conservationists ask why it has proven to be so frustratingly difficult to convince the masses to care about nature. A nearly endless list of explanations has been proffered, but the challenge to change popular priorities remains as daunting as ever. Here, I propose for contemplation that humans themselves are not precisely the driving force of environmental impacts, but instead function more like pawns in the game of evolutionary ecology. It is evident that selective forces are at play in our behavior and since "nothing succeeds like success," we continue to use the same strategies that have allowed us to dominate nearly every landscape we have encountered. Part of our domination has come through the fact that we function essentially as allelopathic organisms (Swing 2014).

Generally speaking, environmentalists blame human selfishness, multiplied by our vast numbers, for most of the devastating impacts we see around us (Terborgh 1999; Wilson 2003). However, from an evolutionary perspective, we should recognize that the human race is simply winning the game of competition (Mayr 1970). Selection pressures favor species that out-compete rivals (Darwin 1859). After all, why should things be any different for us? Our population growth, particularly in the last half century, tells us that we have competed exceedingly well; our huge numbers reinforce the idea that the positive evolutionary feedback loop has favored behavior that pushes all lesser competitors toward extinction. Gause (1934) first illustrated this phenomenon using protozoans within a fishbowl set-up, but conditions are quite similar for us humans within our limited biosphere. We are winning the game, so we are encouraged to continue using the winning strategy—hoard resources, eliminate competitors, occupy more space, etc. And if we are indeed winning, why would we stop now?

As a part of this process, humans (or modern human societies) have evolved to be allelopathic. Allelopathy is typically defined as the production of chemical compounds that result in excluding or eliminating competitors (Willis 2007). Classically, the production of allelopathic compounds allows some level of monopolization of resources (starting with space perhaps but including various resources within that space) by disallowing survival or encroachment of others that have similar requirements. The phenomenon is typically seen as a naturally occurring form of chemical warfare. Classroom treatments of the subject traditionally refer to plants such as black walnuts and creosote bushes as easily recognized examples, but many benthic and planktonic marine organisms exemplify similar strategies (Graneli and Pavia 2006).

For the longest time, we referred to the active toxins in these scenarios as "secondary compounds" because they are not considered direct requirements for maintenance, growth, or reproduction. Moreover, we saw them as chemical by-products that, purely by chance, provided some benefit to the organism producing them. The benefit between plants of various species is usually achieved by essentially elbowing neighbors out through the use of biochemistry; the benefit for organisms (plants or animals) facing ingestion might come from the production and storage of toxins that cause their consumers to sicken or die. Early along in evolutionary history, such acquired advantages resulted in powerful selection pressure to continue to do this even more, by rewarding the accumulation of greater quantities of these by-products. Through differential survival and reproduction, allelopathy was favored and developed in more species in many ecosystems (Willis 2007).

Whether human behaviors and cultural characters are subject to precisely the same mechanisms as genetically based evolution or how memes or exaptations (Gould and Lewontin 1979) might be transmitted between individuals or generations has been amply discussed in the literature (Dawkins 1976; Lynch 1996). Specific details may be controversial at some level, but behavior is undoubtedly subject to selective processes as much as other aspects of all living organisms, including humans.

And now, at least partially through the same mechanism, the human race has come to dominate the planet. During most of our evolutionary history, we could only impact small, defined areas primarily through physical interventions, but during the last few centuries, especially advancing with the Industrial Revolution and then lurching forward after WWII, we have come to use our "chemical by-products" to aggressively push aside nearly everything else, thereby giving us the opportunity to monopolize more space every day. Rachel Carson (1962) was one of the first to sound the alarm as to where this could lead. A portion of what we do on a chemical level is entirely intentional; the use of pesticides, herbicides, fungicides, and antibiotics are all considered justifiable in our struggle to survive and thrive. In many cases, the ultimate goal is precisely to eradicate another species altogether (usually weedy plants, crop-destroying insect pests, invasive-introduced species, or infectious diseases), at least on a local scale. Although the term has not been previously applied to humans, we have indeed been using chemicals in precisely the same way that formally recognized allelopathic organisms do, but on a broader scale. The only real difference is that instead of glandular tissue within our bodies, we utilize laboratories and industry to produce the chemicals noxious to other beings in far greater quantities and varieties; we also use globalized transportation systems to deliver them on vast geographic scales.

So far, the feedback loop has been mostly positive for us, so we keep right on poisoning everything else. In many instances, this may have been no more intentional than it is for a eucalyptus tree. Nonetheless, the results are similar—we, as a species, end up with more and everything else ends up with less. On the economic scale, or any other for that matter, increasing access to more resources, through any means, is expected to provide incentives for the winner to continue using the same methods (Diamond 1999, 2005). If we are benefited by certain characters, physical or behavioral, we should expect selection pressures, social or evolutionary, to favor those specific attributes (Darwin 1859), and on a proportional basis (Malthus 1798; Mayr 1970).

In a short-sighted, superficial, and selfish world, it becomes clear that the elimination of competitors, intra- and inter-specific, means more of everything for us. Following this line of thinking, societies have decimated one another through warfare of every kind (Diamond 1999), including the use of toxins and even radioactivity as offensive weapons. Fishermen have killed sharks and sea lions so as to have more fish for themselves; some gardeners kill deer to avoid their feeding on flowers and vegetable crops, and opportunistically consume the competitors as well. As a corollary, we also target large mammals like bears and wolves because they could eat us or our children. Whether the threat is perceived or real is inconsequential. Evolutionary feedback loops continue to push us along this path of decreasing our competitors, so we go right on creating, producing, and dispersing more "secondary" compounds into the environment as well as directly killing anything in our way (Diamond 2005). Short-term benefits are quite evident in most cases, so selection pressures continue to enforce this behavior.

Predictably in this scenario, humankind races forward supplanting or eliminating almost everything within reach, blindly (for the most part) feeding into the vicious cycle that is driving the current pulse of extinction. Often, without any defined targets, chemical pollutants, true by-products originating in factories, have eliminated many species around and downstream from the source. The same is true when we consider the quantities of household waste and sewage released into the environment, especially in light of the fact that 95 % of the developing world employ no sewage treatment whatsoever (World Future Fund). As a default management strategy, "dilution is the solution to pollution" is reaching its limits. Public health is strongly tied to environmental health, and although some humans in some places are already highly impacted, we have yet to exceed the threshold that might provoke widespread alarm and reaction. Jared Diamond (2005) masterfully illustrated how this story has turned out repeatedly on many geographical scales. As our reach is extended through population growth, concepts of equilibrium (Malthus 1798) tell us that we will someday surpass a breaking point when long-term negative impacts override the short-term positive feedback of our greed even at the planetary scale (Diamond 2005; Gause 1934; Malthus 1798; Mayr 1970; Murdoch et al. 2003; Terborgh 1999; Wilson 2003). But because evolution cannot predict the future (Mayr 1942, 1970; Dawkins 2013), we must depend upon our intellect to overcome the positive feedback loop that competitive exclusion (Gause 1934; Hardin 1960) has yielded for ourselves—if indeed we are to survive well into the future. Although our intellect is undoubtedly one more product of evolution, this particular offspring may not be powerful enough to overcome its progenitor's *modus operandi* with sufficient lead time to avoid serious outcomes (Margulis 1997). While we are certainly subject to evolutionary pressures, intelligence should give us an alternative to simply bowing to them.

As our numbers increase within a finite setting with a relatively finite set of resources, all the limitations that affected the *Paramecium* species in Gause's aquaria as well as all our monocultural crops will eventually come to present stronger negative feedback on our own population. Although eucalyptus and black walnut trees have tremendous capacities in the realm of allelopathy, they have not independently taken over the world due to limitations imposed by both biotic and abiotic factors. For ourselves, we have been able to use our wit to minimize many of those same kinds of limitations for some time, but we do not have the capacity to eliminate them altogether.

In the last century, humans have evolved to become behavioral allelopaths. Recognizing that evolutionary selection, a force possibly even greater than our base selfishness, is integral to how we interact with nature does not resolve any of the challenges associated with our growing numbers (ongoing conversion of the surrounding landscape and overuse of natural resources), but it can give us better context for understanding how and why we have arrived at current conditions without reacting in a more profound and consequential manner. Recognizing the nature of this mechanism does not provide any excuse for our actions, but should give us greater opportunities to manage our behavior so as to increase the longevity of humankind within the confines of our blue planet fishbowl. We've been able to control much of the nature around us. Will we now be able to control our own nature?

6.6 A Sense of Entitlement Due to Religious Beliefs

Since we started to notice that human impacts on nature are indeed substantial, the scientific community has logically made efforts to gather data, analyze the numbers, and thereby develop arguments in order to make a change in the way the general public views purported unsustainable harvest, habitat loss, species extirpations, and extinctions. After a half century of communicating related messages, we have accomplished little beyond cementing the support of those who were already interested in nature and associated causes (Brown 1988, Norton et al. 1995). For the masses, campaigns to bring attention to concerns about emblematic or commercially important species have even fueled mistrust between environmentalists and those who desire to make money from natural resources or simply need to feed their families. Instead of using the strategy of singling out "guilty" parties and looking to punish them (financially in most countries) as a method of bringing them in line, there must be a better way to recruit a greater proportion of the world's population into the ranks of those who genuinely care about their surroundings, resources, and sustainability—even if purely for purposes of self-interest.

As it turns out, most people on the planet are affiliated with some religion—about half belong to the Judeo-Christian-Mohammedan lineages alone. Nearly every religion or traditional belief system, whether organized on the global scale or practiced only by tiny indigenous groups in scant remote villages, includes a creation story (Renard 2002). Typically, this involves special intervention by a supreme being (or beings) who, through a series of miraculous acts, has brought into existence all the plants and animals of the Earth. A nearly universal part of all these cosmologies is extraordinary respect for and worship of this creator, undoubtedly out of awe for such almighty power. According to traditional teachings, subsequent to the creation event(s), members of the favored kind (all humanity or some specific group of humans) were granted access to everything across the planet, giving rise to the mindset that all that exists was brought forth specifically for human use or consumption. It has been pointed out, however, that said "dominion over nature" should not be interpreted as a license to damage or destroy but rather as a contract of stewardship (Cobb 1988).

Since Darwin's time, in the Western World at least, there has been a particularly wide and growing schism between science and religion (Dunn 2010). Ironically, since before the Renaissance, it was often religious scholars who had made most observations about nature and tried to provide understanding of natural phenomena for the masses. In modern times, however, scientists have come to essentially discount the perspectives of religious figures and, as if in reciprocation, some sects in some regions have grown more skeptical of the teachings of science with the passage of time. It is no news that these two arenas have come to have a rather adversarial relationship (Thiele 2011); this friction is definitely not compatible with overcoming the situation or even managing it.

Everyone, religious or not, occupies space and consumes resources. Another fact is that the human race is responsible directly or indirectly for the loss of most species that are being driven toward extinction (Simberloff 1986; Wilson 1992). In the modern world, with ever-increasing concern for the resources that we require and for the species that are disappearing all around us, the time is ripe to bring all these believers into the fold that works with the scientific community toward sustainability and conservation. Considering that for billions of religious individuals, all species created through divine intervention represent the tangible works of a "higher power", each and every one should be, by default, considered worthy of respect and protection (Heltne 2008). In turn, the elimination of any species would be seen as a sign of disrespect or even an insult to its creator. To make an analogy, if we like an artist or value his/her works, we make very special efforts to maintain those pieces in the best possible condition for as long as possible. The pieces that do not get curated in order to guarantee their survival, protection, and longevity are simply not considered worthy. By the same token, if we allow species to disappear, or actually drive them to extinction, are we insulting the works themselves or their creator? Given this scenario, one might logically conclude that truly religious individuals should be conservationists of the greatest fervor. And since such a large proportion of Earth's human inhabitants categorize themselves as being devoted to one religion or another, a reminder of this situation should result in the decrease of habitat degradation, over-exploitation, and extinctions.

As it turns out, however, religion has been a major factor in the civilization of the world which means conversion of wilderness into exploited lands. The influence of religion at the level of work ethics has played a major role in our behavior and has been an important part of the driving force that has allowed humankind to dominate the planet (Weber 1905; Boltanski and Chiapello 2005). This sets up a serious conflict of interests between basic desires to continually improve quality of life, which requires greater consumption of resources and occupation of space, and any respect for the original producer of those resources (commonly known as the creator of everything).

Keeping in mind that the best estimates for species loss at the planetary level are tens of thousands per year, conservation efforts in every corner of the globe require all the support that can be mustered (Raven and Wilson 1992). No matter how a person thinks or believes species showed up originally, the current pulse of extinctions cannot be controlled or reversed without broader acceptance of culpability and the involvement of a large portion of the world's religious contingent, without any need to target specific denominations.

During several months in mid-2015, many important religious leaders have sequentially stepped forward to make statements about human responsibility in relation to everything from global climate change to the sixth mass extinction. Pope Francis, advised by a prestigious team of scientists, published an extensive encyclical (2015) that provided some accounting of our destruction as well as ample instruction as to what our next steps should be. In short order, the Dalai Lama joined him with a related message. Numerous Jewish leaders, a council of hundreds of rabbis, soon followed suit and, by mid-August, Islamic leaders had put forth a similar plea to all Muslims of the world.

Science has a choice: continue to struggle against mass extinction nearly single-handedly in direct conflict with the greedy and the needy or recruit the masses to become part of the fight based on existing widespread beliefs and the associated implied responsibilities.

6.7   Conclusion

Given so much inextricable dependence, if we cannot understand or justify optimal management and protection for the world ecosystem, if we do not exhibit the necessary forethought and actions to sustain that which sustains all earthly biota, including our own species, what can be realistically expected for the future of the planet or humankind itself?

Edward Abbey (1975) spoke out about loss of wilderness and even advocated sabotage of construction equipment and infrastructure to avoid facilitating access to untouched areas. In the present, use of the word "sabotage" should probably be applied amply to what humans are doing to the natural world and to their own future.

References

Abbey E (1975) The monkey wrench gang. Penguin, London
Baker S, Clapham PJ (2004) Modeling the past and future of whales and whaling. Trends in Ecology and Evolution 19(7):365–371
Boltanski L, Chiapello E (2005) The new spirit of capitalism. Verso, London
Brown LR (1988) And today we're going to talk about biodiversity... That's right, biodiversity. In: Peter FM, Wilson EO (eds) Biodiversity. National Academy Press, Washington, DC, pp 446–449
Cairns J Jr (1988) Increasing diversity by restoring damaged ecosystems. In: Peter FM, Wilson EO (eds) Biodiversity. National Academy Press, Washington, DC, pp 333–343
Carson RL (1962) Silent spring. Houghton Mifflin, New York
Cobb JB Jr (1988) A Christian view of biodiversity. In: Peter FM, Wilson EO (eds) Biodiversity. National Academy Press, Washington, DC, pp 481–485

Daily GC (1997) Nature's services: societal dependence on natural ecosystems. Island Press, Washington, DC

Dawkins R (1976) The selfish gene. Oxford University Press, Oxford

Dawkins R (2013) Commentary in "The unbelievers". A Black Chalk Film. http:// unbelieversmovie.com/. Accessed 20 October 2015

Diamond J (1999) Guns, germs and steel; the fates of human societies. Norton, New York

Diamond J (2005) Collapse: how societies choose to fail or succeed. Penguin, New York

Gause GF (1934) The struggle for existence. Williams & Wilkins, Baltimore

Gibbons W (1993) Keeping all the pieces: perspectives on natural history and the environment. Smithsonian Institution, Washington, DC

Gleason HA (1922) On the relationship between species and area. Ecology 3:158–162

Gould SJ, Lewontin RC (1979) The spandrels of San Marco and the Panglossian paradigm: a critique of the adaptationist programme. Proc R Soc Lond B Biol Sci 205:581–598

Hardin G (1960) The competitive exclusion principle. Science 131(3409):1292–1297

Heltne PG (2008) Imposed ignorance and humble ignorance—two worldviews. In: Vitek B, Jackson W (eds) The virtues of ignorance: complexity, sustainability and the limits of knowledge. University of Kentucky Press, Lexington, pp 135–150

Louv R (2005) Last child in the woods. Algonquin Books, Chapel Hill

Lovejoy TE (1988) Diverse considerations. In: Wilson EO, Peter FM (eds) Biodiversity. National Academy Press, Washington, DC, pp 421–427

Lynch A (1996) Thought Contagion: how belief spreads through society. The new science of memes. Basic Books, New York

MacArthur RH, Wilson EO (1967) The theory of island biogeography. Princeton University Press, Princeton

MacDonald GM (2003) Biogeography: introduction to space, time and life. Wiley, New York

Malthus TR (1798) An essay on the principle of population. Johnson, London

Margulis L (1997) A pox called Man. In: Margulis L, Sagan D (eds) Slanted truths: essays on Gaia, symbiosis, and evolution. Springer, New York, pp 247–261

Mayr EW (1942) Systematics and the origin of species. Columbia University Press, New York

Mayr EW (1970) Populations, species and evolution. Harvard University Press, Boston

Miller JD (2002) Civic scientific literacy: a necessity in the 21st century. Journal of the Federation of American Scientists 55(1), http://www.fas.org/faspir/2002/v55n1/scilit.htm. Accessed 13 August 2015

Mittermeier C (2014) Enoughness. TEDxVailWomen. https://www.youtube.com/watch? v=Xw8U5LXaItM. Accessed 20 July 2015

Murdoch WM, Briggs CJ, Nisbet RM (2003) Consumer-resource dynamics. Princeton University Press, Princeton

Norton BG, Hutchins M, Stevens EF, Maple TL (eds) 1995. Ethics on the Ark. Smithsonian Institution Press, Washington, DC. Pew research: religion and public life project: the global religious landscape. http://www.pewforum.org/2012/12/18/global-religious-landscape-exec/. Accessed 15 June 2015

Phys.org website (2012) 5 October. The sad state of biodiversity. http://phys.org/news/2012-10-sad-state-biodiversity.html. Accessed 15 June 2015

Pitman NCA, Terborgh J, Silman MR, Núñez P (1999) Tree species distributions in an upper Amazonian forest. Ecology 80:2651–2661

Pitman NCA, Widmer J, Jenkins CN, Stocks G, Seales L, Paniagua F, Bruna E (2011) Volume and geographical distribution of ecological research in the Andes and the Amazon, 1995-2008. Tropical Conservation Science 4(1):64–81

Pope Francis (2015) Laudato Si. Vatican Press. www.novusordowatch.org/wire/laudato-sii-francis.htm. Accessed 20 June 2015

Quinn D (1992) Ishmael. Bantam, New York

Randall A (1988) What mainstream economists have to say about the value of biodiversity. In: Wilson EO, Peter FM (eds) Biodiversity. National Academy Press, Washington, DC, pp 217–223

Renard J (2002) The handy religion answer book. Visible Ink Press, Canton

Reuth RG (1990) Goebbels: the life of Joseph Goebbels, the Mephistophelian genius of Nazi propaganda. Harcourt Brace, New York, p 471

Seuss D (1971) The Lorax. Random House, New York

Simberloff D (1986) Are we on the verge of a mass extinction in tropical rain forests? In: Elliot DK (ed) Dynamics of extinction. Wiley, New York

Suárez E (2010) La falacia del desarrollo sustentable. Polémika 2(5):102–109

Swing K (2014) Evolution in relation to environmental impacts and extreme species loss. In: Trueba G (ed) Why does evolution matter? Cambridge Scholars, Newcastle, pp 77–86

Terborgh J (1999) Requiem for nature. Island Press, Washington, DC

Wallace S (2013) Rain Forest for Sale. National Geographic, January: 82-119

Weber M (1905) The Protestant ethic and "The Spirit of Capitalism". Roxbury, Los Angeles, Translated by Stephen Kalberg (2002)

Westwood JO (1833) On the probable number of species of insects in the Creation; together with descriptions of several minute Hymenoptera. Magazine of Natural History VI:116–123

Willis RJ (2007) The history of allelopathy. Springer, New York

Wilson EO (1992) The diversity of life. Norton, New York

Wilson EO (2000) A global biodiversity map. Science 289:2279

Wilson EO (2003) The future of life. Vintage, New York

World Future Fund. The safe water crisis in the third world, http://www.worldfuturefund.org/wffmaster/Charts-HTML/wff-sanitation.htm. Accessed 14 May 2015

Chapter 7
Biogeochemistry in the Scales

S.A.F. Bonnett, P.J. Maxfield, A.A. Hill, and M.D.F. Ellwood

Abstract Global environmental change is challenging our understanding of how communities as a whole interact with their physical environment. Ideally, we would model the impacts of global environmental change at a global level. However, in order to mathematically model the sheer functional diversity of Earth's dynamic ecosystems, we need to integrate the scales at which these processes operate. Traditionally, studies of ecosystem function have focused on singular ecological, evolutionary or biogeochemical process within an environment. Such studies have contributed much more to the development of our understanding of ecosystem function than those focused on the interactions between biotic and abiotic factors. Ultimately, the productivity of most ecosystems is controlled by the concentration, molecular form and stoichiometry of the macronutrients thereby highlighting the importance of biogeochemical modelling for dynamic ecosystem models across molecular, habitat, landscape and global scales. But as we face unprecedented rates of habitat degradation and species extinctions, few traditional theories can predict in detail how ecosystems will respond to perturbations such as environmental disturbance or shifting weather patterns. To be both statistically and ecologically informative, future ecosystem and biogeochemical models must address complex interactions from atoms to ecosystems. Unless ecological processes are modelled explicitly, significant feedbacks, thresholds and constraints will be missed. The aim of this chapter is to review the state of the art in the use of such models, and suggest new approaches for ecologists, biogeochemists and mathematicians to work together to model the inputs and outputs of entire ecosystems rather than as a series of individual interactions.

Keywords Biogeochemistry • Modelling • Ecosystems • Scales • Global environmental change • Functional diversity

S.A.F. Bonnett (✉) • P.J. Maxfield • A.A. Hill • M.D.F. Ellwood
Department of Applied Sciences, University of the West of England,
Frenchay Campus, Bristol BS16 1QY, UK
e-mail: sam.bonnett@uwe.ac.uk

© Springer International Publishing Switzerland 2017
J.N. Furze et al. (eds.), *Mathematical Advances Towards Sustainable Environmental Systems*, DOI 10.1007/978-3-319-43901-3_7

7.1   Introduction

Biogeochemistry is a multidisciplinary subject dealing with chemical, physical, geological and biological processes that drive the flow of matter and energy within the Earth system. In particular, the biogeochemical cycling of water, carbon (C), nitrogen (N) and phosphorus (P) between abiotic and biotic pools are considered essential to the sustainability of ecosystem functions and the biosphere as a whole. That productivity of most ecosystems is in part controlled by the concentration, molecular form and stoichiometry of the macronutrients highlights the importance of biogeochemical modelling for dynamic ecosystem models across molecular, habitat, landscape and global scales.

Global environmental change is challenging our understanding of how communities as a whole interact with their physical environment. Soils, plants and microbial communities respond to global change perturbations through coupled, non-linear interactions (Sistla et al. 2014) that complicate projecting how global change disturbances will influence ecosystem processes, such as carbon (C) storage and greenhouse gas climate feedback loops. Ideally, we would model the impacts of global environmental change at a global level. However, in order to mathematically model the sheer functional diversity of Earth's dynamic ecosystems, we need to integrate the scales at which these processes operate. Biogeochemical cycles operate at a landscape or global scale but are composed of countless microbial functions that have been experimentally examined using an array of distinct laboratory and field-based experiments at differing scales with differing degrees of explanatory power and system representativeness. Issues of scale are also apparent in mathematical models of complex systems that rely on either a top-down or bottom-up approach. Top-down models can detect general trends at a larger scale but cannot produce specific, detailed predictions such as stability of ecological systems. Bottom-up models require detailed micro-scale process equations that have been evaluated through empirical studies and are therefore inherently complex and impractical for ecological systems. Most models therefore incorporate both top-down and bottom-up into what is known as Middle-out modelling. However, this approach still lacks the resolution from a true 'bottom-up' approach. In ecology, natural microcosms are used as model communities for testing links between biodiversity and ecosystem function. Issues in this field range from a lack of standardization between models, or artificiality in experiments using laboratory-assembled communities, to problems of complexity with experiments based on entire ecosystems. However, carefully chosen natural microcosms offer ecologists a way forward. To be both statistically and ecologically informative, future ecosystem and biogeochemical models must address complex interactions from atoms to ecosystems. Unless ecological processes are modelled explicitly, significant feedbacks, thresholds and constraints will be missed.

The aim of this chapter is to review the state of the art in the use of such models, and suggest new approaches for ecologists, biogeochemists and mathematicians to work together to model the inputs and outputs of entire ecosystems across scales rather than as a series of individual interactions. Problems of scale in biogeochemistry are similar to those in ecology and mathematics. In this chapter, we will use examples from disciplines where scale has been more fully addressed to the field of biogeochemistry.

7.2 Loose Definitions and the Problems of Scale

7.2.1 Views of Experimental Scale Across Scientific Disciplines

The most accurate way to parameterize a biogeochemical process is to study it at its most irreducible scale. Biogeochemical cycles at the micro-scale are driven by microbial functions (see Sect. 7.3.2), and the establishment of microcosm experiments to investigate the environmental controls over specific biogeochemical processes offer significant explanatory power. Due to the sheer number of different micro-scale processes that are operating within just one biogeochemical cycle, and the vast range of microcosm experiments that have been established to study these processes, it is beyond the scope of this chapter to give a comprehensive overview of the range of studies that have been performed. Some of the most significant microcosms that have been utilized in biogeochemical modelling are discussed below. It should be noted that whilst this approach is most accurate it is by no means necessarily the most effective, as the heterogeneity of the environment can render microcosm manipulation experiments meaningless if sufficient care is not taken to embed them within a wider scientific context. Furthermore, the isolation of microcosms from the wider environment can cause significant problems when interpreting and re-integrating experimental results into their wider context.

Within certain disciplines, mesocosms are favoured as more representative experimental systems. It is difficult to clearly differentiate between microcosms and mesocosms based on a simple definition as there are differences of opinion on what constitutes a microcosm or a mesocosm. A range of different criteria have been used to differentiate mesocosms from microcosms. The three main factors include size, location (field vs. laboratory) and isolation (whether the system is open or closed to external exchange). In addition, these definitions can vary greatly between different scientific areas of study from a definition based purely on scale (e.g. freshwater hydrology) (Cooper and Barmuta 1993) to a definition that also incorporates experimental setting (e.g. soil ecology) and as such, many experimental systems labelled as microcosms in certain scientific disciplines would be considered as mesocosms in other areas of research. A good example is that under certain definitions the Biosphere 2 project in the Arizona desert during the 1980s, which explored the potential use of closed biospheres for Mars colonization, has been categorized as a microcosm despite its size as it was a fully enclosed system (Odum 1984). In the field of soil ecology, Kampichler et al. (2001)

attempted to clarify the definition of mesocosms as systems within the realm of microcosms or 'enclosed model ecosystems' that were a special type of microcosm which allowed for controlled exchange between the model system and the wider environment (Fig. 7.1a). As such mesocosms are therefore the type of microcosm that was most representative of the real world (or macrocosm). One problem with this definition is that it negates the effect of the size of a model system; for example, the interaction between invertebrates and two different plant species within a greenhouse could be considered a mesocosm, whereas the entire greenhouse

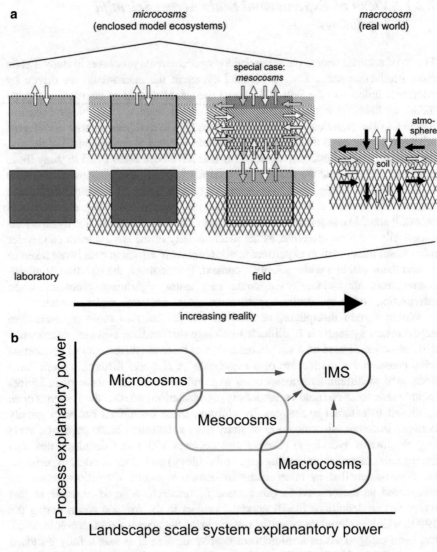

Fig. 7.1 (a) Mesocosms as systems. (b) Explanatory power of an Idealized Model System

would be classified as a microcosm. In reality, despite attempts to clarify the definition of a mesocosm, the term is used very loosely and is often applied to model systems that are larger in size than those classified as microcosms.

As indicated by the above proposal, a macrocosm is representative of the real-world system and as such is not enclosed and places no restriction on the free movement of nutrients and organisms (as would be the case with microcosms, and to a degree mesocosms). In terms of scale, experimental results from macrocosm experiments are far easier to relate to global biogeochemical cycles. However, in terms of explanatory power, the precision of macrocosm experiments to explain changes to discreet processes within complex systems can be limited, a fact not clearly recognized by Kampichler et al. (2001).

Figure 7.1a shows mesocosms as systems within the realm of microcosms or 'enclosed model ecosystems' that are a special type of microcosm which allows for controlled exchange between the model system and the wider environment (Kampichler et al. 2001); Fig. 7.1b shows the explanatory power of an idealized model system (IMS).

7.2.2 Problems of Experimental Scale

Irrespective of the precise definition of microcosm, mesocosm and macrocosm, it is apparent that manipulation experiments within microcosms offer the highest degree of explanatory power for discreet biogeochemical processes, whereas experimental macrocosms are far more powerful and representative of biogeochemical cycles at the landscape scale. Thus, an idealized model system (IMS) would encapsulate the explanatory power of both of these experimental settings (Fig. 7.1b).

The challenge is to identify ideal model systems (whether micro-, meso- or macrocosms) that can best bridge the gap between accurately explaining specific biogeochemical processes, whilst encapsulating these within a broader suite of interconnected processes that is more representative of the operation of a biogeo-chemical cycle at the landscape scale.

7.3   Mathematical Modelling Approaches

7.3.1   Top-Down and Bottom-Up Modelling

Traditionally, two alternative modelling methodologies based on the top-down and bottom-up approaches, respectively, have been used in describing complex systems cf. Noble (2003). In the top-down methodology, the approach begins with a high-level functional model of the entire system and then successively replaces each functional block with a model of the mechanism that implements it. The obvious advantage of this approach is the ability to detect general trends in the system at every stage without requiring every possible interaction to be modelled. In the context of ecosystems, this had been a widely implemented approach that has led to a substantial increase in our understanding of the spatial and temporal scales of biophysical interaction at both global and regional scales, e.g. Moore et al. (2004), Soberón (2010). Mathematical modelling on an ecological macro-scale usually relies on modern and stochastic control theory (Astrom 2006), optimization theory (Lasdon 2011) and parameter estimation (Aster 2011). In essence, these approaches attempt to encapsulate a generalized system through macro-scale data, where the exact model is not known. However, due to its inherent 'black box' approach which averages out the mechanisms of lower level components, a top-down methodology does not tend to produce specific, detailed predictions. Another potential issue lies in its ability to accurately model the stability of ecological systems in a changing environment as a top-down approach tends to create a self-regulating model which relies heavily on recreating macro-scale data, and thus tends to focus on stability to environmental perturbation, removing the effects of potential destabilizing lower level components.

In the bottom-up methodology, the global state of all the components is assumed to be impossible to obtain, so the approach begins with the properties of the component parts and their interactions, and higher level processes are constructed by assembling these detailed components. Micro-scale modelling relies heavily on mean-field techniques such as statistical mechanics, master equations and dynamical systems theory (Gaspard 2005). In essence, these approaches recreate individual physical interactions through well-defined models (i.e. basic first principles), and then combine them to attempt to encapsulate higher level interactions. Although it is clear that this approach has the potential to include all interactions of a system, it is usually impractical to manage the computational complexity of a fully encompassing, bottom-up approach for most ecological systems.

7.3.2 Middle-Out Modelling

Mathematical modelling on the meso-scale adopts a 'middle-out' approach, which starts at an intermediate level of scale and reaches out to link with larger and smaller scale components. This can be implemented using a combination of mathematical approaches form both top-up and bottom-down methodologies and can potentially encapsulate the lower level effects of a system, without suffering from over generality, see e.g. Allen and Fulton (2010), Creutzig et al. (2012).

At any scale, whether macro, micro or meso, mathematical models rely on conceptually derived equations, which are parameterized and evaluated through empirical studies. Intuitively, to maximize the effectiveness of a model in addressing the hypothesis being tested, it should be constructed at an appropriate level of complexity and data availability to support it, cf. Noble (2003). Existing models can certainly be used (in any of the approaches), but they should always be assessed as to their usefulness in their current state, and be pliable to revision. 'Therefore rather than slaving ourselves to a single approach, we should consider a balanced combination of all three (top-down, bottom-up and middle out) approaches and focus on the following criteria; a good model should be, descriptive (represents the available data), integrative (demonstrate how elements interact), explanatory (provide biological insight)' (Allen and Fulton 2010).

It is important to note that to develop a truly representative mathematical model of any ecological system, the dialogue between the modellers and practitioners needs to be substantial. Without context and statistical data for verification, any model would have no validity, and similarly, without modelling expertise to construct a model at the most applicable level, context and statistical data will provide limited predictive benefits.

7.3.3 Example of Biogeochemical Integration of Top-Down, Bottom-Up and Middle-Out Modelling

Peatland C dynamics are a subject of great current interest as they are intrinsically linked with the atmospheric concentration of methane (CH_4), a potent greenhouse gas. Due to the heterogeneity and widespread distribution of natural wetlands, accurately modelling C fluxes from peatlands has been problematic. Net soil C flux is predominantly controlled by soil hydrology, and thus approaches to integrate and model the global impact of peatland C emissions have been based on a top-down modelling approach linked with the soil hydrological cycle.

Whilst this approach has proven effective at integrating global C flux, there are a number of uncertainties linked with a lack of knowledge of the geographical distribution and interannual variability of CH_4 emissions from natural wetland ecosystems. Unfortunately, many of the processes governing CH_4 exchange between ecosystems and the atmosphere are poorly constrained on the global scale. Measurements of localized C emissions have indicated a number of both temporal and areal discrepancies with observations. There are a range of reasons for these inaccuracies, but one major factor is the lack of accurate vegetational model components, which are not captured by top-down approaches. Typically, biogeo-chemical processes linked with CH_4 exchange have been considered as one process linked with wetland emissions, when in reality a number of processes contribute to the net flux and estimates of global emissions, which have only been calculated from simple parameterizations (e.g. Kaplan (2002)).

Bottom-up or inversion modelling approaches have been far more widely applied in dynamic global vegetational models, which provide a versatile platform for studying the interactions between vegetation and C cycling. In recent years, attempts have been made to formulate similar 'bottom-up' emission models for natural wetlands (Zhuang et al. 2004; Wania et al. 2010), incorporating soil thermal dynamics, peatland hydrology, and peatland-specific plant functional types into a dynamic global vegetation model. It is hoped that the development of these models will increase the explanatory power of global models of peatland C flux, but many are still at a relatively early developmental stage.

Spahni and co-workers (2011) have attempted an intermediate model of global wetlands through combining a 'bottom-up' model of the biogeochemical cycle of the land biosphere with two 'top-down' models of atmospheric inversions, incorporating atmospheric chemistry and CH_4 emission transport. To constrain this 'middle-out' approach, both local emissions (flux data) and global observations (satellite data) were incorporated and estimated outputs correlated with real-world observations (Spahni et al. 2011). Whilst this approach offers improvements over previous estimates, which have broadly characterized all natural terrestrial CH_4 sources as 'wetlands', the model still lacks the necessary resolution that would be obtained from a true 'bottom-up' approach.

7.4 How Do We Model Complex Ecosystems?

7.4.1 Biodiversity

Biodiversity is also a multidisciplinary subject, dealing with many species, a wide range of evolutionary relationships, and a multitude of ecological interactions. Ecologists attempting to study biodiversity must therefore by definition consider more than one species at a time. To do this, ecologists often use natural microcosms (Srivastava et al. 2004) as model communities for testing the effects of metacommunity theory or the links between biodiversity and ecosystem function (Ellwood and Foster 2004). Such microcosm studies combine the advantages of small size, short generation times, contained structure and hierarchical spatial arrangement with the authenticity of field studies: natural environmental variance, and realistic species combinations with shared evolutionary histories (Srivastava et al. 2004). An important distinction between biology and ecology is that—for ecologists—the model system is often the environment, or the community, rather than a specific organism.

Biologists make extensive use of organisms as model systems. Classic examples include the bacteria *Escherichia coli* for molecular biology, the pipid frog *Xenopus* for developmental biology, the fruit fly *Drosophila melanogaster* for animal genetics and the thale cress *Arabidopsis thaliana* for plant genetics. Such biological models have proven to be powerful research tools because they are tractable, general and realistic (Levins 1984). With the notable exceptions of *Tribolium* beetles for population ecology, or Darwin's finches for evolutionary ecology, ecologists have overcome their lack of model systems by developing their own model ecological communities by using natural microcosms. Ecologists have also had to overcome the problem of many of their model systems meeting some but not all of the criteria necessary for future experiments to effectively standardize and utilize previous results. For example, laboratory-assembled communities of protozoa which enable rapid, precise and replicated experiments (Lawler 1998; Jessup et al. 2004) have been criticized for their artificiality (Carpenter 1996). Experiments based on entire ecosystems represent the other extreme: whilst entire natural communities are highly realistic, their large size and vague dimensions usually necessitate poorly replicated, long duration, overly simplified studies. This has led to heated debate among ecologists focused on the issue of replication versus realism (Schindler 1998). However, carefully chosen natural microcosms offer ecologists a way forward; in other words, natural microcosms allow ecologists to study biodiversity in a similar manner to the way biologists study genetics or developmental biology using model organisms. However, ecological model systems (natural microcosms) must satisfy the same criteria as model organisms which, although they do not correspond completely to the systems that they represent, must be similar enough to be useful. For example, the nematode *Caenorhabditis elegans* is clearly different from humans. However, both have eukaryotic genomes, meaning that the small nematode genome could be a useful model for understanding how the larger human genome functions. Again this raises the issue of scale: in biology, where model systems are used for theory testing and development, the theory may need to be parameterized for other systems (e.g. evolution theory developed based on experiments with *Drosophila* needs to be adjusted for selection coefficients and modes of inheritance before it can be applied to other organisms (Srivastava et al. 2004)). Likewise, rates and processes observed in ecological model systems may need to be scaled up to describe the natural ecosystem.

7.4.2 Biogeochemistry

Terrestrial ecosystem biogeochemistry models have been used to simulate the productivity and carbon storage of a diversity of ecosystems (Baisden and Amundson 2003). Models have been used to evaluate ecosystem management (Riley and Matson 2000) and to predict or understand the response of ecosystems to global changes including nitrogen deposition (Holland et al. 1997), CO_2 fertilization (Paustian et al. 1996), land-use change (Parton et al. 1996) and global climate change (Schimel et al. 1996). Traditionally, process-based and data-driven biogeochemical models have been conceptualized, parameterized, calibrated and validated using datasets from monitoring and experimental networks (e.g. Rothamsted Experimental Station established in 1843, the International Biological Program (1970–1975), Hubbard Brook Ecosystem Study began in 1964, Long Term Ecological Research Network (LTER) started in 1980, the global network of flux towers (FluxNet) (Evrendilek 2012)). The related literature contains numerous process-based (mechanistic) biogeochemical models such as BIOME-BGC (BioGeochemistry Cycles) (Running and Hunt 1993), CENTURY (Parton et al. 1987), TEM (Terrestrial Ecosystem Model) (Raich et al. 1991) and DNDC (Denitrification-Decomposition) (Li et al. 1992) as well as data-driven (empirical) models such as multiple regression models, artificial neural networks and geostatistical models. The Global Change and Terrestrial Ecosystems Soil Organic Matter Network (GCTE-SOMNET) database (online at http://saffron.rothamsted.bbsrc.ac.uk/cgi-bin/somnet/) contains metadata on over 30 current operational soil organic matter (SOM) models (see Falloon and Smith 2009).

An SOM model simulates either the decomposition of SOM as whole or as soil organic carbon (SOC), N or other nutrient and may be part of a larger model package, e.g. RothC (Coleman et al. 1997) and CENTURY models (Falloon and Smith 2009). Most models are process based as they focus on the processes mediating the movement and transformations of matter or energy and usually assume first-order rate kinetics (Falloon and Smith 2009). Food web models simulate C and N transfers through a food web of soil organisms (Falloon and Smith 2009) explicitly accounting for different trophic levels or functional groups of biota in the soil. Many studies have used SOM models to assess SOM dynamics at the regional, national and global scales such as in the post-Kyoto Protocol debate on the ability of terrestrial biosphere to store carbon (IPCC 2000). The RothC SOM model is concerned only with soil processes and is therefore not linked to plant production. The CENTURY SOM model, however, is part of a larger ecosystem model that simulates crop, grass and tree growth and the effects of different management practices on both plant production and SOM (Falloon and Smith 2009). Both models have been adapted to simulate N and S dynamics, but only CENTURY simulates P dynamics. RothC has been applied to databases of soil and land use and climate in the UK (Falloon et al. 2006). Upscaling the results from process-based soil–plant models to assess regional SOC change and sequestration potential is a great challenge due to the lack of detailed spatial information, particularly soil properties (Luo et al. 2013). Meta-modelling can be used to simplify and summarize process-based models and significantly reduce the demand for input data and thus could be easily applied on regional scales. For example, Luo et al. (2013) used the pre-validated Agricultural Production Systems SIMulator (APSIM) to simulate the impact of climate, soil and management on SOC at reference sites across Australia's cereal-growing regions under a continuous wheat system. They developed a simple meta-model to link the APSIM-modelled SOC change to primary drivers and used the meta-model to assess SOC sequestration potential and the uncertainty associated with the variability of soil characteristics. Sistla et al. (2014) provide an example of an ecosystem-scale model (SCAMPS—Stoichiometrically Coupled, Acclimating Microbe–Plant–Soil model) that simulates the dynamic feedbacks between above-ground and below-ground communities that affect their shared soil environment. This model has shown that incorporating dynamically interacting microbial and plant communities into ecosystem models might increase the ability to link ongoing global change field observations with macro-scale projections of ecosystem biogeochemical cycling in systems under change.

7.4.3 Potential Solutions (Principle of Model Systems in Ecology)

Whilst ecological model systems (natural microcosms) may not always represent direct analogues of other ecological systems, in the way that *C. elegans* can model the human genome, they can serve as scaled analogues (Petersen et al. 2003; Schneider 2001). In this approach, rates are scaled to system size: for example, marine ecologists have derived empirical relationships relating productivity to mesocosm size. When scaled up to coastal systems, the mesocosms provide remarkably good predictions for productivity (Petersen et al. 2003). Natural microcosms may also be used for screening potential hypotheses before resources are committed to experiments in larger, slower systems such as forests or lakes. To this end, only rough concordance between the model and target systems is needed, such that effects seen in the model system (e.g. rock pools, pitcher plants or epiphytes) are likely to be worth investigating in the target system (e.g. lakes or forests). However, as well as screening hypotheses, natural microcosms have been used extensively as systems for testing and developing theory.

7.5 A Way Forward for Integrating Biogeochemical and Ecosystem Models Using Natural Microcosms

7.5.1 Mathematical Models

The key issue when modelling complex ecosystems is not necessarily the number of components to be modelled, but how they interact. Systems which can be successfully modelled by considering averaged effects can be entirely represented using a bottom-up approach. However, most ecosystems contain non-linear interactions, making it impossible to simply reduce the system's behaviour to the sum of its parts (combining separately modelled biogeochemical processes for example). Feedback loops can play a significant role in these non-linear interactions (Kell and Knowles 2006), where the outputs of the system impact onto its inputs. An understanding of such loops and their effects is central to building and understanding models of complex systems (Milo et al. 2002).

The potential way forward to address this crucial issue for the mathematical modelling of ecosystems is to explore natural microcosms as a simulation for the larger system. By modelling a microcosm, for which the inputs and outputs can be controlled experimentally, an understanding of its interaction mechanisms can be developed. The possibility of experimental control allows for the significant testing of model accuracy, which would not be possible in larger uncontrollable systems. Once a model is constructed, substantial hypothesis testing can then be performed, allowing one to analyse the effects of manipulating experimental conditions in the model without having to perform multiple experiments. Even if the microcosm is a

Table 7.1 Modelling considerations for simulating microcosms

Dimension or feature	Possible choices	Comments
Stochastic or deterministic	*Stochastic*: Monte Carlo methods, or statistical distributions	Phenomena are not of themselves either stochastic or deterministic; large-scale, linear systems can be modelled deterministically, whilst a stochastic model is often more appropriate when nonlinearity is present
	Deterministic: equations, e.g. ordinary differential equations (ODEs), partial differential equations (PDEs)	
Fully quantitative versus partially quantitative versus qualitative	*Qualitative*: direction of change modelled only, or on/off states (Boolean network)	Reducing the quantitative accuracy of the model can reduce complexity greatly and many phenomena may still be modelled adequately
	Partially quantitative: fuzzy models	
	Fully quantitative: ODEs, PDEs	
Predictive versus exploratory/ explanatory	*Predictive*: specify every variable that could affect outcome	If a model is being used for precise prediction or forecasting of a future event, all variables need to be considered. The exploratory approach can be less precise but should be more flexible, e.g. allowing different control scenarios to be tested
	Exploratory: only consider some variables of interest	

heavily idealized version of the larger ecosystem, it can suggest areas of investigation (such as regions of instability) that would not have been predictable utilizing a system which is uncontrollable experimentally.

When deciding on the type of mathematical model for the microcosm, there are several strategic choices to be made when deciding what may be most appropriate to simulate the physical system. Table 7.1 (based on Table (1.2) of Kell and Knowles (2006)) suggests the three main mathematical considerations which should be made for this simulation.

When applying these approaches to complex systems, the key factor is a model system that enables the delineation of specific interactive processes in order that the interactions can be characterized. We will consider the possibilities through examples taken from ecological and biogeochemical studies.

7.5.2 Ecology Models

Natural microcosms, the model systems of ecology, are small, discrete habitats containing natural communities of organisms (Srivastava et al. 2004). Examples include the animal communities of aquatic *Phytotelmata* (e.g. pitcher plants, bromeliads) and of non-aquatic epiphytes such as birds nest ferns (Ellwood and Foster 2004; Fig. 7.2). Other natural microcosms include the microarthropod communities of moss patches and crustacean communities in rock pools. Natural

Fig. 7.2 Bird's nest ferns are islands of biodiversity in a sea of canopy

microcosms are, by definition, small habitats, facilitating replicated experimentation and allowing for robust statistical analysis. Another feature of natural microcosms is that they are contained habitats with clearly delineated arenas for species interactions. Compared with food webs and communities in continuous habitats, natural microcosms represent ideal experimental units allowing the precise delineation of communities, without problems of artificial inclusion or exclusion of animals (Krebs 1996). On the contrary, the physical boundaries between natural microcosms and their surrounding habitat—for example, air or water—can aid in the removal or addition of species (Kneitel and Miller 2002; Fincke et al. 1997). In some cases, entire communities can be assembled from scratch (Ellwood et al. 2009; Miller et al. 2002a; Srivastava and Lawton 1998; Bengtsson 1989; Levine

2001). However, natural microcosms are not closed systems; eventually, all experimental manipulations will be altered by, for example, the emergence of adult insects, oviposition of eggs and colonization by microorganisms (Srivastava et al. 2004). However, these dispersal processes can be valuable in their own right for answering questions about metacommunity dynamics.

Owing to the relatively small size of natural microcosms, the organisms inhabiting them are also small: insects and smaller arthropods (e.g. amphipods, mites and collembola), annelids, protozoa (e.g. rotifers, ciliates and flagellates) and bacteria. Small size (<5 mm) and rapid generation times allow studies of long-term effects to be completed in a matter of weeks or months (Ellwood et al. 2009; Bengtsson 1989; Miller et al. 2002b). In contrast, even when studies are conducted over several years, experiments with larger organisms often capture only transient dynamics (Tilman 1989).

Understanding the scale and importance of ecological processes is a key goal in ecology. The usefulness of natural microcosms will therefore depend on their physical, chemical and biological properties—in relation to the theoretical questions being asked. In terms of scale and process, natural microcosms are ideal for testing two very active areas of ecological theory: the effects of neighbouring communities on species richness and the effect of declining diversity on ecosystem function. It is this second area which is most relevant to the current discussion. In the face of mass extinctions and global environmental change, it is critical that we understand the effects of species loss on the stability of ecosystem functions. Previous studies have focused on manipulation of monotrophic communities such as grassland plants in 1 m^2 plots; whilst such studies are useful for developing theory, we must move beyond these singular interactions to multitrophic food webs of coevolved species experiencing real patterns of species loss (Duffy 2003; Srivastava 2002). It is here that ecology's model system (the natural microcosm) has potential to play a particularly important role in diversity–function research programmes. Local extinctions can be easily (and ethically) induced by changes in the habitat; and the effects of changes to species diversity on ecosystem function can be monitored over multiple generations and through multiple trophic levels.

An example of local extinction being experimentally induced is by the fragmentation of moss patches before tracking the effects on microarthropod biomass (Gonzalez and Chaneton 2002). This study noted a lag effect between species extinctions and declining microarthropod biomass because it was the rare microarthropod species that were the first to disappear, which had minimal impact on community biomass. Community biomass was most affected by the declining abundance of common species as they neared extinction. This highlights the fact that real extinctions can occur long before the effects are observed in the functioning of ecosystems (Gonzalez and Chaneton 2002). Rock pools offer another natural microcosm for diversity–stability studies. An open question in ecology is whether increased species diversity can reduce variability in ecosystems (Kampichler et al. 2001). By using natural variation in the faunal diversity of different rock pools, positive diversity–stability relationships have been shown at the levels of both populations and communities. These relationships were, however, only observed when the confounding effects of environmental variation (Kaplan et al. 2002) or habitat specialization (Kell and Knowles 2006) were removed. Tropical epiphytes, another example of natural microcosms, have been used to tease apart the effects of environmental gradients and ecological succession on the relative importance of interspecific competition versus stochastic processes instructuring rainforest arthropod communities (Ellwood et al. 2009). In all of these examples, natural microcosms have provided genuine ecosystems in which to test theory developed from mathematical models or synthetic communities, revealing how theory can be modified to incorporate real patterns in extinctions or community assembly.

Possibly, the ecology of natural microcosms will be dominated by non-equilibrium dynamics caused by dispersal, local extinction and spatial structure. For example, drought and treefalls have catastrophic effects on the fauna of individual treeholes (Kneitel and Miller 2002) but minor effects on the larger forest ecosystem. Diurnal variation in the temperature of rock pools exceeds that in large bodies of water (Krebs 1996). However, such processes are important in all communities, regardless of size, and most taxa in natural microcosms have short generation times. Once disturbance and dispersal are scaled by generation times, there may be little difference between natural microcosms and larger systems (Lawler 1998). Spatial effects could similarly be scaled by body size of organisms. This type of biological scaling has rarely been attempted but is a fertile area for future research (Srivastava et al. 2004).

7.5.3 Biogeochemical Models

Whilst biological model systems are relatively widespread, model systems have received less emphasis in ecological and biogeochemical studies. As discussed above, a 'model system' is a system—which could be a gene and its regulators, an organism or an ecosystem—that displays a general process or property of interest, in a way that makes it understandable. Oceanic islands are a good example of such a system that is relatively large in scale. They are useful as analogues of model systems as they are identical to natural ecosystems in many ways but are (naturally) clearly bounded and therefore simpler than other similar systems. Even though oceanic islands are not as controllable as would be in an ideal model system, islands can be used a scaled analogue of the wider ecosystem, similar to the way C. Elegans can represent a scale analogue of the human genome (as discussed in Sect. 7.3.1). Thus, an island is an integrated, natural example of a larger complex system where the natural functioning of the island ecosystem can shed light on the functioning of much more complicated systems, whilst retaining a full suite of natural functions. Whilst islands simplify the wider natural environment (similarly to a microcosm), realistic processes and dynamics are retained because they are complete, real systems. In contrast, artificial microcosms are not always authentic as they can fail to include many ecosystem processes. Furthermore, the use of

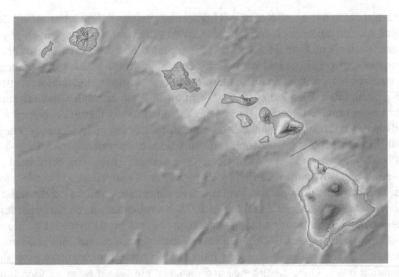

Fig. 7.3 Hawaiian archipelago

islands as analogues of model systems is that their unique setting can isolate specific variables, which can enable unique insights to be made.

One of the most well-known island systems used in biogeochemical models is the Hawaiian archipelago (Fig. 7.3). A range of ecological studies within these bounded tropical forests have revealed a number of scientific principles that are also relevant to larger comparative ecosystems, which can commonly be hidden by the added complexity of larger systems. The main factors that ensure Hawaii is an effective model system include unique geographical isolation, and the constancy and predictability of a number of environmental variables. In the case of Hawaii, a well-studied aspect of the island's geological setting is the effect that substrate age (from 300 to 4.1 Myr) has on the island's ecology and habitats. Thus, process studies in this environment offer the rare opportunity for naturally isolating and characterizing specific biogeochemical processes within a real complex ecosystem setting. Vitousek (2002) used this approach to identify the crucial limiting nutrients within the system, and the inputs and potential influence of external nutrient sources including marine aerosols and of Aeolian dust transported from central Asia. Crucially, it is not that Hawaii uniquely receives more of these external inputs than other tropical forests, but that the model system allows a much clearer characterization of specific phenomena within an ecosystem that inform in general settings much more widely (e.g. old, deeply leached soils depend on arid ecosystems hundreds to thousands of miles away for nutrient inputs).

The approach ecologists have taken (see previous section) for modelling biodiversity across landscapes using natural mesocosms is a good example for biogeochemists to follow. Benton et al. (2007) recently argued that small-scale experiments using 'model organisms' in microcosms or mesocosms can be a useful approach for apparently intractable global problems, such as ecosystem responses to climate change or managing biodiversity through the design of nature reserves.

Increasing evidence from microcosm studies suggest that both the rate and magnitude of important below-ground processes, such as decomposition of organic matter and resulting nutrient release, depend not only on the biomass of the decomposer community but are also influenced by the detrital food web (Setälä et al. 2005). However, evidence at the species level in microcosms and field experiments shows that there is considerable functional redundancy in soil species, suggesting little effect on rates of general decomposition processes with loss of species (Hunt and Wall 2002; Wall et al. 2005). As long as there is a functional group available to perform a particular role in a given ecosystem function, it may not matter whether there are many or a few species within the functional group. This contrasts markedly with experimental results from above-ground (plant) communities (Tilman et al. 1996). However, key species may sometimes strongly affect a range of soil-based ecosystem processes, including decomposition pathways, carbon and nutrient cycling, hydrologic pathways and the maintenance of soil structure, all of which interact in biogeochemical cycles and models (Wall et al. 2005). However, few studies exist in which the relationship between species diversity and system level process has been explicitly studied (Setälä et al. 2005). In an example of a microcosm study, Setälä and McLean (2004) manipulated the species richness of soil saprophytic fungi and showed that decomposition activity (CO_2 production) was only weakly related to the number of fungal species.

Whilst there is evidence suggesting the fundamental importance of structural complexity and diversity of detrital food webs in affecting below-ground processes, and ultimately, net primary production, species diversity per se may not be a significant factor controlling soil processes and plant growth. Most ecosystem-level models ignore microbial community dynamics, yet still do an adequate job of describing large-scale flows of carbon and nitrogen (Schimel 2001). Estimates indicate that less than 5 % of microbial species or less than 1 % of operational taxonomic units (OTUs) in soils are described, and we are therefore only beginning to connect identity to function (Wall et al. 2005). Indeed, an individual species may have an identifiable effect on function in microcosm experiments but at the mesocosm or macrocosm scale, species richness results in multiple indirect interactions making observations impossible. Evidence of significant horizontal gene transfer between prokaryotes has brought into question the definition and existence of bacterial species (O'Donnell et al. 2005). Horizontal gene transfer in prokaryotes and eukaryotes is correlated with gene function with marked differences in observed transfer frequencies between 'informational genes' (e.g. the 16S rRNA signature molecule for diversity) and 'operational genes' (e.g. those involved in amino acid synthesis). This 'complexity hypothesis' implies that sequence data from 'informational genes' may not be truly representative for studying diversity–function relationships in soil where function (phenotype expression) is largely coded for by operational genes (O'Donnell et al. 2005). However, the advent of next-generation sequencing, microarrays and other environmental genomic, metabolomic and proteomic technologies is improving our understanding of diversity and function relationships. A better understanding of functional redundancy in natural mesocosms could lead to simplification and advancement of complex landscape and global biogeochemical models.

7.6 Summary

Ecosystems are in part controlled by the concentration, molecular form and stoichiometry of the macronutrients, highlighting the importance of biogeochemical modelling for dynamic ecosystem models across molecular, habitat, landscape and global scales. Global environmental change is challenging our understanding of how communities as a whole interact with their physical environment. In order to mathematically model the functional diversity of Earth's dynamic ecosystems, we need to integrate the scales at which these processes operate. Microbial functions have been experimentally examined using an array of distinct micro, meso and macrocosm laboratory and field-based experiments with differing degrees of explanatory power and system representativeness for given scales. Issues of scale are also apparent in mathematical models of complex systems that rely on either a top-down or bottom-up approach. In ecology, natural microcosms are used as model communities for testing links between biodiversity and ecosystem function but through careful selection, natural mesocosms may offer a new way forward for biogeochemical modelling as scaled analogues or mathematical simulations that can also be used for testing and developing theory. In conclusion, the natural microcosm has potential to play a particularly important role in diversity–function research programmes linking biogeochemistry, ecology and mathematics across multiple scales.

References

Allen JJ, Fulton EA (2010) Top-down, bottom-up or middle-out? Avoiding extraneous detail and over-generality in marine ecosystem models. Progr Oceanogr 84:129–133
Aster R (2011) Parameter estimation and inverse problems. Academic, Waltham
Astrom KJ (2006) Introduction to stochastic control theory. Dover, New York
Baisden WT, Amundson R (2003) An analytical approach to ecosystem biogeochemistry modelling. Ecol Appl 13:649–663
Bengtsson J (1989) Interspecific competition increases local extinction rate in a metapopulation system. Nature 340:713–715
Benton TG, Solan M, Travis JMJ, Sait SM (2007) Micocosm experiments can inform global ecological problems. Trends Ecol Evol 22:516–521
Carpenter SR (1996) Microcosm experiments have limited relevance for community and ecosystem ecology. Ecology 77:677–680
Coleman K, Jenkinson DS, Crocker GJ et al (1997) Simulating trends in soil organic carbon in long-term experiments using RothC-26.3. Geoderma 81:29–44
Cooper SD, Barmuta LA (1993) Field experiments in biomonitoring. In: Rosenberg DM, Resh VH (eds) Freshwater biomonitoring and Benthic macroinvertebrates. Chapman & Hall, New York, pp 399–441
Creutzig F, Popp A, Plevin R, Luderer G, Minx J, Edenhofer O (2012) Reconciling top-down and bottom-up modelling on future bioenergy deployment. Nat Clim Change 2:320–327
Duffy JE (2003) Biodiversity loss, trophic skew and ecosystem functioning. Ecol Lett 6:680–687
Ellwood MDF, Foster WA (2004) Doubling the estimate of invertebrate biomass in a rainforest canopy. Nature 429:549–551. doi:10.1038/nature02560

Ellwood MDF, Manica A, Foster WA (2009) Stochastic and deterministic processes jointly structure tropical arthropod communities. Ecol Lett 12:277–284. doi:10.1111/j.1461-0248. 2009.01284.x

Evrendilek F (2012) Modeling of spatiotemporal dynamics of biogeochemical cycles in a changing global environment. J Ecosyst Ecogr 2, e113. doi:10.4172/2157-7625.1000e113

Falloon P, Smith P (2009) Modelling soil carbon dynamics. In: Kutsch WL, Bahn M, Heinemeyer A (eds) Soil carbon dynamics. Cambridge University Press, Cambridge

Falloon P, Smith P, Bradley RI et al (2006) RothC UK: a dynamic modelling system for estimating changes in soil C at 1-km resolution in the UK. Soil Use Manag 22:274–288

Fincke OM, Yanoviak SP, Hanschu RD (1997) Predation by odonates depresses mosquito abundance in water-filled tree holes in Panama. Oecologia 112:244–253

Gaspard P (2005) Chaos, scattering and statistical mechanics. Cambridge University Press, Cambridge

Gonzalez A, Chaneton EJ (2002) Heterotroph species extinction, abundance and biomass dynamics in an experimentally fragmented microecosystem. J Anim Ecol 71:594–602

Holland EA, Braswell BH, Lamarque JF, Townsend A, Suleman J, Muller JF, Denterer F, Brasseur G, Levy H, Penner JE, Roelofs GJ (1997) Variations in the predicted spatial distribution of atmospheric nitrogen deposition and their impact on carbon uptake by terrestrial ecosystems. J Geophys Res Atmos 102(D13):15849–15866

Hunt HW, Wall DH (2002) Modelling the effects of loss of soil biodiversity on ecosystem function. Glob Chang Biol 8:33–50

IPCC (2000) Land use, land-use change, and forestry. A special report of the IPCC. Cambridge University Press, Cambridge

Jessup CM, Kassen R, Forde SE, Kerr B, Buckling A, Rainy PB, Bohannan BJM (2004) Big questions, small worlds: microbial model systems in ecology. Trends Ecol Evol 19:189–197

Kampichler C, Bruckner A, Kandeler E (2001) Use of enclosed model ecosystems in soil ecology: a bias towards laboratory research. Soil Biol Biochem 33:269–275

Kaplan JO (2002) Wetlands at the Last Glacial Maximum: distribution and methane emissions. Geophys Res Lett 29(6):3-1–3-4

Kell DB, Knowles JD (2006) The role of modeling in systems biology. In: Szallasi Z, Stelling J, Periwal V (eds) System modeling in cellular biology: from concepts to nuts and bolts. MIT Press, Cambridge, pp 3–18

Kneitel JM, Miller TE (2002) Resource and top-predator regulation in the pitcher plant (Sarracenia purpurea) inquiline community. Ecology 83:680–688

Krebs CJ (1996) Population cycles revisited. J Mammal 77:8–24

Lasdon LS (2011) Optimization theory for large systems. Dover, New York

Lawler SP (1998) Ecology in a bottle: using microcosms to test theory. In: Resetarits WJ Jr, Bernando J (eds) Experimental ecology: issues and perspectives. Oxford University Press, New York, pp 236–253

Levine JM (2001) Local interactions, dispersal, and native and exotic plant diversity along a California stream. Oikos 95:397–408

Levins R (1984) The strategy of model building in population biology. In: Sober E (ed) Conceptual issues in evolutionary biology. Cambridge University Press, Cambridge, pp 18–27

Li C, Frolking S, Frolking TA (1992) A model of nitrous oxide evolution from soil driven by rainfall events: 1. Model structure and sensitivity. J Geophys Res Atmos 97:9759–9776

Luo Z, Wang E, Bryan BA, King D, Zhao G, Pan X, Bende-Michl U (2013) Meta-modelling soil organic sequestration potential and its application at regional scale. Ecol Appl 23:408–420

Miller TE, Horth L, Reeves RH (2002a) Trophic interactions in the phytotelmata communities of the pitcher plant, Sarracenia purpurea. Community Ecol 3:109–116

Miller TE, Kneitel JM, Burns JH (2002b) Effect of community structure on invasion success and rate. Ecology 83:898–905

Milo R, Shen-Orr S, Itzkovitz S, Kashtan N, Chklovskii D, Alon U (2002) Network motifs: simple building blocks of complex networks. Science 298:824–827

Moore JK, Doney SC, Lindsay K (2004) Upper ocean ecosystem dynamics and iron cycling in a global three-dimensional model. Glob Biogeochem Cycles 18:4028

Noble D (2003) The future: putting Humpty-Dumpty together again. Biochem Soc Trans 31:156–158

O'Donnell AG, Colvan SR, Malosso E, Supaphol S (2005) Twenty years of molecular analysis of bacterial communities in soils and what have we learned about function? In: Bardgett RD, Usher MB, Hopkins DW (eds) Biological diversity and function in soils. Cambridge University Press, New York, pp 44–56

Odum EP (1984) The mesocosm. BioScience 34(9):558–562

Parton WJ, Schimel DS, Cole CV, Ojima DS (1987) Analysis of factors controlling soil organic matter levels in Great Plains Grasslands. Soil Sci Soc Am J 51:1173–1179

Parton WJ, Ojima DS, Schimel DS (1996) Models to evaluate soil organic matter storage and dynamics. In: Carter MR, Stewart BA (eds) Structure and organic matter storage in agroecosystems. Advances in soil science. CRC, New York, pp 421–448

Paustian K, Elliott ET, Peterson GA, Killian K (1996) Modelling climate, CO_2 and management impacts on soil carbon in semi-arid agroecosystems. Plant and Soil 187:351–365

Petersen JE et al (2003) Multiscale experiments in coastal ecology: improving realism and advancing theory. Bioscience 53:1181–1197

Raich JW, Rastetter EB, Melillo JM, Kicklighter DW, Steudler PA, Peterson BJ, Grace AL, Moore B, Vörösmarty CJ (1991) Potential net primary productivity in South America: application of a global model. Ecol Appl 1:399–429

Riley WJ, Matson PA (2000) NLOSS: a mechanistic model of denitrified N_2O and N_2 evolution from soil. Soil Sci 165:237–249

Running SW, Hunt ER (1993) Generalization of a forest ecosystem process model for other biomes, BIOME-BGC, and an application for global-scale models. In: Ehleringer JR, Field CB (eds) Scaling physiological processes: leaf to globe. Academic, San Diego, pp 141–158

Schimel JP (2001) Biogeochemical models: implicit vs. explicit microbiology. In: Schulze ED, Harrison SP, Heimann M et al (eds) Global biogeochemical cycles in the climate system. Academic, San Diego, pp 177–183

Schimel DS, Braswell BH, Makeown R, Ojima DS, Parton WJ, Pulliam W (1996) Climate and nitrogen controls on the geography and timescales of terrestrial biogeochemical cycling. Global Biogeochem Cycles 10:677–692

Schindler DW (1998) Replication versus realism: the need for ecosystem-scale experiments. Ecosystems 1:323–334

Schneider DC (2001) Spatial allometry: theory and application to experimental and natural aquatic ecosystems. In: Gardner RH et al (eds) Scaling relations in experimental ecology. Columbia University Press, New York, pp 113–153

Setälä H, McLean MA (2004) Decomposition rate of organic substrates in relation to the species diversity of soil saprophytic fungi. Oecologia 139:98–107

Setälä H, Berg MP, Jones TH (2005) Trophic structure and functional redundancy in soil communities. In: Bardgett RD, Usher MB, Hopkins DW (eds) Biological diversity and function in soils. Cambridge University Press, New York, pp 236–249

Sistla SA, Rastetter EB, Schimel JP (2014) Responses of a tundra system to warming using SCAMPS: a stoichiometrically coupled, acclimating microbe-plant-soil model. Ecol Monogr 84:151–170

Soberón JM (2010) Niche and area of distribution modeling: a population ecology perspective. Ecography 33:159–167

Spahni R, Wania R, Neef L, van Weele M, Pison I, Bousque P, van Velthoven P (2011) Constraining global methane emissions and uptake by ecosystems. Biogeosciences 8 (6):1643–1665. doi:10.5194/bg-8-1643-2011

Srivastava DS (2002) The role of conservation in expanding biodiversity research. Oikos 98:351–360

Srivastava DS, Lawton JH (1998) Why more productive sites have more species: an experimental test of theory using tree-hole communities. Am Nat 152:510–529

Srivastava DS, Kolasa J, Bengtsson J, Gonzalez A, Lawler SP, Miller TE, Munguia P, Romanuk T, Schneider DC, Trzcinski MK (2004) Are natural microcosms useful model systems for ecology? Trends Ecol Evol 19:379–384

Tilman D (1989) Ecological experimentation: strengths and conceptual problems. In: Likens GE (ed) Long-term studies in ecology. Springer, New York, pp 136–157

Tilman D, Wedin D, Knops J (1996) Productivity and sustainability influenced by biodiversity in grassland ecosystems. Nature 379:718–720

Vitousek PM (2002) Oceanic islands as model systems for ecological studies. J Biogeogr 29:573–582

Wall DH, Fitter AH, Paul E (2005) Developing new perspectives from advances in soil biodiversity research. In: Bardgett RD, Usher MB, Hopkins DW (eds) Biological diversity and function in soils. Cambridge University Press, New York, pp 3–27

Wania R, Ross I, Prentice IC (2010) Implementation and evaluation of a new methane model within a dynamic global vegetation model: LPJ-WHyMe v1.3.1. Geosci Model Dev 3 (2):565–584. doi:10.5194/gmd-3-565-2010

Zhuang Q, Melillo JM, Kicklighter DW, Prinn RG, McGuire AD, Steudler PA, Felzer BS, Hu S (2004) Methane fluxes between terrestrial ecosystems and the atmosphere at northern high latitudes during the past century: a retrospective analysis with a process-based biogeochemistry model. Global Biogeochem Cycles 18, GB3010

Chapter 8
Plant Metabolites Expression

H.A. Hashem and R.A. Hassanein

Abstract Plant metabolic engineering is an emerging discipline that promises to create new opportunities in agricultural research, environmental applications, production of chemicals, and even medicine. Crucial to the success of this technology is the enhanced understanding of plant metabolite production. This chapter reviews recent research on plant metabolite production and aims to illustrate the link between different elements in the metabolism regulation process. Considerable attention is given to the environmental factors affecting metabolite production in plants. Hormone metabolism has also emerged as a key factor in regulating the plant stress response and recent developments in defining the functional and genetic basis of plant hormones in regulating stress response will be addressed. Finally, this chapter will discuss both primary and secondary metabolism in plants and will highlight transcriptional regulation of secondary metabolite biosynthesis, including that of flavonoids, alkaloids, and terpenoids.

Keywords Metabolites • Transcriptional regulation • Stress • Genome • Plant hormones • Secondary metabolism • Signal transduction

8.1 Introduction

The ability to control metabolic processes in response to changes in internal or external environmental conditions is an indispensable attribute of living cells. This adaptability is necessary for maintaining a stable intracellular environment. Which is—in turn—essential for maintaining an efficient functional state (Plaxton and Mcmanus 2006). Although biochemists frequently employ the terms "regulation" and "control" interchangeably, the need to discriminate between these terms has been emphasized (Fell 1997). Metabolic control refers to adjusting the output of a metabolic pathway in response to an external stimulus. In contrast, metabolic

H.A. Hashem (✉) • R.A. Hassanein
Botany Department, Faculty of Science, Ain Shams University,
Khalifa El-Maamon st, Abbasiya sq., Cairo 11566, Egypt
e-mail: hashem.hanan@gmail.com

© Springer International Publishing Switzerland 2017
J.N. Furze et al. (eds.), *Mathematical Advances Towards Sustainable Environmental Systems*, DOI 10.1007/978-3-319-43901-3_8

regulation occurs when an organism maintains some internal conditions as constant over time, despite fluctuations in external conditions.

The gene is the basic hereditary unit in all living organisms and is defined as a segment of deoxyribonucleic acid (DNA) involved in the production of a polypeptide chain. Each gene is composed of a coding region along with regions preceding and following this coding region that are involved in regulating its expression. Genes may also contain intervening noncoding sequences that are located within the coding region. The coding region of a given gene provides information about the amino acid sequence of a specific protein and is also referred to as an open reading frame (ORF). Regulatory sequences, which usually precede the ORF, generally consist of specific DNA sequences that act as recognition sites for various proteins involved in ribonucleic acid (RNA) synthesis. In this way, regulatory sequences control the rate and amount of RNA that is synthesized. Transcription and translation are two major genetic events during which control of gene expression can be asserted.

Transcription is the process by which RNA is synthesized using a single strand of DNA as the template. The DNA sequence (genetic code) comprises triplet base coding units called codons. Once transcribed, each codon in the resulting mRNA (messenger RNA) strand directs the incorporation of a particular amino acid into the nascent protein primary sequence by the translational apparatus. Translation of mRNAs into their corresponding proteins occurs on ribosomes located in the cytoplasm or on the rough endoplasmic reticulum. Alternatively, genes located within the mitochondrial or chloroplast genomes are translated by ribosomes within these semiautonomous organelles (Lefebvre and Gellatly 1997; Kuhlemeier et al. 1987).

The advent of genomics, proteomics, and metabolomics has revolutionized the study of plant development and metabolism (Plaxton and Mcmanus 2006). Proteomics can be defined as the systematic analysis of the proteome, the protein complement of genome (Pandey and Mann 2000; Patterson and Aebersold 2003; Phizicky et al. 2003). Proteomics allows for the global analysis of gene products in various tissues and cell physiological states. With the sequencing of a growing number and variety of genomes and the development of analytical methods for protein characterization, proteomics has become an integral component of the field of functional genomics. The initial objective of proteomics was the large-scale identification of all protein species in a given cell or tissue. The applications have since expanded to include analysis of protein–protein interactions, posttranslational modifications, protein activities, and structures (Park 2004). New advances in proteomics include the use of highly efficient modern techniques such as bidimensional electrophoresis, multidimensional chromatography, mass spectrometry and bioinformatics tools, along with second-generation technologies for the analysis of polypeptides and proteins at tissue, organ, organelle, and membrane levels.

Metabolomics can be defined as the comprehensive identification and quantification of all the metabolites of an organism (Fiehn 2002). Along with proteomics, metabolomics contributes to our understanding of the complex molecular interactions that occur in biological systems (Hall et al. 2002). The biochemical response of an organism to a conditional perturbation can be characterized by the differential accumulation of individual metabolites (Raamsdonk et al. 2001).

In the last few years, significant advances in the study of plant metabolism have been made as enzyme gene families are elucidated, and new protein regulators are identified. Enzyme activity is the major factor influencing the magnitude of metabolic fluxes in any cell. Metabolic control may occur at several levels, beginning with gene transcription and proceeding through the various stages of protein synthesis and turnover. More rapid alterations in metabolic flux occur through activation and inhibition of key enzymes along the major metabolic pathways, particularly by mechanisms such as reversible covalent modification and by the actions of allosteric effectors molecules that reflect the cell's adenylate energy charge, oxidation/reduction potential, and/or the accumulation of metabolic end products (Plaxton and Mcmanus 2006). Discoveries concerning plant metabolic control continue to be made at a rapid rate, particularly in the field of signal transduction. Each new discovery adds to the assessment that plant signal transduction and metabolic control networks display remarkable intricacy. This chapter focuses on the links between different elements in the metabolism regulation process, with particular emphasis on the environmental factors affecting metabolites production in plants.

8.2   Metabolic Regulation

Plants obtain the major elements that make up the plant body—carbon, oxygen, hydrogen, and nitrogen—mainly as carbon dioxide, water, and nitrate. They also take up and use many other minerals and elements, albeit in much smaller quantities. This means of nutrition, from inorganic compounds, is known as autotrophy ("self-feeding"). Organisms that obtain their carbon and nitrogen only from organic compounds—that is, ultimately, from plants—have a form of nutrition known as heterotrophy ("other-feeding") (Smith et al. 2010).

Plants show huge genetic variation in metabolism, the multitude of interrelated biochemical reactions that maintain plant life. Metabolites are compounds synthesized by plants for both essential functions, such as growth and development (primary metabolites), and specific functions, such as pollinator attraction or defense against herbivore and abiotic environmental factor (secondary metabolites). Primary metabolites comprise many different types of organic compounds, including, but not limited to, carbohydrates, lipids, proteins, and nucleic acids. They are found universally in the plant kingdom because they are the components or products of fundamental metabolic pathways or cycles such as glycolysis, the Krebs cycle, and the Calvin cycle. Because of the importance of these and other primary pathways in enabling a plant to synthesize, assimilate, and degrade organic compounds, primary metabolites are essential. On the other hand, secondary metabolites are largely fall into three classes of compounds: alkaloids, terpenoids,

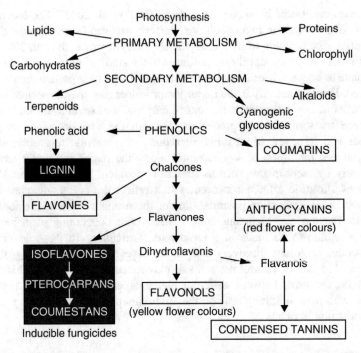

Fig. 8.1 Interrelationships between the primary and secondary metabolism in plants (Giada 2013)

and phenolics (Wink 2010). However, these classes of compounds also include primary metabolites, so whether a compound is a primary or secondary metabolite is a distinction based not only on its chemical structure but also on its function and distribution within the plant kingdom.

Many thousands of secondary metabolites have been isolated from plants, and many of them have powerful physiological effects in humans and are used as medicines (Wink 2008). It is only since the late twentieth century that secondary metabolites have been clearly recognized as having important functions in plants. Research has focused on the role of secondary metabolites in plant defense. This is discussed at the end of this chapter with reference to those variously distributed in the plant kingdom, and their functions are specific to the plants in which they are found (Fig. 8.1).

8.2.1 Complexity of Metabolism

Plant metabolic networks are more complex than those of other organisms. There are several interlinked aspects of plant life: sessile, ectothermic, and autotrophic. Plants have chemical repertories and a high degree of subcellular compartmentation, with the chloroplast being the most prominent compartment exclusive to plants. The metabolic complexity is a result of separate enzymatic reactions which make up the pathways constituting metabolism (Plaxton 1997).

The metabolic networks responsible for the assimilation and utilization of carbon, nitrogen, and other elements are highly regulated. Regulation allows the integration and coordination of processes, as well as rapid responses to environmental changes that directly influence plant metabolism, such as temperature, light intensity, and water availability. For example, enzymes of the metabolic pathway that assimilates carbon dioxide during photosynthesis (the Calvin cycle) are regulated by a host of factors that allow the rate of assimilation and the fate of the immediate products to respond rapidly and sensitively to change—in the availability of carbon dioxide inside the leaf, in the supply of energy from light-driven processes, and in the demand for organic compounds by the nonphotosynthetic cells of the plant.

8.2.2 Metabolic Control by Compartmentalization

Broadly speaking, metabolic processes in plant cells are controlled at two levels: by compartmentation of metabolic pathways into different organelles and by mechanisms that modulate the activities of individual enzymes or blocks of enzymes that catalyze the reactions of these pathways. We consider first the importance of compartmentation in increasing the potential for metabolic diversity in the cell, and then the ways in which metabolic processes are integrated and controlled by modulation of enzyme activities.

Compartmentalization of metabolism and the sequestration of metabolic pathways into membrane-bound organelles occur in all eukaryotes but are most apparent in the cells of higher plants. It is a means by which pathways can be regulated independently from one another and is a mechanism of controlling the flux of carbon through the pathways in response to the demands of cellular metabolism. In addition, the division of the cell into smaller volumes can serve to concentrate the intermediates of a pathway and may prevent futile cycling by opposing reactions. It may also provide specialized environments more favorable to certain reactions.

The major organelles found in plant cells are the nucleus, mitochondria, Golgi apparatus, microbodies, peroxisome, glycosomes, vacuoles, and plastids. In addition, it should not be forgotten that the cytoplasm is also a compartment in its own right. In some cases, the sequence of reactions that comprise a metabolic pathway may be located in more than one organelle.

The best known example in plant for metabolites coordination between cytosol and plastids is the metabolism of trehalose-6-phosphate (T6P) (Fig. 8.2). T6P is needed for reproductive development in maize (Satoh-Nagasawa et al. 2006) and *Arabidopsis* (Eastmond et al. 2002), and is also important for seed germination (Gómez et al. 2006) and sugar signaling (Kolbe et al. 2005). T6P is synthesized in the cytosol and is strongly correlated with sucrose levels and the rate of starch synthesis in the plastid (Lunn et al. 2006). Supplying T6P to isolated chloroplasts

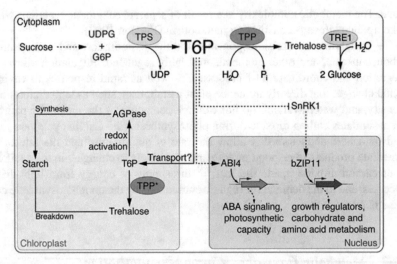

Fig. 8.2 Trehalose biosynthesis and its role in carbohydrate metabolism Trehalose-6-phosphate (T6P) is synthesized from UDP-glucose (UDPG) and glucose-6-P (G6P) by the activity of trehalose-6-phosphate synthase (TPS) and subsequently converted to trehalose by trehalose-6-phosphate phosphatase (TPP). trehalase1 (TRE1) hydrolyzes trehalose into two molecules of glucose. T6P plays a central role in regulating sugar metabolism in plants. The precursors of T6P are derived from the sucrose metabolism. It has been suggested that T6P is transported by an unknown mechanism into plastids where it induces starch synthesis via thioredoxin-mediated activation of AGPase. T6P might be converted into trehalose, which has been shown to regulate starch breakdown in plastids. Several TPPs (marked with an asterisk) have been predicted to localize to plastids, but this still needs to be confirmed experimentally. SnRK1, which represses plant growth, is inhibited by T6P. A regulatory loop, which involves T6P, SnRK1, and bZIP11, and that is thought to control sucrose availability and utilization, has been proposed.(Ponnu et al. 2011)

promotes redox activation of AGPase (Kolbe et al. 2005) although the molecular mechanism is not fully understood. The direct effects of T6P on signaling/target proteins still need to be characterized in detail. T6P may act allosterically through either SnRK1 (Sucrose nonfermenting protein kinase) or NADP-thioredoxin reductase C, which regulates AGPase (Michalska et al. 2009). T6P may act as a secondary messenger of carbon status between the cytosol and the chloroplasts (Kolbe et al. 2005).

8.2.3 Metabolic Control by Regulation of Enzyme Activities

The magnitude of metabolic fluxes through any metabolic pathway will depend upon the activities of individual enzymes involved. Two basic mechanisms can potentially be used by the cell to vary the reaction velocity of a particular enzyme. The first type, known as "coarse control," determines the amount of enzyme protein present in a cell. Coarse control operates through changes in the relative rates of synthesis and degradation (turnover) of the enzyme protein. As with any protein, the rate of synthesis depends on the rate of transcription of the gene encoding the enzyme (transcriptional control), the turnover of the messenger RNA (mRNA), and the rate of translation of the mRNA (translational control). Coarse control can be regarded as setting the capacity of the cell to catalyze a particular reaction or metabolic pathway. Generally speaking, it brings about changes in amounts of enzymes in hours or days rather than minutes. Coarse control can be applied to one or all of the enzymes in a particular pathway, and most frequently comes into play during tissue differentiation or long-term environmental (adaptive) changes. The other type of regulation, known as "fine control," works on a much shorter time scale to modulate the activities of enzymes within the framework set by coarse control. Thus, coarse control determines how many molecules of enzyme protein are present in the cell, while fine control determines the activity of these molecules (Ap Ress and Hill 1994).

Trehalose-6-phosphate (T6P) is synthesized from UDP-glucose (UDPG) and glucose-6-P (G6P) by the activity of trehalose-6-phosphate synthase (TPS) and subsequently converted to trehalose by trehalose-6-phosphate phosphatase (TPP). Trehalase1 (TRE1) hydrolyzes trehalose into two molecules of glucose. T6P plays a central role in regulating sugar metabolism in plants. The precursors of T6P are derived from the sucrose metabolism. It has been suggested that T6P is transported by an unknown mechanism into plastids, where it induces starch synthesis via thioredoxin-mediated activation of AGPase. T6P might be converted into trehalose, which has been shown to regulate starch breakdown in plastids. Several TPPs (marked with an asterisk) have been predicted to localize to plastids, but this still needs to be confirmed experimentally. SnRK1, which represses plant growth, is inhibited by T6P. A regulatory loop, which involves T6P, SnRK1, and bZIP11, and that is thought to control sucrose availability and utilization, has been proposed (Ponnu et al. 2011).

8.3 Role of Biotic and Abiotic Stresses in Plant Metabolite Expression

Stress in plants could be defined as any change in growth condition(s) that disrupts metabolic homeostasis and requires an adjustment of metabolic pathways in a process that is usually referred to as acclimation. External stress can have a biotic or abiotic origin. Biotic stress may be caused by the attack of herbivores and pathogen. Abiotic stress refers to the physical or chemical changes in the environment of the individual (Madlung and Comai 2004). Among the most common abiotic stresses are those related to drought, excess salt in the soil, extremes of temperature, and the presence of toxins contaminating the environment (Bhatnagar-Mathur et al. 2008). Plant are sessile beings, so the lack of mechanisms to escape from adverse conditions has fostered, through evolution, the development of unique and sophisticated responses to environmental stress. The chain of events that culminate in a response begins with the perception of a specific signal. This perception and signaling lead to gene expression changes, protein expression changes, the activation of new biochemical pathways, and repression of others which are characteristic of the unstressed state (Agrawal et al. 2010). These protective metabolic changes represent nothing less than the effort of this sessile organism to overcome the stress situation, maintain homeostasis and adapt (Hazen et al. 2003).

Stress inducers from abiotic as well as biotic factors have some common signal and response pathways in plants (Leakey et al. 2006 and Humphreys et al. 2006) and thereby have the potential to moderate each other's effects through cross-talking (Soltis and soltis 2003; Hongbo et al. 2005). However, apart from specific stress responses at the gene or metabolic level, there is a common signal transduction pathway model for stress, which is shared by many higher plants. This model proceeds through the perception of the environmental signal, and subsequently the production of a secondary messenger (such as inositol phosphates and reactive oxygen species— ROS), which will regulate the endogenous level of Ca^{2+}. From this point, a chain of events occurs that affects protein phosphorylation, reaching proteins linked to the protection of cellular structures or transcription factors controlling specific sets of stress-regulated genes. These genes are related to the production of regulatory molecules such as the plant hormones abscisic acid (ABA), ethylene, jasmonic acid (JA), and salicylic acid (SA). Some of these regulatory molecules can, in turn, initiate a second round of circulation (Macedo 2012).

In this part, the focus is on primary metabolites adjustments in response to abiotic stress (high salt stress, drought, and extreme temperatures) and our current knowledge of signal transduction components that regulate metabolite levels under these stress conditions will be summarized.

8.4 Osmotic Adjustment Imposed by Stress and Metabolic Compensation Mechanisms

Drought, salinity, and low temperature stress impose an osmotic stress that can lead to turgor loss. Membranes may become disorganized, proteins may undergo loss of activity or be denatured, and often excess levels of reactive oxygen species (ROS) are produced leading to oxidative damage. As a consequence, inhibition of photosynthesis, metabolic dysfunction, and damage of cellular structures contribute to growth perturbations, reduced fertility, and premature senescence.

Responses to environmental stresses occur at all levels of organization. Cellular responses to stress include adjustments of the membrane system, modifications of the cell wall architecture, and changes in cell cycle and cell division. In addition, plants alter metabolism in various ways, including production of osmoprotectants, such as sugars (trehalose, sucrose, and fructan), amino acids (tryptophan and proline), and ammonium compounds (polyamines and glycinebetaine) that are able to stabilize proteins and cellular structures and/or to maintain cell turgor by osmotic adjustment, and redox metabolism to remove excess levels of ROS and reestablish the cellular redox balance (Bartels and Sunkar 2005; Valliyodan and Nguyen 2006; Munns and Tester 2008; Janska et al. 2010).

Accumulation of amino acids has been observed in many studies on plants exposed to abiotic stress (Kaplan et al. 2004; Brosche et al. 2005; Zuther et al. 2007; Usadel et al. 2008, Lugan et al. 2010). This increase might stem from amino acid production and/or from enhanced stress-induced protein breakdown. While the overall accumulation of amino acids upon stress might indicate cell damage in some species (Widodo et al. 2009), increased levels of specific amino acids have a beneficial effect during stress acclimation.

For many years, the capacity to accumulate proline has been correlated with stress tolerance (Singh et al. 1972; Stewart and Lee 1974). Proline is considered to act as an osmolyte, an ROS scavenger, and a molecular chaperone stabilizing the structure of proteins, thereby protecting cells from damage caused by stress (Verbruggen and Hermans 2008; Szabados and Savoure 2010). Proline accumulates in many plant species in response to different environmental stresses including drought high salinity high light and UV irradiation, heavy metals, oxidative stress, and in response to biotic stresses (Saradhi et al. 1995, Fabro et al. 2004, Yang et al. 2009). Proline is produced in the cytosol or chloroplasts from glutamate, which is reduced to glutamate-semialdehyde (GSA) by Δ-1-pyrroline-5-carboxylate synthetase (P5CS). GSA can spontaneously convert to pyrroline-5-carboxylate (P5C), which is then further reduced by P5C reductase (P5CR) to proline. Proline is degraded in mitochondria by proline dehydrogenase (ProDH) and P5C dehydrogenase (P5CDH) to glutamate. Stress conditions stimulate proline synthesis while proline catabolism is enhanced during recovery from stress. Over-expression of P5CS in tobacco and petunia led to increased proline accumulation and enhanced salt and drought tolerance (Hong et al. 2000; Yamada et al. 2005), whereas *Arabidopsis* P5CS1 knock-out plants were impaired in stress-induced proline synthesis and were hypersensitive to salinity (Szekely et al. 2008).

Improvement of drought or salt tolerance of crop plants via engineering proline metabolism is an existing possibility and should be explored more extensively. The fact that proline can act as a signaling molecule and influence defense pathways, regulate complex metabolic and developmental processes, offer additional opportunities for plant improvement. Further studies are required to study the feasibility to engineer flowering time or to improve defenses against certain pathogens via the targeted engineering of proline metabolism.

8.4.1   Carbohydrate Metabolism

Starch is composed of glucose polymers arranged into osmotically inert granules. It serves as the main carbohydrate store in most plants and can be rapidly mobilized to provide soluble sugars. Its metabolism is very sensitive to changes in the environment. In addition to diurnal fluctuations in starch levels, salt and drought stress generally leads to a depletion of starch content and to the accumulation of soluble sugars in leaves (Kaplan and Guy 2004; Kempa et al. 2008). Sugars that accumulate in response to stress can function as osmolytes to maintain cell turgor and have the ability to protect membranes and proteins from stress damage (Madden et al. 1985; Kaplan and Guy 2004).

Hydrolysis of starch by the β-amolytic pathway represents the predominant pathway of starch degradation in leaves under normal growth conditions and may also be involved in stress-induced starch hydrolysis. *Arabidopsis sex1* (*starch excess 1*) mutants, that are impaired in glucan-water dikinase (GWD) activity, were compromised in cold-induced malto-oligosaccharide, glucose, and fructose accumulation during the early stages of cold acclimation and *sex1* plants that have been briefly pre-exposed to cold showed a reduced freezing tolerance (Yano et al. 2005). Osmotic stress was shown to enhance total β-amylase activity and to reduce light-stimulated starch accumulation in wild-type *Arabidopsis* but not in *bam1* (*bmy7*) mutants, who appeared to be hypersensitive to osmotic stress (Valerio et al. 2011). Similarly, BMY8 (BAM3) antisense plants accumulated high starch levels were impaired in cold-induced maltose, glucose, fructose, and sucrose accumulation, and showed a reduced tolerance of photosystem II to low temperature stress (Kaplan and Guy 2005). Data by Zeeman et al. (2004) also suggest a role of the phosphorolytic starch degradation pathway during stress.

In the desiccation-tolerant plant *Craterostigma plantagineum*, dehydration induces the conversion of 2-octulose, an eight-carbon sugar, to sucrose (Bianchi et al. 1991). This conversion correlates to increases in the gene expression for sucrose synthase (SUS) and sucrose phosphate synthase (SPS) (Ingram et al. 1997; Kleines et al. 1999), which are considered key enzymes of sucrose synthesis/ metabolism. Under conditions of dehydration/osmotic stress, the expression of genes coding for SUS isoforms is upregulated in several plants (Pelah et al. 1997; Déjardin et al. 1999). Similarly, antisense expression of the SPS coding sequence in potato plants completely suppressed the water stress-induced stimulation of sucrose synthesis (Geigenberger et al. 1999). Thus, SUS and SPS in plants are crucial steps in the acclimation process of dehydration. Highly soluble sugars, such as the polyfructose molecules fructans, are involved in plant and bacterial adaptation to osmotic stress. The introduction of fructosyltransferases to fructan non-accumulating tobacco and rice plants stimulated fructan production associated with enhanced tolerance to low-temperature stress and drought (Li et al. 2007; Kawakami et al. 2008). Fructans can stabilize membranes (Valluru and Van den Ende 2008) and might indirectly contribute to osmotic adjustment upon freezing and dehydration by the release of hexose sugars (Spollen and Nelson 1994; Olien and Clark 1995).

The nonreducing disaccharide trehalose accumulates to high amounts in some desiccation-tolerant plants, for example, *Myrothamnus flabellifolius* (Drennan et al. 1993). In living organisms, several functional properties have been proposed for trehalose: energy and carbon reserve, protection from dehydration, protection against heat, protection from damage by oxygen radicals and protection from cold. At sufficient levels, trehalose can function as an osmolyte and stabilize proteins and membranes (Paul et al. 2008). Transgenic expression of trehalose biosynthesis genes (Trehalose-6-phosphate synthase; TPS and trehalose-6-phosphate phosphatase; TPP) showed that enhanced trehalose metabolism can positively regulate tolerance to abiotic stress, even though only a limited increase in trehalose content could be observed, excluding a direct protective role of trehalose in these plants. Heterologous expression of genes involved in the trehalose pathway from *E. coli* or *Saccharomyces cerevisiae* enhanced tolerance to drought, salt, and low temperature stress in several plant species (Iordachescu and Imai 2008). Over-expression of different isoforms of TPS from rice conferred enhanced resistance to salinity, cold, and/or drought (Li et al. 2011).

8.4.2   The Active Role of Polyols in Protective Mechanisms

Polyols are implicated in stabilizing macromolecules and in scavenging hydroxyl radicals, thereby preventing oxidative damage of membranes and enzymes (Smirnoff and Cumbes 1989; Shen et al. 1997). Accumulation of the straight-chain polyols, mannitol and sorbitol, has been correlated with stress tolerance in several plants species (Stoop et al. 1996). Expression of mannitol-1-phosphate dehydrogenase (mtlD) from *E. coli*, which catalyzes the reversible conversion of fructose-6-phosphate to mannitol-1-phosphate in *Arabidopsis*, tobacco, poplar, and wheat induced mannitol accumulation and enhanced tolerance to salinity and/or water deficit (Abebe et al. 2003; Chen et al. 2005). Similarly, photosystem II was less affected by salinity in persimmon trees that accumulated sorbitol by over-expression of sorbitol-6-phosphate dehydrogenase (S6PDH) from apple (Gao et al. 2001).

The cyclic polyols *myo*-inositol, and its methylated derivatives D-ononitol and D-pinitol, accumulates upon salt stress in several halo-tolerant plant species (Sengupta et al. 2008). L-*myo*-inositol-1-phosphate synthase (MIPS) forms *myo*-inositol-1-P from glucose-6-P, which is dephosphorylated by *myo*-inositol 1-phosphate phosphatase (IMP) to form *myo*-inositol. Inositol *O*-methyltransferase (IMT) methylates inositol to form D-ononitol, which is epimerized to D-pinitol. In line with a stress-protective role, over-expression of MIPS and IMT from halo-tolerant plants increased cyclic polyols levels and salt stress tolerance of tobacco (Majee et al. 2004; Patra et al. 2010).

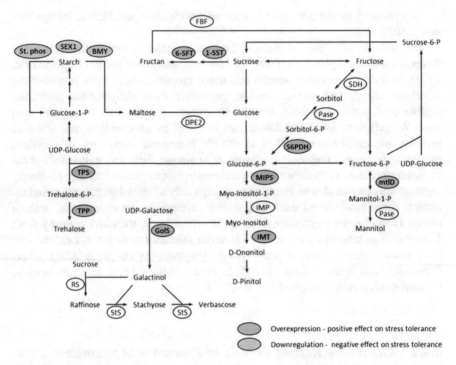

Fig. 8.3 Schematic overview of starch, fructan, sugar, and polyol metabolism (Krasensky and Jonak 2012)

Plants with enhanced or reduced activity of the indicated enzymes (shown here with their Enzyme Commission (EC) numbers indicating reactions they catalyze) show altered tolerance to abiotic stress. SEX1, α-glucan water kinase (EC 2.7.9.4); St.phos, starch phosphorylase (2.4.1.1); BMY, β-amylase (EC 3.2.1.2); DPE2, glucanotransferase (EC 2.4.1.25); 1-SST, sucrose:sucrose 1-fructosyltransferase (EC 2.4.1.99); 6-SFT, sucrose:fructan 6-fructosyltransferase (EC 2.4.1.10); FBF, fructan beta-fructosidase (EC 3.2.1.80); mtlD, mannitol-1-phosphate dehydrogenase (EC 1.1.1.17); S6PDH, sorbitol-6-phosphate dehydrogenase (EC 1.1.1.200); SDH, sorbitol dehydrogenase (EC 1.1.1.14); Pase, unspecific phosphatase; MIPS, inositol-1-phosphate synthase (EC 5.5.1.4); IMP, inositol-1-phosphate phosphatase (EC 3.1.3.25); IMT, inositol methyltransferase (EC 2.1.1.40); GolS, galactinol synthase (EC 2.4.1.123); RS, raffinose synthase (EC 2.4.1.82), StS, stachyose synthase (EC 2.4.1.67); TPS, trehalose-6-phosphate synthase (EC 2.4.1.15); TPP, trehalose-6-phosphate phosphatase (EC 3.1.3.12) (Fig. 8.3).

8.4.3 Amines

Polyamines (PAs) are small (low-molecular-weight), positively charged, aliphatic amines that are found in all living organisms. The major forms of PAs are putrescine (Put), spermidine (Spd), and spermine (Spm) although plants also synthesized a variety of other related compounds. Arginine (Arg) and ornithine (Orn) are the precursors of plant PAs. Ornithine decarboxylase (ODC; EC 4.1.1.17) converts Orn directly into Put. The other biosynthetic route to Put, via arginine decarboxylase (ADC; EC 4.1.1.19), involves the production of the intermediate agmatine (Agm) followed by two successive steps catalyzed by agmatine iminohydrolase (AIH, EC 3.5.3.12) and *N*-carbamoylputrescine amidohydrolase (CPA, EC 3.5.1.53). In animals and fungi, Put is synthesized primarily through the activity of ODC while in plants and bacteria the main pathway involves ADC. Aminopropyl groups, donated by decarboxylated *S*-adenosyl methionine (dcSAM), must be added to convert Put into Spd and Spm in a reaction catalyzed by spermidine synthase (SPDS; EC 2.5.1.16) and spermine synthase (SPMS; EC 2.5.1.22), respectively (Alcázar et al. 2010). Polyamine levels in plants increase under a number of environmental stress conditions, including drought and salinity (Kasinathan and Wingler 2004). Several biological roles were proposed for polyamines action in stress situations; PAs could act as osmoprotectants, as scavengers of active oxygen species (AOS) or by stabilizing cellular structures, such as thylakoid membranes (Alcázar et al. 2011; Hussain et al. 2011). The first reports of transgenic approaches using genes responsible for PA biosynthesis were conducted in two species, tobacco and rice (Capell et al. 2004, Roy and Wu 2002). Recently, PAs were proposed to be components of signaling pathways and fulfill the role of second messengers (Kuznetsov et al. 2006). Studies with ABA-deficient and ABA-insensitive *Arabidopsis* mutants with differential abiotic stress adaptations (Alcázar et al. 2005) support the conclusion that the up-regulation of PA biosynthetic genes and Put accumulation under water stress are mainly ABA-dependent responses.

8.4.4   Glycine Betaine

Glycine betaine (GB) is a quaternary ammonium compound that occurs in a wide variety of plants. *Arabidopsis* and many crop species do not accumulate GB. In plants that produce GB naturally, abiotic stress, such as cold, drought, and salt stress, enhances GB accumulation (Rhodes and Hanson 1993; Chen and Murata 2011).

GB can be synthesized from choline and glycine (Chen and Murata 2011). Introduction of the GB biosynthesis pathway genes into non-accumulators improved their ability to tolerate abiotic stress conditions (Park et al. 2004; Bansal et al. 2011) pointing to the beneficial role of GB in stress tolerance. Targeting GB production to the chloroplasts led to a higher tolerance of plants against stress than in the cytosol (Park et al. 2007). In salt-tolerant plant species, GB can also accumulate to osmotically significant levels (Rhodes and Hanson 1993). It has been suggested that GB protects photosystem II, stabilizes membranes, and mitigates oxidative damage (Chen and Murata 2011).

8.5   Utilizing Functional Genomics Approaches
       to Elucidate Plant Stress Responses

Completions of the genomes of *Arabidopsis* and *Oryza sativa* have added to the
information available on stress physiology. The absolute genome sequence is
nowaday accessible and also has enabled genomewide gene expression profiling
to a variety of abiotic stresses (Kilian et al. 2007). Complete transcriptome analysis
has facilitated the relationships between stress-regulated transcripts, and their
regulatory elements (Weston et al. 2008). High-throughput functional genomics
technologies, such as transcriptomics, metabolomics and proteomics are powerful
tools for investigating the molecular responses of plants to abiotic stress.
The advancement of these technologies has allowed for the identification and
quantification of transcript/metabolites in specific cell types and/or tissues
(Cramer et al. 2011). These platforms can be utilized to improve our ability to
discover the genes and pathways that control specific traits in response to abiotic
stress (Fig. 8.4).

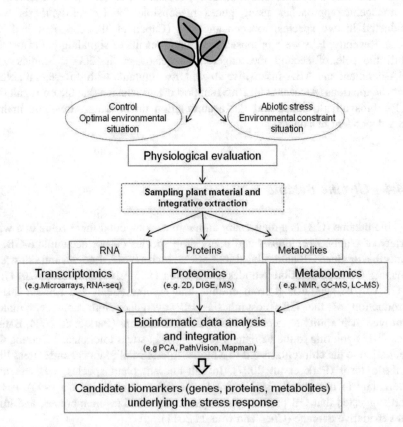

Fig. 8.4 Proposed strategy for the integration of physiology and systems biology (Duque et al.
2013)

Figure 8.4 strategy enables us to gain insights into abiotic stress responses in plants and the future development of abiotic stress-tolerant crops (Duque et al. 2013).

8.5.1   Signal Transduction Involved in Stress-Induced Metabolic Changes

Major components of the regulatory networks underlying environmental stress adaptation, pathogen recognition, and defense include reactive oxygen species (ROS) signaling (Miller et al. 2008), plant hormones (Bari and Jones 2009; Peleg and Blumwald 2011), changes in redox status (Munne-Bosch et al. 2013), and inorganic ion fluxes, such as Ca^{2+} (Martí et al. 2013). A start has been made to link signal transduction with the metabolite response upon drought, salt, and temperature stress conditions including targeted analyses of metabolites and metabolic profiling of mutants in signaling components.

8.6   Plant Hormones Have Pivotal Roles in Plant Stress Signaling

Plant hormones function as central integrators that link and reprogram the complex developmental and stress-adaptive signaling cascades. The phytohormone abscisic acid (ABA) functions as a key regulator in the activation of plant cellular adaptation to drought and salinity and has a pivotal function as a growth inhibitor (Cutler et al. 2010; Raghavendra et al. 2010; Weiner et al. 2010). Additionally, the view of function of ABA as a linking hub of environmental adaptation and primary metabolism is increasingly emerging. The stress responses are integrated and occur in the entire organism. The integration of the response requires the involvement of multiple hormones.

8.6.1   Abscisic Acid

ABA quickly accumulates in response to different environmental stress conditions and ABA-deficient plants have an altered stress response. ABA promotes stomatal closure, inhibits stomatal opening to reduce water loss by transpiration, induces the expression of numerous stress-related genes, and recent studies indicate a role in regulation of stress-induced metabolic adjustments. Comparison of the metabolic profile of *Arabidopsis* plants treated with ABA, or exposed to high soil salinity, revealed ABA-induced and ABA-independent steps of salt stress-induced metabolic rearrangements (Kempa et al. 2008). Both ABA and salt stress led to a depletion of starch and the accumulation of maltose. However, subsequent carbon flux appears to be differentially regulated. While glucose-6-phosphate and fructose-6-phosphate levels, and the glucose/fructose ratio, decreased under salt stress conditions, glucose-6-phosphate and fructose-6-phosphate levels and the glucose/fructose ratio increased in response to ABA treatment. Interestingly, ABA did not induce galactinol and raffinose accumulation (Kempa et al. 2008). In support of an ABA-independent induction of these sugars, *Arabidopsis* mutants deficient in stress-induced ABA accumulation were able to induce galactinol and raffinose under drought conditions (Urano et al. 2009). Similarly, stress-induced accumulation of citrate, malate, succinate (TCA cycle), and GABA is ABA independent (Kempa et al. 2008; Urano et al. 2009). In contrast, accumulation of many osmotic stress-induced proteinogenic amino acids, including proline, were induced and depended on ABA (Kempa et al. 2008; Urano et al. 2009).

8.6.2 Gibberellic Acid

Currently, increasing evidence has been emerging for modulation of ABA-mediated environmental signaling by interaction and competition with hormonal key regulators of plant cellular developmental and metabolic signaling. The complex and divergent endogenous and exogenous signals perceived by plant cells during development and environmental adversity are linked and integrated by distinct and interactive hormonal pathways. Particularly, convergence and functional modulation of ABA signaling by the plant growth regulating phytohormones gibberellic acid (GA) has a key regulatory function in the plant cellular network of stress and developmental signaling (Golldack et al. 2013).

8.6.3 Jasmonates

Recently, interesting evidence has been also provided for a convergence and cross talk of GA and ABA signaling with the developmental regulator jasmonate in plant responses to drought. Jasmonates are membrane lipid-derived metabolites that originate from linolenic acid and have signaling functions in plant growth and biotic stress responses (Wasternack 2007; Wasternack and Hause 2013). Drought-induced transcriptional regulation of the rice JA receptor protein *OsCOI1a* (CORONATINE INSENSITIVE 1) and of key regulators of JA signaling *OsJAZ* (jasmonic acid ZIM-domain proteins) indicate significant integration of JA metabolism and signaling in plant abiotic stress responses (Du et al. 2013).

8.7 Transcriptional Regulation of Secondary Metabolites

A typical trait of plants is the production of a high diversity of secondary metabolites (SM) (the number of identified substances exceeds 100,000 at present). These compounds are synthesized in plants in a tissue-, organ-, and developmental-specific way by specific biosynthetic enzymes (Facchini and De-Luca 2008; Murata et al. 2008). The corresponding genes are regulated accordingly and gene regulation shows all the complexity known for genes encoding enzymes of primary metabolism. Their function in plants is now attracting attention as some appear to have a key role in protecting plants from herbivores and microbial infection, as attractants for pollinators and seed-dispersing animals, as allelopathic agents, UV protectants, and signal molecules in the formation of nitrogen-fixing root modules in legumes. Secondary metabolites are also of interest because of their use as dyes, fibers, glues, oils, waxes, flavoring agents, drugs, and perfumes, and they are viewed as potential sources of new natural drugs, antibiotics, insecticides, and herbicides (Croteau et al. 2000; Dewick 2002; Schmidt et al. 2012). Between 1983 and 1994, 520 new drugs were approved. Of these 39 % were or were derived from natural products. On top of this, 60–80 % of antibacterial and anticancer drugs are derived from natural products (Harvey 2000).

Structurally and biosynthetically, SMs are classified into three major groups: terpenoids, phenolic compounds, and nitrogen-containing compounds. Terpenoids contain one or more C5 units, which are synthesized either via the cytosolic mevalonate pathway or the plastidial methylerythritol phosphate pathway. Phenolic compounds are highly diverse and include phenylpropanoids, coumarins, stilbenes, and flavonoids. Phenylpropanoids contain at least one aromatic ring with one or more hydroxyl groups and are synthesized via the shikimate pathway alone or in combination with the melonate pathway. The nitrogen-containing compounds are also highly diverse and include alkaloids, nonprotein amino acids, and amines. Their biosynthetic pathways usually have multiple routes, frequently starting from amino acids. To this point, more than 36,000 terpenoids, 12,000 alkaloids, and 10,000 flavonoids have been found although this represents only a fraction of what exists in nature (Chen et al. 2007; Fang et al. 2011).

The synthesis and proper accumulation of secondary metabolites are strictly controlled in a spatial and temporal manner and influenced by a number of biotic and abiotic factors. The spatio-temporal transcriptional regulation of metabolic pathways is controlled by a complex network involving many regulatory proteins known as transcription factors (TFs). TFs are sequence-specific DNA-binding proteins that recognize specific *cis*-regulatory sequences in the promoters of target genes and activate or repress their expression in response to developmental and/or other environmental cues. Some TFs do not bind DNA but interact with other cofactors to form complexes that regulate the expression of the target genes (Yang et al. 2012). Recent research has revealed that posttranscriptional and posttranslational mechanisms also play significant roles in the regulation of metabolic pathways (Li et al. 2012; Luo et al. 2012; Maier et al. 2013).

8.7.1 Terpenoids

The terpenes, or terpenoids, constitute the large class of secondary products. The diverse substances of this class are generally insoluble in water. They are biosynthesized from acetyl-CoA or glycolytic intermediates. Terpenes are biosynthesized from primary metabolites in at least two different ways. In the well-studied mevalonic acid pathway, three molecules of acetyle-CoA are joined together to form mevalonic acid (Fig. 8.5). This key six-carbon intermediate is then phosphorylated, decarboxylated, and dehydrated to yield isopentenyl diphosphate (IPP). IPP and its isomer dimethylallyl diphosphate (DMAPP) are the activated five-carbon building block of terpenes. IPP also can be formed from intermediated of glycolysis or the photosynthetic carbon reaction cycle via a separate set of reactions called the methylerythritol phosphate (MEP) pathway that operates in chloroplasts and other plastids (Lichtenthaler et al. 1997). The activities of three prenyltransferases produce the direct precursors of terpenes, the linear prenyl diphosphates geranyl diphosphate (GPP, C_{10}), farnesyl diphosphate (FPP, C_{15}), and geranylgeranyl diphosphate (GGPP, C_{20}). Terpene synthases (TPS) are the primary enzymes responsible for catalyzing the formation of hemiterpenes (C_5), monoterpenes (C_{10}), sesquiterpenes (C_{15}), or diterpenes (C_{20}) from the substrates DMAPP, GPP, FPP, or GGPP, respectively.

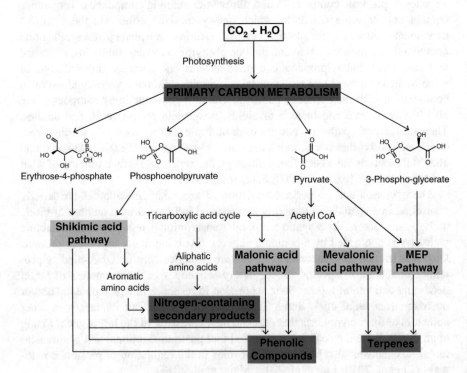

Fig. 8.5 Main pathways leading to secondary metabolites

The production of secondary metabolites is tightly associated with the pathways of primary metabolism, such as glycolysis, shikimate, and production of aliphatic amino acids (Uarrota et al. 2011).

Volatile and nonvolatile terpenes are implicated in the attraction of both pollinators and predators of herbivores, in protection against photo-oxidative stress and in mediating thermotolerance (Pichersky and Gershenzon 2002). Terpenes are toxins and feeding deterrents to many plant-feeding insects and mammals; thus they appear to play important defensive roles in the plant kingdom. For example, the monoterpene esters called pyrethroids that occur in the leaves and flowers of *Chrysanthemum* species show striking insecticidal activity. Both natural and synthetic pyrethroids are popular ingredients in commercial insecticides because of their low persistence in the environmental and their negligible toxicity to mammals. Given the increasing interest of researchers in the biological functions of plant volatiles, many studies are concerned with the role of terpene volatiles in the indirect defense of plants by attracting natural enemies of herbivores and plant parasites. Consequently, several *TPS* genes that are induced in leaves by herbivore attack or spider mite infestation, and which are responsible for the *de novo* biosynthesis of volatile monoterpenes and sesquiterpenes, have been identified in microarray and cDNA analyses of maize, *Medicago*, *Lotus*, and cucumber (Tholl 2006).

Indications of the posttranscriptional or posttranslational regulation of terpene synthases were obtained in studies of sesquiterpene volatile biosynthesis in flowers of *Arabidopsis* ecotypes (Tholl et al. 2005). In addition, analyses of the seasonal variation of isoprene formation in poplar leaves suggested posttranslational modifications of isoprene synthase activity (Mayrhofer et al. 2005). Furthermore, isoprene formation in poplar appears to be regulated at metabolic levels because concentrations of the substrate DMAPP and emissions of isoprene are subject to coordinated diurnal changes without alterations in isoprene synthase activity. In snapdragon flowers, monoterpene biosynthesis is closely correlated with the expression levels of the small subunit of GPP synthase, indicating a tight control of the GPP substrate pool for monoterpene formation during flower development (Tholl et al. 2004).

8.7.2 Alkaloids

Alkaloids are nitrogen-containing molecules mostly derived from amino acids such as tryptophan, tyrosine, phenylalanine, and lysine, as well as ornithine (De Luca and Pierre 2000). Alkaloids, including terpenoid indole alkaloids (e.g., vinblastine, vincristine), tropane alkaloids (cocaine, scopolamine), and purine alkaloids (caffeine), are known to protect plants from microbial or herbivore attack and from UV radiation. In general, alkaloids are infamous as animal toxins and certainly serve mainly as defense chemicals against predators (herbivores, carnivores) and to a lesser degree against bacteria, fungi, and viruses (Wink 2008). Alkaloids and amines often affect neuroreceptors as agonist or antagonists, or they modulate other steps in neuronal signal transduction, such as ion channels or enzymes, which take up or degrade neurotransmitters or second messengers. Since alkaloids often derive from the same amino acid precursor as the neurotransmitters, acetylcholine (ACh), serotonin, noradrenaline (NA), dopamine, gamma-aminobutyric acid (GABA), glutamic acid, or histamine, their structures can frequently be superimposed on those neurotransmitters (Wink 2010).

Large-scale expression analyses have recently begun to provide a broad picture of the gene expression profiles associated with alkaloid biosynthesis. Large-scale transcriptome and metabolite profiling has been performed in opium poppy cell cultures treated with a fungal elicitor (Zulak et al. 2007), and in *Catharanthus roseus* and tobacco cell cultures treated with methyl jasmonate (Goossens et al. 2003). In all cases, researchers observed a coordinated increase in the expression of genes implicated in alkaloid metabolism. Moreover, profound changes in the level of gene transcripts encoding primary metabolic enzymes also occurred. Interestingly, transcripts involved in SAM (S-Adenosylmethionine) recycling increased in all three systems, indicating a high demand for this cofactor in the modification of alkaloid backbone structures. As expected, genes implicated in aromatic amino acid metabolism were also induced in opium poppy and *C. roseus* in response to the increased demand for the precursors of alkaloids biosynthesis. With respect to early signal transduction events, considerable effort has been focused on events associated with the induction of alkaloid metabolism in *Eschscholzia californica* cell cultures, which accumulate benzophenanthridine alkaloids such as sanguinarine in response to treatment with a fungal elicitor. Two different signal transduction pathways were identified. One cascade is jasmonate dependent and responds to high elicitor concentrations. The other is jasmonate independent and is triggered by low elicitor concentrations (Farber et al. 2003). The jasmonate-independent pathway involves Gα proteins that interact and activate phospholipase A2, which leads to a transient proton efflux from the vacuole and a subsequent activation of other cytoplasmic components (Warzecha et al. 2000).

8.7.3 Flavonoids

Flavonoids belong to the large family of phenolic compounds also known as polyphenols. They accumulate in all organs and tissues, at different stages of development, and depending on the environmental conditions. Besides their multiple roles in plant development and adaptation to the environment, these molecules are of major interest for human nutrition and health (Wink 2010). Indeed, they contribute to the organoleptic quality of plant-derived products (color, taste, flavor, etc.) (Hichri et al. 2011). The flavonoid family encompasses at least 10,000 molecules, chiefly divided into phlobaphenes, aurones, isoflavonoids, flavones, flavanols, and anthocyanins. Unlike the other classes of flavonoid compounds, phlobaphenes and isoflavonoids are synthesized almost exclusively by some maize varieties and leguminous plants, respectively. All flavonoids display a C6–C3–C6 skeleton structure, except for the aurones (C6–C2–C6) (Marais et al. 2006). Their classification is based upon the oxidation level of the central C heterocycle, the presence of hydroxyl and methyl substitutions on the A and B rings, and also on supplemental modifications such as glycosylation (glucose, galactose, arabinose, rhamnose, and, to a lesser extent, disaccharides), acylation (notably coumaric and caffeic acids), and polymerization (Kong et al. 2003; Macheix et al. 2005; Aron and Kennedy 2008).

Flavonoids are synthesized in the cytosol and are mainly transported to the vacuole for storage. They can also be found in cell walls, the nucleus, chloroplasts, and even in the extracellular space, depending on the plant species, the tissue, or the stage of development (Feucht et al. 2004; Zhao and Dixon 2010). Flavonoids play a role in the interaction between plants and animals, as exemplified in leaves, where the concentration and nature of phenolic acids (PAs) determine the bitter taste and thus prevent feeding by herbivores (Harborne and Williams 2000, Aron and Kennedy 2008). In addition, flavonoids control pollen fertility, and modulate auxin transport (Peer and Murphy 2007; Thompson et al. 2010). As well as controlling physiological traits of plant development, flavonoids play a protective role against an array of abiotic stresses. Flavones, flavonols, and anthocyanins accumulate in leaf epidermal cells, waxes, and trichomes, where they act as UV-B filters, but can also complex with DNA and protect it from oxidative damage (Harborne and Williams 2000; Dixon 2005; Aron and Kennedy 2008; Albert et al. 2009). Likewise, cold stress induces anthocyanin accumulation in maize (*Zea mays*) and *Arabidopsis thaliana* seedlings (Leyva et al. 1995). Flavonoids, and more generally phenolic compounds, also contribute to defense against biotic stresses (Bhattacharya et al. 2010). They may either be constitutively synthesized or accumulate in response to microbial invasion since most of these compounds exhibit antimicrobial and pesticide properties, by acting as a repellent, and inhibiting growth and development of pests (Chong et al. 2009).

Besides their numerous functions in plants, flavonoids present a plethora of medicinal, pharmaceutical, and nutritional properties and are thus termed "neutraceutical" compounds (Lin and Weng 2006). Understanding the different steps of the flavonoid biosynthetic pathway and their regulation is important to generate and select fruits and vegetables enriched in these compounds, with desirable dietary and medicinal properties.

8.8 Transcription Factors Involved in Secondary Metabolism

In recent years, an increasing number of transcription factors and the underlying mechanisms of plant secondary metabolism regulation have been elucidated. However, other mechanisms regulating specific pathways do exist. Investigation of the network controlling the biosynthesis, transportation, accumulation, and release of secondary metabolites will further our understanding of plant adaptation to changing environments and the interaction among plants and other organisms. Several families of transcription factors have been described to be regulators of plant secondary metabolism. The following are representative transcription factors involved in plant secondary metabolism regulation.

8.8.1 MYB

MYB transcription factors are characterized by varying numbers of the MYB DNA-binding domain, which consists of up to four imperfect repeats of 52 amino acids (Feller et al. 2011). The MYB family proteins can be divided into four families, and several members of the R2R3 family have been reported to be regulators of multiple biosynthetic pathways in various plant species. In *Arabidopsis*, AtMYB113, AtMYB114, AtMYB75, and AtMYB90 are involved in controlling anthocyanin content in vegetative tissues by activating the entire phenylpropanoid pathway (Borevitz et al. 2000; Stracke et al. 2001; Tohge et al. 2005; Gonzalez et al. 2008), members of this family also regulate glucosinolate biosynthesis (Gigolashvili et al. 2008).

8.8.2 bHLH

bHLH transcription factors often interact with MYB family proteins to form a complex, and then regulate the downstream expression of target genes (Feller et al. 2011). A well-characterized example is the transcriptional regulation of anthocyanin, which is mostly studied in genes of the anthocyanin pathway in *Arabidopsis*. The bHLH proteins GL3, eGL3, and TT8 interact with MYB proteins in the presence of a WD40 repeat containing protein TTG1, forming a transcriptional regulation complex which activates anthocyanin biosynthetic genes (Gonzalez et al. 2008; Dubos et al. 2010).

Another important bHLH regulator is MYC2. MYC2 has widely been found to directly or indirectly participate in secondary metabolism in multiple plant species. In *Arabidopsis*, it integrates gibberellin (GA) and jasmonate (JA) signaling pathways to upregulate the expression of sesquiterpene synthase genes in flowers (Hong et al. 2012).

8.9 Conclusion

Many secondary metabolites are highly valuable to humans. However, due to their low concentrations in plant tissues, direct extraction of these metabolites is usually expensive and inefficient. Engineering biosynthetic pathways in microbes or other plant species via a synthetic approach is a recent phenomenon that holds promise. To increase the production of a (group of) compound (s), two general approaches have been followed. Firstly, methods have employed to change the expression of one or a few genes, thereby overcoming specific rate-limiting steps in the pathway, to shut down competitive pathways, and to decrease catabolism of the product of interest. Secondly, attempts have been made to change the expression of regulatory genes that control multiple biosynthesis genes. Successful examples of this approach are still limited, partially due to the complexity of biosynthesis pathways and the frequent toxicity of metabolic intermediates. Unraveling plant secondary metabolites pathways are the challenging way forward for successful applications in the field such as molecular farming, health food, functional food, and plant resistance.

References

Abebe T, Guenzi AC, Martin B, Cushman JC (2003) Tolerance of mannitol-accumulating transgenic wheat to water stress and salinity. Plant Physiol 131:1748–1755

Agrawal GK, Jawa N-S, Lebrun M-H, Job D, Rakwal R (2010) Plant secretome: unlocking secrets of the secreted proteins. Proteomics 10:799–827

Albert NW, Lewis DH, Zhang H, Irving LJ, Jameson PE, Davies KM (2009) Light-induced vegetative anthocyanin pigmentation in Petunia. J Exp Bot 60:2191–2202

Alcázar R, Garcia-Martinez JL, Cuevas JC, Tiburcio AF, Altabella T (2005) Overexpression of ADC2 in Arabidopsis induces dwarfism and late-flowering through GA deficiency. Plant J 43:425–436

Alcázar R, Planas J, Saxena T, Zarza X, Bortolotti C, Cuevas J, Bitrian M, Tiburcio AF, Altabella T (2010) Putrescine accumulation confers drought tolerance in transgenic *Arabidopsis* plants over-expressing the homologous arginine decarboxylase 2 gene. Plant Physiol Biochem 48:547–552

Alcázar R, Cuevas JC, Planas J, Zarza X, Bortolotti C, Carrasco P, Salinas J, Tiburcio AF, Altabella T (2011) Integration of polyamines in the cold acclimation response. Plant Sci 180:31–38

Ap Ress T, Hill SA (1994) Metabolic control analysis of plant metabolism. Plant Cell Environ 17:587–599

Aron PM, Kennedy JA (2008) Flavan-3-ols: nature, occurrence and biological activity. Mol Nutr Food Res 52:79–104

Bansal KC, Goel D, Singh AK, Yadav V, Babbar SB, Murata N (2011) Transformation of tomato with a bacterial codA gene enhances tolerance to salt and water stresses. J Plant Physiol 168:1286–1294

Bari R, Jones JD (2009) Role of plant hormones in plant defense responses. Plant Mol Biol 69:473–488

Bartels D, Sunkar R (2005) Drought and salt tolerance in plants. Crit Rev Plant Sci 24:23–58

Bhatnagar-Mathur P, Vadez V, Sharma KK (2008) Transgenic approaches for abiotic stress tolerance in plants: retrospect and prospects. Plant Cell Rep 27:411–424

Bhattacharya A, Sood P, Citovsky V (2010) The roles of plant phenolics in defence and communication during Agrobacterium and Rhizobium infection. Mol Plant Pathol 11:705–719

Bianchi G, Gamba A, Murelli C, Salamini F, Bartels D (1991) Novel carbohydrate metabolism in the resurrection plant *Craterostigma plantagineum*. Plant J 1:355–359

Borevitz JO, Xia Y, Blount J, Dixon RA, Lamb C (2000) Activation tagging identifies a conserved MYB regulator of phenylpropanoid biosynthesis. Plant Cell 12:2383–2394

Brosche M, Vinocur B, Alatalo ER et al (2005) Gene expression and metabolite profiling of *Populus euphratica* growing in the Negev desert. Genome Biol 6:101

Capell T, Bassie L, Christou P (2004) Modulation of the polyamine biosynthetic pathway in transgenic rice confers tolerance to drought stress. Proc Natl Acad Sci U S A 101:9909–9914

Chen TH, Murata N (2011) Glycinebetaine protects plants against abiotic stress: mechanisms and biotechnological applications. Plant Cell Environ 34:1–20

Chen XM, Hu L, Lu H, Liu QL, Jiang XN (2005) Overexpression of mtlD gene in transgenic *Populus tomentosa* improves salt tolerance through accumulation of mannitol. Tree Physiol 25:1273–1281

Chen AX, Lou YG, Mao YB, Lu S, Wang LJ, Chen XY (2007) Plant terpenoids: biosynthesis and ecological functions. J Integr Plant Biol 49:179–186

Chong J, Poutaraud A, Hugueney P (2009) Metabolism and roles of stilbenes in plants. Plant Sci 177:143–155

Cramer GR, Urano K, Delrot S, Pezzotti M, Shinozaki K (2011) Effects of abiotic stress on plants: a systems biology perspective. BMC Plant Biol 11:163. doi:10.1186/1471-2229-11-163

Croteau R, Kutchan TM, Lewis NG (2000) Natural products (secondary metabolites). In: Buchnnan BB, Gruissem W, Jones RL (eds) Biochemistry and molecular biology of plant. American Society of Plant Physiologists, Rockville, pp 1250–1318

Cutler SR, Rodriguez PL, Finkelstein RR, Abrams SR (2010) Abscisic acid: emergence of a core signaling network. Annu Rev Plant Biol 61:651–679. doi:10.1146/annurev-arplant-042809-112122

De Luca V, Pierre B (2000) The cell and developmental biology of alkaloid biosynthesis. Trends Plant Sci 5:168–173

Déjardin A, Sokolov LN, Kleczkowski LA (1999) Sugar/osmoticum levels modulate differential abscisic acid-independent expression of two stress responsive sucrose synthase genes in *Arabidopsis*. Biochem J 344:503–509

Dewick PM (2002) Medicinal natural products. A biosynthetic approach, 2nd edn. Wiley, Chichester

Dixon RA (2005) Engineering plant natural product pathways. Curr Opin Plant Biol 8:329–336

Drennan PM, Smith MT, Goldsworthy D, Van Staden J (1993) The occurrence of trehalose in the leaves of the desiccation-tolerant angiosperm *Myrothamnus flabellifolius* Welw. J Plant Physiol 142:493–496

Du H, Liu H, Xiong L (2013) Endogenous auxin and jasmonic acid levels are differentially modulated by abiotic stresses in rice. Front Plant Sci 4:397. doi:10.3389/fpls.2013.00397

Dubos C, Stracke R, Grotewold E, Weisshaar B, Martin C, Lepiniec L (2010) MYB transcription factors in *Arabidopsis*. Trends Plant Sci 15:573–581

Duque AS, Martinho de Almeida A, Bernardes da Silva A, Marques da Silva J, Farinha AP, Santos D, Fevereiro P, Araújo S (2013) Abiotic Stress Responses in Plants: Unraveling the Complexity of Genes and Networks to Survive. In: Vahdati K, Leslie C (eds). Abiotic Stress - Plant Responses and Applications in Agriculture, pp 49-101.

Eastmond PJ, Van Dijken AJH, Spielman M, Kerr A, Tissier AF, Dickinson HG, Jones JDG, Smeekens SC, Graham IA (2002) Trehalose-6-phosphate synthase 1, which catalyses the first step in trehalose synthesis, is essential for Arabidopsis embryo maturation. Plant J 29:225–235

Fabro G et al (2004) Proline accumulation and AtP5CS2 gene activation are induced by plant-pathogen incompatible interactions in *Arabidopsis*. Mol Plant Microbe Interact 17:343–350

Facchini PJ, De-Luca V (2008) Opium poppy and Madagascar periwinkle: model non model systems to investigate alkaloid biosynthesis in plant. Plant J 54:763–784

Fang X, Yang CQ, Wei YK, Ma QX, Yang L, Chen XY (2011) Genomics grand for diversified plant secondary metabolites. Plant Div Res 33:53–64

Farber K, Schumann B, Miersch O, Roos W (2003) Selective desensitization of jasmonate-and pH-dependent signaling in the induction of benzophenanthridine biosynthesis in cells of Eschscholzia californica. Phytochemistry 62:491–500

Fell D (1997) Understanding the control of metabolism (frontiers in metabolism). Portland Press, London

Feller A, Machemer K, Braun EL, Grotewold E (2011) Evolutionary and comparative analysis of MYB and bHLH plant transcription factors. Plant J 66:94–116

Feucht W, Treutter D, Polster J (2004) Flavanol binding of nuclei from tree species. Plant Cell Rep 22:430–436

Fiehn O (2002) Metabolomics-the link between genotypes and phenotypes. Plant Mol Biol 48:155–171

Gao M, Tao R, Miura K, Dandekar AM, Sugiura A (2001) Transformation of Japanese persimmon (*Diospyros kaki* Thunb.) with apple cDNA encoding NADP-dependent sorbitol-6-phosphate dehydrogenase. Plant Sci 160:837–845

Geigenberger P, Reimholz R, Deiting U et al (1999) Decreased expression of sucrose phosphate synthase strongly inhibits the water stress-induced synthesis of sucrose in growing potato tubers. Plant J 19:119–129

Giada MLR (2013) Food phenolic compounds: main classes, sources and their antioxidant power. In: Morales-González JA (ed) Oxidative stress and chronic degenerative diseases-role of antioxidants. InTech Publisher, Rijeka

Gigolashvili T, Engqvist M, Yatusevich R, Muller C, Flugge UI (2008) HAG2/MYB76 and HAG3/MYB29 exert a specific and coordinated control on the regulation of aliphatic glucosinolate biosynthesis in *Arabidopsis thaliana*. New Phytol 177:627–642

Golldack D, Li C, Mohan H, Probst N (2013) Gibberellins and abscisic acid signal crosstalk: living and developing under unfavorable conditions. Plant Cell Rep 32:1007–1016

Gómez LD, Baud S, Gilday A, Li Y, Graham IA (2006) Delayed embryo development in the ARABIDOPSIS TREHALOSE-6-PHOSPHATE SYNTHASE 1 mutant is associated with altered cell wall structure, decreased cell division and starch accumulation. Plant J 46:69

Gonzalez A, Zhao M, Leavitt JM, Lloyd AM (2008) Regulation of the anthocyanin biosynthetic pathway by the TTG1/bHLH/Myb transcriptional complex in *Arabidopsis* seedlings. Plant J 53:814–827

Goossens A, Hakkinen ST, Laakso I, Seppanen-Laakso T, Biondi S et al (2003) A functional genomics approach toward the understanding of secondary metabolism in plant cells. Proc Natl Acad Sci U S A 100:8595–8600

Hall R, Beale M, Fiehn O, Hardy N, Summer L, Bino R (2002) Plant metabolomics as a missing link in functional genomics strategies. Plant Cell 14:1437–1440

Harborne JB, Williams CA (2000) Advances in flavonoid research since 1992. Phytochemistry 55:481–504

Harvey A (2000) Strategies for discovering drugs from previously unexplored natural products. Drug Discov Today 5(7):294–300

Hazen SP, Wu Y, Kreps JA (2003) Gene expression profiling of plant responses to abiotic stress. Funct Integr Genomics 3:105–111

Hichri I, Barrieu F, Bogs J, Kappel C, Delrot S, Lauvergeat V (2011) Recent advances in the transcriptional regulation of the flavonoid biosynthetic pathway. J Exp Bot 62:2465–2483

Hong Z, Lakkineni K, Zhang Z, Verma DP (2000) Removal of feedback inhibition of delta(1)-pyrroline-5-carboxylate synthetase results in increased proline accumulation and protection of plants from osmotic stress. Plant Physiol 122:1129–1136

Hong GJ, Xue XY, Mao YB, Wang LJ, Chen XY (2012) *Arabidopsis* MYC2 interacts with DELLA proteins in regulating sesquiterpene synthase gene expression. Plant Cell 24 (6):2635–2648

Hongbo S, Zongsuo L, Mingan S, Bochu W (2005) Impacts of PEG-6000 pretreatment for barley (Hordeum vulgare L.) seeds on the effect of their mature embryo in vitro culture and primary investigation on its physiological mechanism. Colloids Surf B Biointerfaces 41:73–77

Humphreys MW, Yadav RS, Cairns AJ, Turner LB, Humphreys J, Skøt L (2006) A changing climate for grassland research. New Phytol 169:9–26

Hussain SS, Ali M, Ahmad M, Siddique KH (2011) Polyamines: natural and engineered abiotic and biotic stress tolerance in plants. Biotechnol Adv 29:300–311

Ingram J, Chandler J, Gallagher L et al (1997) Analysis of cDNA clones encoding sucrose-phosphate synthase in relation to sugar interconversions associated with dehydration in the resurrection plant *Craterostigma plantagineum* Hochst. Plant Physiol 115:113–121

Iordachescu M, Imai R (2008) Trehalose biosynthesis in response to abiotic stresses. J Integr Plant Biol 50:1223–1229

Janska A, Marsik P, Zelenkova S, Ovesna J (2010) Cold stress and acclimation: what is important for metabolic adjustment? Plant Biol 12:395–405

Kasinathan V, Wingler A (2004) Effect of reduced arginine decarboxylase activity on salt tolerance and on polyamine formation during salt stress in *Arabidopsis thaliana*. Physiol Plant 121:101–7

Kaplan F, Guy CL (2004) beta-Amylase induction and the protective role of maltose during temperature shock. Plant Physiol 135:1674–1684

Kaplan F, Guy CL (2005) RNA interference of *Arabidopsis* beta-amylase8 prevents maltose accumulation upon cold shock and increases sensitivity of PSII photochemical efficiency to freezing stress. Plant J 44:730–743

Kaplan F, Kopka J, Haskell DW, Zhao W, Schiller KC, Gatzke N, Sung DY, Guy CL (2004) Exploring the temperature-stress metabolome of *Arabidopsis*. Plant Physiol 136:4159–4168

Kawakami A, Sato Y, Yoshida M (2008) Genetic engineering of rice capable of synthesizing fructans and enhancing chilling tolerance. J Exp Bot 59:793–802

Kempa S, Krasensky J, Dal Santo S, Kopka J, Jonak C (2008) A central role of abscisic acid in stress-regulated carbohydrate metabolism. PLoS One 3, e3935

Kilian J, Whitehead D, Horak J, Wanke D, Weinl S (2007) The At gene expression global stress expression data set: protocols, evaluation and model data analysis of UV-B light, drought and cold stress responses. Plant J 50:347–363

Kleines M, Elster RC, Rodrigo MJ et al (1999) Isolation and expression analysis of two stress-responsive sucrose synthase genes from the *Craterostigma plantagineum* (Hochst). Planta 209:13–24

Kolbe A, Tiessen A, Schluepmann H, Paul M, Ulrich S, Geigenberger P (2005) Trehalose 6-phosphate regulates starch synthesis via posttranslational redox activation of ADP-glucose pyrophosphorylase. Proc Natl Acad Sci U S A 102:11118–11123

Kong J-M, Chia L-S, Goh N-K, Chia T-F, Brouillard R (2003) Analysis and biological activities of *anthocyanins*. Phytochemistry 64:923–933

Krasensky J, Jonak C (2012) Drought, salt, and temperature stress-induced metabolic rearrangements and regulatory networks. J Exp Bot 63(4):1593–1608

Kuhlemeier C, Green PJ, Chua NH (1987) Regulation of gene expression in higher plants. Annu Rev Plant Physiol 38:221–257

Kuznetsov V, Radyukina NL, Shevyakova NI (2006) Polyamines and stress: biological role, metabolism, and regulation. Russ J Plant Physiol 53:583–604

Leakey ADB, Uribelarrea M, Ainsworth EA, Naidu SL, Rogers A, Ort DR, Long SP (2006) Photosynthesis, productivity, and yield of maize are not affected by open-air elevation of CO_2 concentration in the absence of drought. Plant Physiol 140:779–790

Lefebvre DD, Gellatly KS (1997) Fundamentals of gene structure and control. In: Dennis D, Turpin DH, Lefebvre DD, Layzell DB (eds) Plant metabolism. Addison Weskey Lonhman, Reading

Leyva A, Jarillo JA, Salinas J, Martinez-Zapater JM (1995) Low temperature induces the accumulation of phenylalanine ammonia-lyase and chalcone synthase mRNAs of *Arabidopsis thaliana* in a light-dependent manner. Plant Physiol 108:39–46

Li HJ, Yang AF, Zhang XC, Gao F, Zhang JR (2007) Improving freezing tolerance of transgenic tobacco expressing sucrose: sucrose 1-fructosyltransferase gene from Lactuca sativa. Plant Cell Tiss Org Cult 89:37–48

Li HW, Zang BS, Deng XW, Wang XP (2011) Overexpression of the trehalose-6-phosphate synthase gene OsTPS1 enhances abiotic stress tolerance in rice. Planta 234:1007–1018

Li YY, Mao K, Zhao C, Zhao XY, Zhang HL, Shu HR, Hao JY (2012) MdCOP1 ubiquitin E3 ligases interact with MdMYB1 to regulate light-induced anthocyanin biosynthesis and red fruit coloration in apple. Plant Physiol 160:1011–1022

Lichtenthaler HK, Schwender J, Disch A, Rohmer M (1997) Biosynthesis of isoprenoids, in higher plant chloroplasts proceeds via a mevalonate-independent pathway. FEBS Lett 400:271–274

Lin JK, Weng MS (2006) Flavonoids as nutraceuticals. In: Grotewold E (ed) The science of flavonoids. Springer, Berlin, pp 213–238

Lugan R, Niogret MF, Leport L, Guegan JP, Larher FR, Savoure A, Kopka J, Bouchereau A (2010) Metabolome and water homeostasis analysis of Thellungiella salsuginea suggests that dehydration tolerance is a key response to osmotic stress in this halophyte. Plant J 64:215–229

Lunn JE, Feil R, Hendriks JHM, Gibon Y, Morcuende R, Osuna D, Scheible WR, Carillo P, Hajirezaei M-R, Stitt M (2006) Sugar-induced increases in trehalose 6-phosphate are correlated with redox activation of ADPglucose pyrophosphorylase and higher rates of starch synthesis in Arabidopsis thaliana. Biochem J 397:139

Luo QJ, Mittal A, Jia F, Rock CD (2012) An autoregulatory feedback loop involving PAP1 and TAS4 in response to sugars in Arabidopsis. Plant Mol Biol 80:117–129

Macedo A (2012) Abiotic stress responses in plants: metabolism to productivity. In: Ahmad P, Prasad M (eds) Abiotic stress responses in plants: metabolism, productivity and sustainability. Springer, New York, pp 41–61

Macheix JJ, Fleuriet A, Jay-Allemand C (2005) Les composés phénoliques des végétaux. Un exemple de métabolites secondaires d'importance économique. Presses polytechniques et universitaires romandes

Madden TD, Bally MB, Hope MJ, Cullis PR, Schieren HP, Janoff AS (1985) Protection of large unilamellar vesicles by trehalose during dehydration: retention of vesicle contents. Biochim Biophys Acta 817:67–74

Madlung A, Comai L (2004) The effect of stress on genome regulation and structure. Ann Bot 94:481–495

Maier A, Schrader A, Kokkelink L, Falke C, Welter B, Iniesto E, Rubio V, Uhrig JF, Hulskamp M, Hoecker U (2013) Light and the E3 ubiquitin ligase COP1/SPA control the protein stability of the MYB transcription factors PAP1 and PAP2 involved in anthocyanin accumulation in Arabidopsis. Plant J 74:638–651

Majee M, Maitra S, Dastidar KG, Pattnaik S, Chatterjee A, Hait NC, Das KP, Majumder AL (2004) A novel salt-tolerant L-myo-inositol-1-phosphate synthase from Porteresia coarctata (Roxb.) Tateoka, a halophytic wild rice: molecular cloning, bacterial overexpression, characterization, and functional introgression into tobacco-conferring salt tolerance phenotype. J Biol Chem 279:28539–28552

Marais JPJ, Deavours B, Dixon RA, Ferreira D (2006) The stereochemistry of flavonoids. In: Grotewold E (ed) The science of flavonoids. Springer, Berlin, pp 1–46

Martí MC, Stancombe MA, Webb AAR (2013) Cell- and stimulus type-specific intracellular free Ca^{2+} signals in Arabidopsis. Plant Physiol 163:625–634

Mayrhofer S, Teuber M, Zimmer I, Louis S, Fischbach RJ, Schnitzler RP (2005) Diurnal and seasonal variation of isoprene biosynthesis-related genes in Grey poplar leaves. Plant Physiol 139:474–484

Michalska J, Zauber H, Buchanan BB, Cejudo FJ, Geigenberger P (2009) NTRC links built-in thioredoxin to light and sucrose in regulating starch synthesis in chloroplasts and amyloplasts. Proc Natl Acad Sci U S A 106:9908–9913

Miller G, Shulaev V, Mittler R (2008) Reactive oxygen signaling and abiotic stress. Physiol Plant 133:481–489. doi:10.1111/j.1399-3054.2008.01090.x

Munne-Bosch S, Queval G, Foyer CH (2013) The impact of global change factors on redox signaling underpinning stress tolerance. Plant Physiol 161:5–1619

Munns R, Tester M (2008) Mechanisms of salinity tolerance. Annu Rev Plant Biol 59:651–681

Murata J, Roepke J, Gordon H, De Luca V (2008) The leaf epidermone of *Catharanthus roseus* reveals its biochemical specification. Plant Cell 20:524–542

Olien CR, Clark J (1995) Freeze-induced changes in carbohydrates associated with hardiness of barley and rye. Crop Sci 35:496–502

Pandey A, Mann M (2000) Proteomics to study genes and genomes. Nature 405:837–846

Park OJ (2004) Proteomic studies in plants. J Biochem Mol Biol 37(1):133–138

Park EJ, Jeknic Z, Sakamoto A, DeNoma J, Yuwansiri R, Murata N, Chen TH (2004) Genetic engineering of glycinebetaine synthesis in tomato protects seeds, plants, and flowers from chilling damage. Plant J 40:474–487

Park EJ, Jeknic Z, Chen TH, Murata N (2007) The codA transgene for glycinebetaine synthesis increases the size of flowers and fruits in tomato. Plant Biotechnol J 5:422–430

Patra B, Ray S, Richter A, Majumder AL (2010) Enhanced salt tolerance of transgenic tobacco plants by co-expression of PcINO1 and McIMT1 is accompanied by increased level of myo-inositol and methylated inositol. Protoplasma 245:143–152

Patterson SD, Aebersold RH (2003) Proteomics: the first decade and beyond. Nat Genet 33:311–323

Paul MJ, Primavesi LF, Jhurreea D, Zhang YH (2008) Trehalose metabolism and signaling. Annu Rev Plant Biol 59:417–441

Peer WA, Murphy AS (2007) Flavonoids and auxin transport: modulators or regulators? Trends Plant Sci 12:556–563

Pelah D, Wang W, Altman A et al (1997) Differential accumulation of water-stress related protein, sucrose synthase and soluble sugars in *Populus* species that differ in their water stress response. Physiol Plant 99:153–159

Peleg Z, Blumwald E (2011) Hormone balance and abiotic stress tolerance in crop plants. Curr Opin Plant Biol 14:290–295

Phizicky E, Bastiaens PI, Zhu H et al (2003) Protein analysis on a proteomic scale. Nature 422:208–215

Pichersky E, Gershenzon J (2002) The formation and function of plant volatiles: perfumes for pollinator attraction and defense. Curr Opin Plant Biol 5:237–243

Plaxton W (1997) Metabolic control. In: Dennis D, Turpin DH, Lefebvre DD, Layzell DB (eds) Plant metabolism. Addison Wesley Longman, Reading, pp 50–68

Plaxton W, McManus M (eds.)(2006) Control of primary metabolism in plants. Annual Plant Reviews Blackwell Publishing Ltd

Ponnu J, Wahl V, Schmid M (2011) Trehalose-6-phosphate: connecting plant metabolism and development. Front Plant Sci 2:70

Raamsdonk LM, Teusink B, Broadhurst D, Zhang N, Hayes A, Walsh MC, Berden JA, Kell DB et al (2001) A functional genomics strategy that uses metabolome data to reveal the phenotype of silent mutations. Nat Biotechnol 19:45–50

Raghavendra AS, Gonugunta VK, Christmann A, Grill E (2010) ABA perception and signaling. Trends Plant Sci 15:395–401. doi:10.1016/j.tplants.2010.04.006

Rhodes D, Hanson AD (1993) Quaternary ammonium and tertiary sulfonium compounds in higher-plants. Annu Rev Plant Physiol Plant Mol Biol 44:357–384

Roy M, Wu R (2002) Over-expression of S-adenosylmethionine decarboxylase gene in rice increases polyamine levels and enhances sodium chloride stress tolerance. Plant Sci 163:987–992

Saradhi PP et al (1995) Proline accumulates in plants exposed to UV radiation and protects them against UV induced peroxidation. Biochem Biophys Res Commun 209:1–5

Satoh-Nagasawa N, Nagasawa N, Malcomber S, Sakai H, Jackson D (2006) A trehalose metabolic enzyme controls inflorescence architecture in maize. Nature 441:227–230

Schmidt TJ, Khalid SA, Romanha AJ, Alves TM, Biavatti MW, Brun R, Da Costa FB et al (2012) The potential of secondary metabolites from plants as drugs or leads against protozoan neglected diseases—part II. Curr Med Chem 19(14):2176–2228

Sengupta S, Patra B, Ray S, Majumder AL (2008) Inositol methyl tranferase from a halophytic wild rice, Porteresia coarctata Roxb. (Tateoka): regulation of pinitol synthesis under abiotic stress. Plant Cell Environ 31:1442–1459

Shen B, Jensen RG, Bohnert HJ (1997) Increased resistance to oxidative stress in transgenic plants by targeting mannitol biosynthesis to chloroplasts. Plant Physiol 113:1177–1183

Singh TN, Aspinal D, Paleg LG (1972) Proline accumulation and varietal adaptability to drought in barley: potential metabolic measure of drought resistance. Nat New Biol 236:188–190

Smirnoff N, Cumbes QJ (1989) Hydroxyl radical scavenging activity of compatible solutes. Phytochemistry 28:1057–1060

Smith AM, Coupland G, Dolan L, Harberd N, Jones J et al (eds) (2010) Metabolism. In: Plant biology. Garland Science, New York, pp 167–296

Soltis DE, Soltis PS (2003) The role of phylogenetics incomparative genetics. Plant Physiol 132:1790–1800

Spollen WG, Nelson CJ (1994) Response of fructan to water-deficit in growing leaves of tall fescue. Plant Physiol 106:329–336

Stewart GR, Lee J (1974) Role of proline accumulation in halophytes. Planta 120:279–289

Stoop JHM, Williamson JD, Pharr DM (1996) Mannitol metabolism in plants: a method for coping with stress. Trends Plant Sci 1:139–144

Stracke R, Werber M, Weisshaar B (2001) The R2R3-MYB gene family in Arabidopsis thaliana. Curr Opin Plant Biol 4:447–456

Szabados L, Savoure A (2010) Proline: a multifunctional amino acid. Trends Plant Sci 15:89–97

Szekely G, Abraham E, Cseplo A et al (2008) Duplicated P5CS genes of Arabidopsis play distinct roles in stress regulation and developmental control of proline biosynthesis. Plant J 53:11–28

Tholl D (2006) Terpene synthases and the regulation, diversity and biological roles of terpene metabolism. Curr Opin Plant Biol 9(3):297–304

Tholl D, Kish CM, Orlova I, Sherman D, Gershenzon J, Pichersky E, Dudareva N (2004) Formation of monoterpenes in Antirrhinum majus and Clarkia breweri flowers involves heterodimeric geranyl diphosphate synthases. Plant Cell 16:977–992

Tholl D, Chen F, Petri J, Gershenzon J, Pichersky E (2005) Two sesquiterpene synthases are responsible for the complex mixture of sesquiterpenes emitted from Arabidopsis flowers. Plant J 42:757–771

Thompson EP, Wilkins C, Demidchik V, Davies JM, Glover BJ (2010) An Arabidopsis flavonoid transporter is required for anther dehiscence and pollen development. J Exp Bot 61:439–451

Tohge T, Matsui K, Ohme-Takagi M, Yamazaki M, Saito K (2005) Enhanced radical scavenging activity of genetically modified Arabidopsis seeds. Biotechnol Lett 27:297–303

Uarrota VG, Severino RB, Maraschin M (2011) Maize Landraces (Zea mays L.): a new prospective source for secondary metabolite production. Int J Agric Res 6:218–226

Urano K, Maruyama K, Ogata Y et al (2009) Characterization of the ABA-regulated global responses to dehydration in Arabidopsis by metabolomics. Plant J 57:1065–1078

Usadel B, Blasing OE, Gibon Y, Poree F, Hohne M, Gunter M, Trethewey R, Kamlage B, Poorter H, Stitt M (2008) Multilevel genomic analysis of the response of transcripts, enzyme activities and metabolites in Arabidopsis rosettes to a progressive decrease of temperature in the non-freezing range. Plant Cell Environ 31:518–547

Valerio C, Costa A, Marri L, Issakidis-Bourguet E, Pupillo P, Trost P, Sparla F (2011) Thioredoxin-regulated beta-amylase (BAM1) triggers diurnal starch degradation in guard cells, and in mesophyll cells under osmotic stress. J Exp Bot 62:545–555

Valliyodan B, Nguyen HT (2006) Understanding regulatory networks and engineering for enhanced drought tolerance in plants. Curr Opin Plant Biol 9:189–195

Valluru R, Van den Ende W (2008) Plant fructans in stress environments: emerging concepts and future prospects. J Exp Bot 59:2905–2916

Verbruggen N, Hermans C (2008) Proline accumulation in plants: a review. Amino Acids 35:753–759

Warzecha H, Gerasimenko I, Kutchan TM, Stockigt J (2000) Molecular cloning and functional bacterial expression of a plant glucosidase specifically involved in alkaloid biosynthesis. Phytochemistry 54:657–666

Wasternack C (2007) Jasmonates: an update on biosynthesis, signal transduction and action in plant stress response, growth and development. Ann Bot 100:681–697. doi:10.1093/aob/mcm079

Wasternack C, Hause B (2013) Jasmonates: biosynthesis, perception, signal transduction and action in plant stress response, growth and development. An update to the 2007 review in Annals of Botany. Ann Bot 111:1021–1058

Weiner JJ, Peterson FC, Volkman BF, Cutler SR (2010) Structural and functional insights into core ABA signaling. Curr Opin Plant Biol 13:495–502

Weston DJ, Gunter LE, Rogers A, Wullschlerger DD (2008) Connecting genes, coexpression modules, and molecular signatures to environmental stress phenotypes in plants. BMC Syst Biol 2:16

Widodo P, Newbigin JH, Tester E, Bacic M, Roessner U (2009) Metabolic responses to salt stress of barley (Hordeum vulgare L.) cultivars, Sahara and Clipper, which differ in salinity tolerance. J Exp Bot 60:4089–4103

Wink M (2008) Ecological roles of alkaloids. In: Fattorusso E, Taglialatela-Scafati (eds) Modern alkaloids. Structure, isolation, synthesis, and biology. Wiley, Weinheim, pp 3–24

Wink M (2010) Functions and biotechnology of plant secondary metabolites. Annu Plant Rev 39:1–16

Yamada M, Morishita H, Urano K, Shiozaki N, Yamaguchi-Shinozaki K, Shinozaki K, Yoshiba Y (2005) Effects of free proline accumulation in petunias under drought stress. J Exp Bot 56:1975-1981.

Yang SL et al (2009) Hydrogen peroxide-induced proline and metabolic pathway of its accumulation in maize seedlings. J Plant Physiol 166:1694–1716

Yang CQ, Fang X, Wu XM, Mao YB, Wang LJ, Chen XY (2012) Transcriptional regulation of plant secondary metabolism. J Integr Plant Biol 54:703–712

Yano R, Nakamura M, Yoneyama T, Nishida I (2005) Starch-related alpha-glucan/water dikinase is involved in the cold-induced development of freezing tolerance in Arabidopsis. Plant Physiol 138:837–846

Zeeman SC, Thorneycroft D, Schupp N, Chapple A, Weck M, Dunstan H, Haldimann P, Bechtold N, Smith AM, Smith SM (2004) Plastidial alpha-glucan phosphorylase is not required for starch degradation in Arabidopsis leaves but has a role in the tolerance of abiotic stress. Plant Physiol 135:849–858

Zhao J, Dixon RA (2010) The 'ins' and 'outs' of flavonoid transport. Trends Plant Sci 15:72–80

Zulak KG, Cornish A, Daskalchuk TE, Deyholos MK, Goodenowe DB et al (2007) Gene transcript and metabolite profiling of elicitor-induced opium poppy cell cultures reveals the coordinate regulation of primary and secondary metabolism. Planta 225:1085–1106

Zuther E, Koehl K, Kopka J (2007) Comparative metabolome analysis of the salt response in breeding cultivars of rice. In: Jenks MA, Hasegawa PM, Jain SM (eds) Advances in molecular breeding toward drought and salt tolerant crops. Springer, Netherlands, pp 285–315

Chapter 9
Tools from Biodiversity: Wild Nutraceutical Plants

S. Kumar and P.K. Jena

Abstract Food security and medicine have always been challenging issues. Strategic efforts from 1989 include DeFelic's term "Nutraceutical" combining "Nutrition" and "Pharmaceutical" uses. Several species of the genus *Dioscorea*, abundant in Similipal Biosphere Reserve forest, Odisha, India, are locally used as medicine for abdominal pain, birth control, diarrhoea, labour pain (during delivery), abdominal worm, as antidotes for scorpion bite, and in treatment of skin infections. Tubers of *Dioscorea* species have high proportions of carbohydrate, starch and fibre. Starchy components (nutritional) make them optional food during critical periods of drought and famine among the tribal communities of the region, whereas "anti-nutritional" components give a bitter taste. Anti-nutritional factors may be neutralized using molecular technique. These plants have bioactive compounds including steroids saponin and diosgenin. Recently, a series of pre-clinical and mechanistic studies have been independently conducted to understand beneficial roles of diosgenin against metabolic diseases (hypercholesterolemia, dyslipidemia, diabetes, and obesity), inflammation, and cancer. Diosgenin is also used in commercial synthesis of Cortisone, Progesterone, and other similar products. In experimental models of obesity, diosgenin decreases plasma and hepatic triglycerides and improves glucose homeostasis, by promoting adipocyte differentiation and inhibition of inflammation in adipose tissues. Expression studies are required to ascertain regulatory modulation of genes and to silence bitterness producing genes. Potentially, *Dioscerea* are new horticultural crop in the farm, of both nutritional and pharmaceutical value, and for increased synthesis of diosgenin and other active secondary metabolites, formulating new drugs as well. Genes involved in biochemical synthesis and pathways are cooperatively expressed to ensure function. Information on co-expression/regulation is the key in understanding biological value at the molecular level. Quantitative genetic/mathematic studies help to dissect the pathway.

Keywords *Dioscorea* species • Ethnobotany • Browning values • Nutraceuticals • Diosgenin • Metabolite expression

S. Kumar (✉) • P.K. Jena
Department of Botany, Ravenshaw University, Cuttack, Odisha 753003, India
e-mail: sanjeet.biotech@gmail.com

© Springer International Publishing Switzerland 2017
J.N. Furze et al. (eds.), *Mathematical Advances Towards Sustainable Environmental Systems*, DOI 10.1007/978-3-319-43901-3_9

9.1   Introduction

Food and medicine are pivotal needs for a healthy life (Norman and Denis 1993). Achieving food security and availability of appropriate medicine continue to be summons for the technically sound world today (Albert 2012). Researchers and technocrats struggle to combine the needs properly. Food and medicine are principally important for third-world countries; secondly, focus on new challenges like antibiotics resistance and other side effects may lead to administration of inappropriate dosages (Sosa et al. 2010), with have equal deleterious effect on the populace of any country. Recent published research emphasizes the need for searching novel foods and medicines from natural resources and gives sound methods of scientific research fighting against harmful diseases (Odingo 1981; Wack and Baricas 2002). Situations in today's society emphasize the desideratum for science to achieve sustainable development. There are renewed interests in green chemistry and sustainable use of natural products. These tendencies are evident in the search for new pharmaceutical and biotechnological possibilities, to traverse new sources of food and medicines in nature (Sylla and Ferroud 2014). Multidisciplinary research, for example directed on different organisms, can provide us with a more complete picture of the complex functioning of the Earth's biome (McRae et al. 2007). Several countries in the tropics including India, endure famines, food shortage, malnutrition, endemic lethal diseases, antibiotic resistance, and other disorders. They have the potential to produce adequate nutritional bases and appropriate medicines for their colossal increasing populations (Bohin et al. 2010).

Despite several policies and programs on food and medicine for self-sufficiency at state and national levels, India has failed to attain food and medicine security at the household level, particularly in remote areas. Considerable proportions of rural and tribal populations are under-nourished and lack appropriate health care (Mittal 2013). People living near forest areas cannot produce ample food grains to meet daily requirements, nor can they afford to purchase required medicines. A large mass of such populace meet their food and medicinal requirements through non-conventional means. They make use of wild plants and animal resources available in biodiversity, especially in critical and off agricultural should be periods, for food and medicinal needs. Collection, consumption, and trading of forest products are skills for a group of people coping and adapting to poverty, meeting the growing food demand and attaining seasonal food security (Cordain et al. 2005; Albert 2012). Indigenous forest foods and medicines are of social significance to the tribal population in India. Forest foods and medicinal plants are neglected and underutilized due to lack of awareness, poor research attention, insufficient commercialization, and inadequate policy formulation. All these stand in the way of harnessing plants' 'de facto' potential. Food and medicines from biodiversity and traditional practice to utilize them can allocate the baseline for future substitution by which third-world countries may fight against food scarcity and serious health problems. Appropriate effort should be given to screen plants; efforts result in finding "Nutraceutical food" (Jain and Ramawat 2013; Yadav and

Dugaya 2013). During the last few years, attempts have been made to document the rich indigenous knowledge on medicinal food plants. There are about 1532 wild edible food species available in India (Reddy et al. 2007). Among indigenous nutraceutical food, root and tubers are the most important wild foods after grains (Edison et al. 2006). They have the highest dry matter production per day and are major calorie contributors to rural and tribal populace. Roots and tubers play vital roles in supplementing staple foods with micronutrients to people of remote areas through preserved and stored starchy foods. These supplement requirements during times of food shortage and lean agricultural seasons, acting as a "safety net" for aboriginals (Edison et al. 2006; Behera et al. 2010).

India has rich genetic diversity of tropical root and tuberous plants such as yams, aroids, and several others like ginger, arrowroot, zedoary, ginger lily, wild turmeric, and some orchids (Misra et al. 2013; Kumar et al. 2013a, b). The hot spots of global biodiversity such as Western Himalayas, North-East regions, humid parts of Western Ghats and Eastern Ghats including East-Coastal region are rich in wild tubers and wild relatives of tropical root and tubers (Reddy et al. 2007). Similipal Biosphere Reserve (SBR) is the major part of Eastern Ghats in Odisha having a bounteous diversity of nutraceutical plants (Misra et al. 2013). It is inhabited by various tribal communities with unique traditional skills allowing consumption of wild foods. Among the tuberous edible medicinal plants available in SBR, *Dioscorea* species or yams are common in all landscapes (Kumar et al. 2012). The genus *Dioscorea* belongs to family Dioscoreaceae, representing more than about 600 species worldwide (Coursey 1967). The Dioscoreales are believed to be among the earliest angiosperms that originated in Southeast Asia, but followed divergent evolution in three continents separated by the formation of the Atlantic Ocean and desiccation of the Middle East (Hahn 1995). Yams are monocotyledonous climbers. Stem twining is the key for identification of species. They are very rarely erect, herbaceous or suffruticose, usually forming a tuberous rootstock or hard rhizome with fleshy and tuberous roots. Leaves are opposite or alternate, entire lobed or digitately 3–5 foliolate, palminerved with reticulate venation between; they are petiolate often angular and twisted at the base. The present study reports that sometimes leaves are 7 foliolate in *D. pentaphylla*. Flowers are regular, small or minute, usually dioecious, rarely hermaphrodite, spicate, racemed, or panicled (Table 9.1). Fruits are 3-valved capsulate or baccate. Seeds are flat or globose with small embryos included in hard albumen. Their habitat is mainly tropics or sub-tropical areas of Africa, America, Asia, and Polynesia (Coursey 1967; Saxena and Brahmam 1995). Most *Dioscerea* species possess starchy tubers/rhizomes (Alexander and Coursey 1969). Tubers importantly have adequate amounts of nutritional composition as well as bioactive compounds, which make them ideal nutraceutical food. The starchy and other energy supplemented components of *Dioscorea* tubers have added to their food value. Tubers rich in phenolic compounds and/or their derivatives are applaudable pharmacological agents. *Dioscorea* species provide an ideal radix of genetic components for plant breeders and biotechnologists; new varieties and breeds could have desired characters with effective energy components as well as bioactive compounds (Bhandari et al. 2003;

Table 9.1 Morphological variations of selected *Dioscerea* species

Specimen	Synonyms	Habitat	Twining	Leaf	Stem	Flower	Fruit	Tuber
D. pentaphylla	*D. communis* Burkill.	Climber	Left	3–5 foliolate	Stem with prickles	Axillary racemes	Oblonged capsule	Clavate proceeding from base of stem
D. puber	*D. anguina* Roxb.	Pubscented climber	Right	Opposite	Tomentose and woody stem	Axillary or shortly panicled	Subcordate capsule	Long cylindric tuber
D. bulbifera	*D. sativa* L.	Climber	Left	Alternate or rarely opposite	Smooth and sub-alate	Axillary and panicled spikes	Quadrately oblong Capsule	Subglobosed without defind stalk
D. hispida	*D. daemona* Roxb.	Weakly prickled climber	Left	3-foliolet	Stems with weak prickles	Short oblong spikes	Quadrately oblong capsule	Irregular shaped tubers
D. alata	*D. globosa* Roxb.	Stout twining climber	Right	Opposite and alternate at base, 9-costae	Stem with scattered prickles, 4-angled	axillary and solitary	Capsule obcordate	Large, shallow without long stalk

Samanta and Biswas 2009; Arinathan et al. 2009). The use of *Dioscorea* can be increased by biotechnological tools such as metabolite expression, gene silencing, and molecular and biotechnological intervention. This chapter attempts to highlight the food and medicinal values of *Dioscorea* species collected from SBR, and to lay emphasis on gene expression and manipulation for production of primary and secondary metabolites to desirable limits.

9.2   Plant Metabolite Expression

Modus operandi of using wild plants for food and simultaneously for treating diseases dates back to pre-historic times. The cumulative knowledge of tribal practitioners and millions of 'old-age' people in each generation flows into the main-stream, emerging as traditional medico-food systems (Misra et al. 2012). The heritable system provides a solid platform to screen the nutraceutical foods. Edible medicinal food plants rich in primary and secondary metabolites offer future horticultural potential. Detoxification, if required, is carried out by ethnic practices. There is a requirement for permanent detoxification in order to produce a safe domesticated variety from wild-types. It is necessary to study that the nutritional metabolite expression required molecular techniques to bring such a change (Penna 2001; Fuller et al. 2014).

Plant nutritional metabolites expression propounds a top-hole future for obtaining adequate food and medicines from plants. Aside from producing primary metabolites, plants also synthesize a vast range of secondary metabolites. While the definition of primary metabolites and secondary metabolites is not entirely clear, for certain compounds primary metabolites essentially represent substances that are ubiquitously produced by all plant species and other organisms. Primary metabolites include principally the universal and essential building blocks of sugar, amino acids, nucleotides, lipids, and energy sources. Metabolic pathways of primary metabolites and the actual compounds produced by these fundamental processes are alike, though in some cases not identical (Verpoorte and Memelink 2002; Capell and Christou 2004).

The major classes of secondary metabolites are produced from biosynthetic pathways of primary metabolites, for example, in carbohydrate synthesis after the shikimate pathway, flavonoids are produced. Plant metabolites are important because they change the quality and performance of useful plants. These biochemical features of plants are controlled by complex regulatory networks at multiple levels, by genetic interactions. The genetic transformation of plants by *Agrobacterium tumefaciens* (Gustavo 1998; Herwig and Jutta 2014), metabolic engineering using genetic engineering and molecular biology tools, has modified metabolic pathways to increase the concentration of desired metabolites. Metabolic pathways can be modulated at multiple levels in a controlled manner, producing food with both high nutrient and secondary metabolite levels. However, disturbances in metabolism can affect the whole plant growth system, resulting in unexpected metabolite production (Verpoorte and Memelink 2002). Development of complex and sophisticated strategies involves multiple changes in metabolic pathways. More than one gene may be inserted strategically to complete the production of a metabolite of interest in the metabolic pathway (Penna 2001).

RNA interference (RNAi) gene silencing was adopted as a tool for the manipulation of metabolic pathways by suppressing the expression of target genes. RNAi has been used to modify genetically engineered plants in producing lower levels of natural plant toxins. Thus, we may take advantage of stable and heritable phenotypes present in plant stocks. The discovery of RNAi was preceded by observations of transcriptional inhibition by antisense RNA expressed in transgenic plants, and more directly by reports of unexpected outcomes in experiments performed by plant scientists in the United States and the Netherlands in the early 1990s (Fire et al. 1998). Metabolite expression initially used a single gene insertion to obtain a final product of interest. Initial experiments yielded the first generation of transgenic crops and the discovery of the mechanism of co-suppression (negative regulation). This led to the development of RNAi strategy (Fig. 9.1).

RNAi is carried out by a range of methods and tools including crossing transgenic lines, sequential transformation, co-transformation with standard binary vectors, conventional vectors, high capacity binary vectors (BIBAC or TAC), assisted direct transfer (high capacity transfer), artificial plant chromosome, and operon systems (Liu et al. 1999; Ma et al. 1995; Cosa et al. 2001; Qi et al. 2004; Twyman et al. 2002; Wu et al. 2005; Tang et al. 2007; Carlson et al. 2007; Wada et al. 2009). Metabolite expression involves modification of endogenous pathways, increasing flux towards particular desirable molecules. Aims may be to enhance production of a natural product or to synthesize novel compounds or molecules (Capell and Christou 2004). The gene-silencing mechanism implemented in the RNAi process is activated by double-stranded RNA (dsRNA), with target mRNA complementary strands. The dsRNA induction methods have been highly successful in reducing target gene expression (Napoli et al. 1990; Tang et al. 2007). Strategies for metabolite expression in *Dioscerea* should enhance starchy components and other primary metabolites present in the tuber, bulbils, and other edible parts of *Dioscorea* species, while simultaneously concentrating on regulation of genes responsible for bitter components (dioscorine, didhydrodioscorine, furanoid norditerpenes, sapogenins, diobulbin-A,B,C and D, dioscin, and cyanogens) and anti-nutritional factors like oxalates, tannin, and phytic acid (Kuete et al. 2012).

9.3 *Dioscorea* at Similipal Biosphere Reserve Forest: Indigenous Uses

The concept of the Biosphere Reserve was initiated by UNESCO (United Nations Educational, Scientific and Cultural Organization) in 1970 as a global measure to promote in situ conservation of biological resources for human welfare and

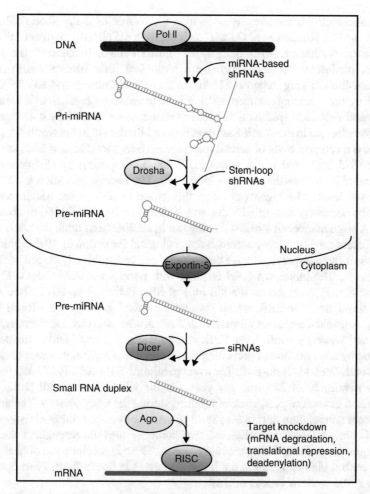

Fig. 9.1 Endogenous RNAi pathways and their use as tools for gene silencing (Fellmann and Lowe 2014)

sustainable development. Representative areas of natural and cultural landscapes, extending over terrestrial and coastal/marine ecosystems, with appropriate zoning pattern, resource base, and management mechanisms, have been designated Biosphere Reserves (Mishra 2010). The approach is an effective means of protecting landscape and biodiversity. To date, 15 Biosphere Reserves have been established in India across different bio-geographical regions. Similipal in Odisha was made the eighth Biosphere Reserve in June 1994, as the representative ecosystem under the Eastern Ghats of India (Mishra et al. 2008a, b). Similipal Biosphere Reserve (SBR) has a unique assemblage of a number of ecosystems (mountains, forests, grasslands, and wetlands) that congregate into a contiguous patch with a range of diverse vegetation types. The biosphere reserve has varied topography, geologic formation, and rich biodiversity. It is also the habitat of many aboriginals. It is known as "The Himalayas of Odisha" as it controls the climatic regime of the state of Odisha, Jharkhand, West Bengal, and Eastern India. It harbours the largest tropical peninsular sal zone, forming a biological link between Northern and Southern India. Lying between 21° 10′ to 22° 12′ N latitude and 85° 58′ to 86° 42′ E longitude, ranging between 300 and 1180 m above sea level, SBR is located in the central part of Mayurbhanj district in Odisha, close to the interstate boundary with West Bengal in the North-East direction and Jharkhand in the North-West. The reserve is a compact mass of natural forest spread over a total area of 5569 km^2 with a core (845 km^2) and buffer zones (2129 km^2) comprising of 16 forest ranges surrounded by a transitional zone (5569 km^2). The average elevation is 559.31 m above sea level. The highest point in this group of hills is the Meghasani Hill (Literally meaning, *Seat of Clouds*), which rises to a height of 1166 m above sea level. A large number of streams flowing out in all directions drain the water of the area. Recent satellite survey confirms the geological formation of SBR is unique. It consists of three impervious huge quartzite bowls, concentrically placed with their interspaces. The innermost bowl is filled with pervious volcanic rocks. Laterite capping is very common on the hill tops of SBR (Misra et al. 2011). Red soil is found throughout the SBR, which favours the sound growth of Sal forest in the region. A tropical monsoon climate with three distinct seasons, i.e. summer, monsoon, and winter, prevails over SBR. It enjoys a warm and humid climate. The topography and direction of prevailing winds result in a general decrease of rainfall from South-West to North-East. The average rainfall of the SBR is 173 mm per year with a maximum of 225 mm per year. On the whole, the rainfall is not well-distributed across the year, most of it falling during the rainy seasons. The average maximum temperature during May is 43 °C and the average minimum temperature is 4 °C during December. High relative humidity prevails throughout the year, which goes up to 90 % during the rainy season. There is precipitation of heavy dew in the central highlands and in the forest clad areas in South-West. Frost occurs in winter in the western valleys of SBR.

The flora of SBR exhibits a rich assemblage of species owing to its diversified hilly topography with lofty mountain crests, deep valleys, abundant springs, and specialized geological formations. The terrain and topography offer a congenial environment for the growth of plants including many that are rare and endangered, restricted to this phyto-geographic region. Since the reserve is located at the junction of four biotic provinces, it forms the agro-ecological link among the geographical regions such as Eastern Ghats, Deccan Plateau, Lower Gangetic plain, and East Coastal zone. Thus, it has a unique biodiversity harbouring a number of endemic, threatened, medicinal, and economically important plants and it is the centre of origin and diversification for a significant number of wild crop plants and their wild relatives. SBR has a mix of vegetation types: Semi-Evergreen forest, Tropical Moist Broadleaf forest, Tropical Moist Deciduous forest, Dry Deciduous hill forest, high level Sal forest with Grasslands, and Savanna. The first attempt to identify the floral biodiversity of SBR is credited to Forester H.H. Haines (1921–1925) who conducted many exploration trips to Similipal area of

Mayurbhanj state, including Meghasini hills. The rich vegetation with mountainous terrain and varied floristic composition with the home of many aboriginal tribes gives immense opportunities to discover variable types of nutraceutical wild food plants in this area (Mishra 1989, 1997; Misra et al. 2003, 2011; Mohanta et al. 2006; Mishra et al. 2008a, b, 2011; Kumar et al. 2013a, b).

The extensive and dense forest-covered hilly tracts of SBR are the home of tribal communities such as Ho, Kolha, Santal, Bathudi, Bhumija, Mahali, Saunti, Munda, Gonda, and Pauri Bhuiyan, which includes two primitive groups, Hill Kharia and Mankirdia. They have unique skills, cultures, and rituals. Their main occupation is food gathering, hunting, collection of forest products, and traditional farming for agriculture. They collect wild products for foods and medicines. The prime wild edible plants available in SBR having medicinal/nutraceutical food values include leaves of *Opilia amentacea* Roxb., *Amaranthus viridis* L., *Cordia oblique* Willd. etc., flowers of *Indigofera cassioides* Rottl.ex DC, *Hibiscus sabdariffa* L., *Moringa oleifera* Lam., vegetables *Coccinia grandis* L. Voigt, *Solanum nigrum* L., *Abelmoschus moschatus* Medic., bulbils of *Dioscorea opositifolia* L. Sp., *Dioscorea pentaphylla* L. Sp., *Dioscorea puber*, fruits of *Cordia oblique* Willd., *Diospyros melanoxylon* Roxb., *Lannea coromandelica* Houtt. Merr. and tubers of *Lasia spinosa* L. Thw., *Dioscorea* species, *Pueraria tuberosa* Willd. DC. Principal medicinal plants available in and around SBR used by local rural and tribal communities are *Rauvolfia serpentina* (L.) Benth., *Celastrus paniculatus* Willd., *Pueraria tuberosa* (Willd.) DC., *Dioscorea bulbifera* L., *Emblica officinalis* Gaertn., *Smilax zeylanica* L., *Curcuma longa* L., *Curcuma aromatica* Salisb., *Cassia fistula* L., *Andrographis paniculata* (Burm. f.) Wall. ex Nees, etc. These plants have ethnobotanical value for aboriginals of SBR. They are used in single and multiple formulations against various common and lethal diseases. All wild food plants, root, and tuber crops in SBR are important because of their easy availability, storage, and as sources of energy-supplemented food. In SBR, *Dioscorea* species is the major wild tuber crop as per availability and consumption palatability. It is locally known as "Bān Aālu" or "Sānga" (Kumar et al. 2012, 2013a, b). Of about 600 species available on the globe, 13 species of *Dioscorea* have been recorded in SBR so far. *D. oppositifolia* L., *D. bulbifera* L., *D. wallichi* H.f., *D. hamiltonii* Hook. f., and *D. spinosa* are found in foot-hills, whereas *D. puber*, *D. pentaphylla*, *D.hispida*, *D. bellophylla*, *D. Glabra* Roxb., *D. belophylla* Voight., and *D. tomentosa* Heyne. are found in moderate and high altitude. *D. alata* L. is cultivated and found in rural and tribal gardens of SBR.

In the present study, five *Dioscorea* species have been selected, *D. alata*, *D. pentaphylla*, *D. bulbifera*, *D. Puber*, and *D. Hispida*, based on the richness of their distribution and their use by local people (Kumar et al. 2012). The enumerations of these selected *Dioscorea* species are cited below with key identification details.

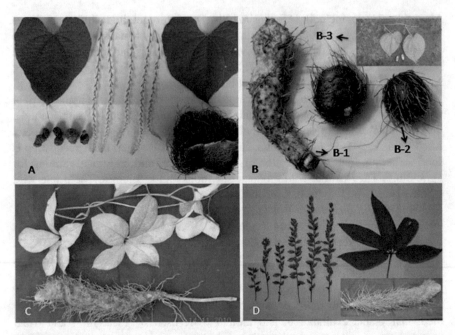

Fig. 9.2 Common *Dioscorea* species of Similipal Biosphere Reserve, Odisha, India. (**a**) leaves, bulbils, flowers, and tuber of *D. bulbifera*; (**b**) tuber of *D. puber* (*B-1*), bulb like tubers of *D. bulbifera* (*B-2*), leaves, and green bulbils of *D. puber* (*B-3*); (**c**) leaves and tuber of *D. hispida*; (**d**): leaf, flowers, and tuber of *D. pentaphylla*

Dioscorea alata L.: A stout twiner having 4-angled, stem twining to the right. Tubers are large, shallow, or deep underground, but without long stalks, bearing edible bulbils. Leaves are mostly opposite and ovate-cordate. Flowers are sub-globose. Fruits are capsules with wings.

Dioscorea bulbifera L.: A climber with stems twining to the left. Tubers are not deep and sub-globose without defined stalks. Bulbils are axillary and warted. Leaves are alternate or rarely opposite. Flowers are purplish/green. Fruits are quadrately oblong capsule with oblong seeds (Fig. 9.2a).

Dioscorea hispida Dennst. : Develops a strong left twining stem bearing a few small, weak prickles. Tubers are sub-globose or irregular covered with root-fibres (Fig. 9.2c). Leaves are 3-foliolate, sometimes prickley, glabrous, or finely pubescent. Flowering branches are small with rounded tips. Fruits are quadrately oblong capsule. Seeds are winged.

Dioscorea pentaphylla L. A prickly vine having stem twining to the left. Tubers are oblong, proceeding direct from the base of the stem. Leaves are 3 to 5-foliolate, and rarely few of the upper leaves are simple. Flowers are shortly pedicelled or sessile above the bract. Fruits are oblong capsule and densely pubescent (Fig. 9.2d).

Dioscorea puber Bl. Enum.: A woody unarmed tomentose climber. Stem twining is
to the right. Tubers are long and cylindrical in shape. Leaves are opposite,
broadly ovate, or sub-orbicular. Spikes are pubescent. Bulbils are axillary and
greenish in colour. Fruits are winged capsule with winged seeds (Fig. 9.2)
(Gamble 1928; Haines 1921–1925; Saxena and Brahmam 1995).

Tubers of all the above contain a good amount of carbohydrate, fibre, and other
nutrients (Kumar et al. 2012). Tribal communities use these tubers as foods such as
chips, vegetables, and salad along with their main meal with rice. The tribal
communities in SBR are often in the grip of food shortage as they hold very
limited/marginal cultivated land (except Mankirdia) as they cannot afford to have
optimum agricultural practice.

Aboriginal agricultural production is insufficient for 3–6 months, during which
they rely on wild plants for their sustenance. The bulk of their current diet is rice
and they obtain tuberous food items including *Dioscorea* species from the forest as
a substitute of regular/part time diet, especially at times of food scarcity. Regarding
health treatment, they have no facilities for modern treatment and no knowledge of
allopathic medicines.

The tribes of the SBR have developed indigenous knowledge of using plants in
their surroundings as cures for diseases. Over the years, they have identified
medicinal edible roots, tubers, rhizomes, corms, fruits, leaves, flowers, and other
plant parts. They use these plants or their parts as medicine against diseases caused
by microbial pathogens. Tubers have sound ethno-botanical values among rural and
tribal communities of SBR. They frequently suffer from serious health problems
like endemic malaria seen prevalent among Mankirdia tribe living in temporary
huts known as "*Kumbhas*". Malnutrition is more prevalent in tribes like Kolha,
Munda, Santal, Bathudi, and Munda. Tuberculosis is frequent among Santhal,
Bathudi, and Munda. Skin infections like eczema, leprosy, and other fungal diseases are common among most tribal communities. The tribe Hill Kharia is famous
for unique medicinal practice in that, they collect medicinal plants raw, store some
for therapeutic use, and market the rest for their livelihood. Aboriginals of SBR
have been using *Dioscorea* species against most of common diseases (Kumar et al.
2013a, b). They use *Dioscorea* species against tuberculosis, poor appetite, and
diabetes. They use the plant or its product in different forms such as the juice of
D. Puber, which is used as a tonic; the paste of *D. bulbifera* is externally applied
against syphilis as well as skin infections. The present study incorporates the
ethnomedicinal values of *Dioscorea* species collected through questionnaires of
rural and tribal communities of SBR (Tables 9.2 and 9.3).

Table 9.2 Details of informants (users) belonging to Similipal Biosphere Reserve and adjoining areas

Name of informants	Sex	Age	Races	Village	Zone
Dimini Makirdia	F	40	Mankidia[a]	Durdura	Transitional zone
Ramdhenu bamudia	M	48	Bathudi		
Topo Mankirdia	M	45	Mankidia[a]		
Ajmal mankardia	M	44	Mankidia[a]		
Bhima Singh	M	55	Munda		
Sokhan Majhi	M	35	Santhal	Gurguria	Buffer
Kusal Ho	M	45	Ho	Handipuan	Transitional zone
Adan marandi	M	33	Santhal		
Sankar munda	M	34	Santhal	Jashipur	Transitional zone
Gopeswar munda	M	45	Santhal		
Sagar munda	M	48	Santhal	Kalikaprasad	Transitional zone
Sarma alda	M	32	Kolho	Kolha	Buffer
Gunjeram Badra	M	30	Kolho		
Baburam Soi	M	59	Munda	Dudhiani	Core
Rama Ho	M	40	Ho	Padampur	Transitional zone
Bnamali Ho	M	41	Ho		
Balram Munda	M	55	Munda		
Sagar Munda	M	48	Santhal		
Janki Munda	F	42	Munda		
Sanu futi	M	50	Bathudia	Sana Baraha Kamuda	Core
Budhi Ho	M	40	Kharia[a]	Sanuski	Buffer
Akhil Naik	M	32	Kolho		
Jambira Bari	M	45	Kolho		
Sahu Bari	M	40	Kolho		
Charan Lohar	M	42	Lohar	Ratuda	Buffer
Sukaral Badra	M	53	Kolho	Kendumundi	Transitional zone
Korma Ho	M	38	Ho		
Parmodh Majhi	M	33	Santhal	Bisoi	Transitional zone
Ansari Soren	M	40	Santhal		
Kusal Ho	M	22	Ho		

[a]Primitive Tribal Group

9.4 *Dioscorea* Species: Future Food and Medicine

In developing countries like India, major populations cannot afford adequate food and medicines to meet daily requirements. People suffer from nutrient deficiency and other diseases (FAO 1994). India has faced a series of famines and major food shortages since before the 1940s. National food grain production was 50.82 million tons during 1950–1951, which has gone up to 264.38 million tons in 2012–2013. While the production of food grain is increasing exponentially, the population of the country has also increased in geometric progression. There has been the need to

Table 9.3 Ethnobotanical values of *Dioscerea* species (tubers) among rural and tribal communities of Similipal Biosphere Reserve

Plant name	Local name (s)	Collection site	Races	Uses against	Mode of use(s)
D. pentaphylla	Panja Sanga	Padampur	Santhal	Skin infections	Macerated tuber paste is applied externally on lesions
		Hatibadi	Bathudi		Tuber paste is used externally for 6–8 days to cure septic and wounds
		Jashipur haat	Mankardia	Cold	Approx 250 g tuber is boiled with about 1 L of water and juice is prepared. One cup of juice with salt is taken thrice a day to remove cough
		Bisoi	Ho	Constipation	One year old tubers are left overnight in running water and this tuber is used as chips to cure stomach pain and constipation problems
		Karanjia	Santhal	Poor appetite	After successive boiling, the tubers are eaten as vegetables and to reduce poor appetite twice a week
		Haatibaadi	Santhal	Against cut	Approx 200 g of fresh tuber is crushed with water and made into paste, which is used externally on cut and other similar wounds thrice a day till cure
D. bulbifera	Pita Aalu	Angarpada	Ho	Cooling agent	Matured tuber is crushed with water and made a light juice with salt. One glass of juice is taken in day time to reduce body heat during working time in summer season
		Durdura	Bathudi	Birth control	Immature bulb/tubers are eaten raw as contraceptive food

(continued)

Table 9.3 (continued)

Plant name	Local name (s)	Collection site	Races	Uses against	Mode of use(s)
		Padampur	Santhal	Skin infections	The tubers are peeled out and the juice is rubbed on the infected skin as antiseptic
		Angarpada	Santhal	Skin infections in cattle	About 100 g of dried tuber powder is mixed with Karanaj oil and rubbed as ointment on infected skin of cattle
D. alata	Khamba Aalu	Sanuski	Kolho	Cooling agent	The tuber juice is used as cooling agents during summer
		Ratuda	Lohar	Diarrhoea	Raw tuber is taken with equal amount of sugar candy twice a day to cure dysentery
D. puber	Kukuai Sanga	Mirginali	Bathudi	Skin infection	One year old tuber is macerated with water and Karanja oil. Paste is applied twice a day to cure wounds and skin infections
		Kalikaparsad	Santhal	Stomach pain	Raw tubers are crushed with water and made light decoction. About 100 mL of decoction is taken during stomach pain to reduce ache
		Dudhiani	Munda	Gastric problems	Matured tuber is dried and made into powder. About 2 teaspoonful powder is taken in empty stomach with 2–5 seeds of black piper to cure gastric problems
		Gurguria	Santhal	Labour pain	Dried tubers are chopped in water overnight and boiled with rhizome of *Curcuma officinalis*. ½ glass of it is taken during delivery to reduce labour pain

(continued)

Table 9.3 (continued)

Plant name	Local name (s)	Collection site	Races	Uses against	Mode of use(s)
D. hispida	Bayan Aalu	Handipuan	Ho	Fish poison	Matured tubers are dug out and cut into small pieces and immediately thrown into the pond during summer to kill or paralyse the fishes to make easy to catch them
		Kiadungri	Ho	Skin infections	Tuber is crushed with water and made paste. Paste is applied on the infected area of skin for 2–3 days as an antiseptic
		Sana Baraha Kamuda	Bathudia	Stomach worm	Fresh tubers are crushed with water and made into juice with salt. 1 cup of juice is taken twice a day continuously for 5 days to kill stomach worm

increase the nutritional status of the community, particularly that of young children and nursing mothers in rural, tribal, and even in urban areas. The Government of India and non-governmental organizations are carrying out a large number of awareness programmes to address these issues, but still malnutrition and adequate medicines are challenges to rural and tribal communities of India (FAO 1998).

Food self-sufficiency is a basic requirement at the household level, especially in remote areas. People living close to nature/biodiversity do not produce enough food grains to meet annual requirements and run out of food in off agricultural seasons/critical periods resulting in famine. They try to source food and medicines through non-conventional means, by consuming various wild plants and their parts. The nonconventional means are directly concomitant with biodiversity. Plants provide numerous edible-medicinal products such as leaves, flowers, fruits, nuts, berries, stems, and roots/tubers. Among all the phyto-resources, roots and tubers are minor wild crops. Edible roots/tubers not only enrich the diet due to presence of starch and energy-supplemented metabolites along with fibres, but also possess medicinal properties due to the presence of diverse secondary metabolites (Kesavan and Swaminathan 2008). *Dioscorea* species rank as the world's fourth most overriding root and tuber crop after potatoes, cassava, and sweet potatoes (Lev and Shriver

1998). Besides being starchy, *Dioscorea* species possess anti-nutritional components that make them bitter in taste.

Tribal people use traditional skills to remove bitterness by various ways throughout the world. In spite of its bitterness, *Dioscorea* are the main food source for the Mbuti pygmies of Eastern Zaire, the Batek of Pen-insular Malaysia, the Baka pygmies in the forests of Southern Cameroon, and people at Kuk Swamp of Papua New Guinea and in Benin, Nepal, Australia, Malaysia, Colombia, and other parts of the world. They are the staple crop of the Igbo people of Nigeria (Milton 1985; Onyilagha and Lowe 1986; Webster et al. 1984; Hart and Hart 1986; Endicott and Bellwood 1991; Sato 2001; Bhandari et al. 2003; Fullagar et al. 2006; Khare 2007; Faostat 2009; N'danikou et al. 2011). Historical records of West African and African yams in Europe date back to the sixteenth century. *Dioscorea* species were taken to the Americas through pre-colonial Portuguese and Spanish on the borders of Brazil and Guyana, followed by dispersion through the Caribbean. Olhuala is a type of local *Dioscorea* tuber that is a staple food in the Maldives and Oca in New Zealand (Maneenoon et al. 2008). In India, particularly in the Western and Eastern Ghats, tubers of *Dioscorea* species are left overnight in stream water and cooked as vegetables. Aboriginals also use tubers as snacks, roasted and powdered with other food stuffs after removal of bitterness.

The same processes are also used in Himalayan regions and North-Eastern part of India (Sheikh et al. 2013), Orissa (Mishra et al. 2008a), Tamil Nadu (Rajyalakshmi and Geervani 1994), among Palliyar and Kanikkar tribes (Santhilkumar et al. 2014) living in South-Eastern slopes of Western Ghats. Of all the *Dioscorea* species reported, only 10 species of *Dioscorea* are commercially cultivated throughout the world. Rural and tribal communities of SBR have developed unique skills to detoxify the bitter content, as shown in Table 9.4.

Knowledge on *Dioscorea* species as wild tubers is gradually disappearing with increased modernization and migration of rural and tribal people to urban areas. Documentation and revaluation of indigenous knowledge on these valuable crops is urgently needed for conservation sustainability of the species. Table 9.5 highlights the food and medicinal values of *Dioscorea* species. Evaluation of nutritional and anti-nutritional values is required along with a genetic strategy to remove the bitter components in order to enhance their utility. Screening of phytochemicals present in these tuber crops and anti-microbial activity against different human pathogens should be documented for isolation of new anti-microbial compounds, which might help to fight against microbial resistance and to manufacture new drugs to cure different types of fatal infectious diseases.

9.5   Nutraceutical Importance of *Dioscorea* Species

"Let Food be the Medicine", as stated by Hippocrates, the Father of Western Medicine, still embodies a practice common in society today. Nutraceutical foods (NF) are the solution to two unsolved problems for an increasing global population.

Table 9.4 Traditional food systems and palatability of *Dioscorea* species among tribal communities of SBR

Botanical name	Local name(s)	Collection site	Parts used	Palatability	Mode of consumption and culinary uses
D. pentaphylla	Panja Sanga	Sana Baraha Kamuda	Tuber	Good	Tubers are soaked in water overnight and boiled with tamarind, then cooked as vegetables
	Bayan aalu	Hatibadi	Tuber	Good	Tubers are left in running water overnight then boiled with tamarind and cooked as vegetables
	Panja sanga	Gurguria	Tuber	Good	Successive boil with leaves of tamarind and cooked with tuber of *D. alata* as vegetables
	Panja sanga	Durdura	Tuber	Good	Properly washed after peeling the outer hairs and cut into small pieces. Left to dry. Dried pieces are cooked with other vegetables
D. bulbifera	Pita aalu	Manda	Tuber	Good	Tubers are left in running water overnight, then cooked as vegetables
	Pita aalu	Hatibadi	Tuber	Good	Tubers are cut into small pieces, dried, and made powder. Powder is used to make vegetables gravy
	Pita aalu	Jashipur	Tuber	Moderate	Tubers are eaten after about 4–5 times boiling as vegetables
	Pita aalu	Durdura	Tuber	Good	Tubers washed repeatedly then boiled and cut into small pieces and cooked as vegetables
	Pita aalu	Handipuhan	Tuber	Moderate	Newly tubers are burned and consumed as snacks by children in Handipuhan and Padampur village
D. puber	Kasa aalu	Durdura	Tuber	Good	Tubers are cut in slices and dried. Dried sliced are cooked and eaten as energy supplement
D. hispida	Khulu sanga	Hatibadi	Tuber	Rare	Newly tubers are burnt and eaten
	Bayan aalu	Hatibadi	Tuber	Rare	Tubers are boiled after soaking overnight in water and cooked as vegetables with tamarind

(continued)

Table 9.4 (continued)

Botanical name	Local name(s)	Collection site	Parts used	Palatability	Mode of consumption and culinary uses
D.alata	Khamba Alu	Angarpada	Tuber	Good	Tubers are cooked as vegetables as a socio-cultural dish of Odisha "Dalma"
	Desia aalu	Padampur	Tuber	Good	Sliced tubers are cooked with pulses and eaten with rice as meal

Table 9.5 Medicinal values of selected *Dioscerea* species

Botanical name	Plant parts	Ethno-botanical/pharmacological values	Supporting source(s)
D. alata	Tuber	Tuber powder is useful in piles	Jadhav et al. (2011)
		Tubers are used to cure ulcers	Singh et al. (2009)
		Used to kill stomach worm	Samanta and Biswas (2009)
		Cake is prepared from the 10 g dried powder of tubers. One cake is taken twice daily for 15 days in the treatment of piles	Kamble et al. (2010)
D. bulbifera	Tuber	Used in piles, dysentery, syphilis, and ulcers	Nashriyah et al. (2011)
		Used as contraceptive	Swarnkar and Katewa (2008)
		Against rheumatism	Sahu et al. (2010)
		In treatment of tuberculosis	Lila and Bastakoti (2009)
		Tubers are taken raw as contraceptive and for birth control	Misra et al. (2012)
		To cure piles	Behera (2006)
		To cure muscular pain	Sahu et al. (2010)
		To cure hernia and against scorpian bite wound	Nayak et al. (2004)
		To enhance appetite	Mishra et al. (2008a, b)
D. hispida	Tuber	An antidote against arrow poison	Edison et al. (2006)
		Against peeling of skin of feet	Lila and Bastakoti (2009)
		It decreases the blood glucose	Harijono et al. (2013)
	Tendrils	For de-worming	Present study
	Tuber	Water of soaked tuber is used as medicine for eyes	Present study
		Used as fish poison	Present study
		Tubers are roasted, pounded, and its paste is applied on wounds and injuries	Edison et al. (2006)

(continued)

Table 9.5 (continued)

Botanical name	Plant parts	Ethno-botanical/pharmacological values	Supporting source(s)
D. pentaphylla	Tuber	Tubers are applied on swelling of joints and also used as tonic to improve body immunity	Edison et al. (2006)
		Fracture	Present study
		Stomach pain	Choudhury et al. (2008)
		Used in digestive tract problems	Choudhury et al. (2008)
		Crushed mass of tuber is given to cattle when they become sick by eating green leaves of maize	Swarnkar and Katewa (2008)
		Tuber is used as tonic and to cure stomach troubles, rheumatic swellings	Lila and Bastakoti (2009)
		Used as medicine for skin problems	Present study
		Inflorescence is used as vegetables for body weakness	Choudhury et al. (2008)
D. puber	Tuber	To cure weakness	Kumar and Satpathy (2011)
	Bulbil	Bulbils are cooked and taken to cure colic pain	Sheikh et al. (2013)

Biodiversity is the hub of uncountable NF, yet we only make use of a few. Among those indigenous forest plants, *Dioscorea* species play a vital role in supplementing the food requirement of rural and tribal people in sub-terrain areas, as preserved/stored food stuffs in times of food shortage and lean agricultural seasons. Besides being used as food, these plants provide raw materials for preparation of medicines. Recent studies evaluated secondary metabolites Tannin (Sadasivam and Manickam 2010) and Saponin (Obadoni and Ochuko 2001) and estimated total phenol (Sudha et al. 2011) and antioxidant activity (Gouda et al. 2013) in selected *Dioscorea* species collected from SBR. Phenols are aromatic secondary metabolites of the plant kingdom and phenolics are associated with colour, palatability, nutritional and antioxidant properties of food (Leja et al. 2013). The phenolic contents of *Dioscorea* tubers cause browning reactions (Fig. 9.3), which take place when tubers are cut/damaged (Constabel et al. 1996; Tao et al. 2007). The browning nature is related to the presence of Tannin (shown in Fig. 9.4) and total Phenol (Fig. 9.5) in selected *Dioscorea* species. The results revealed that tubers of *Dioscorea* possess activity of antioxidant factors (shown in Fig. 9.6) along with polyphenolic compounds.

Figure 9.6 shows IC50 values (mean ± Standard Deviation, $n = 3$). Antioxidant activity of experimental *Dioscorea* species is directly proportional to the browning properties and total phenol content (Figs. 9.5 and 9.6). This makes them a sound and suitable wild nutraceutical.

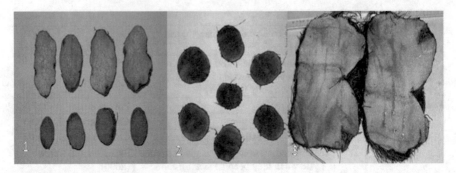

Fig. 9.3 Browning properties of *Dioscorea* species. (**a**) *D. puber*, (**b**) *D. pentaphylla*, (**c**) *D. bulbifera*

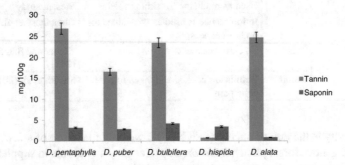

Fig. 9.4 Comparison of two anti-nutritional factors of selected *Dioscorea* species

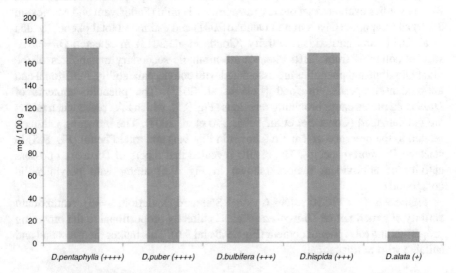

Fig. 9.5 Total phenol of selected *Dioscorea* species and their browning potentials ([+]Browning properties; ++++: Very Fast; +++: Fast, +: Slow)

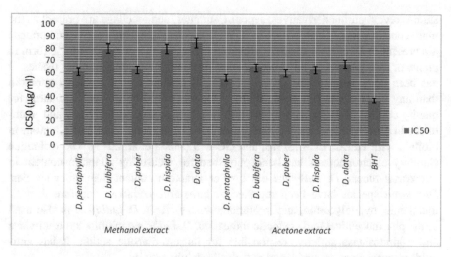

Fig. 9.6 Antioxidant activity (DPPH scavenging activity in IC_{50} values) of selected *Dioscorea* species (DPPH: 2,2-diphenyl-1-picrylhydrazyl; IC_{50}: half maximal inhibitory concentration)

9.6   Active Compounds of *Dioscorea* Species and Pharmacology

Dioscorea species possess diverse primary metabolites and secondary metabolites along with anti-nutritional factors and bitter/toxic components. All these factors make them sound pharmacological agents and good sources for isolation and formulation of new compounds, which can fight against different diseases. Martin (1969) reported a bioactive compound Diosgenin, a sapogenin used in the synthesis of steroidal drugs. Diosgenin is the primary active ingredient of *Dioscorea* species. It is structurally similar to cholesterol. After oral administration, it is metabolized in the liver and eliminated via the bile. Estrogenic and anti-inflammatory effects of diosgenin have been hypothesized due to its structural similarity to estrogen precursors.

D. *deltoidea* is the major species exploited in India for diosgenin production from its rhizomes. Research reports the presence of maximum diosgenin in *D. puber* followed by *D. spicata*, *D. Hispida,* and *D. hamiltonii* (Asha and Nair 2005). *D. alata* possess the phenolic compounds Cyanidin-3-glucoside and the procyanidin dimers B-1 and B-3 (Ozo et al. 1984). Studies reveal that dioscorins present in *Dioscorea* species may exhibit carbonic anhydrase and trypsin inhibitor activity (Hou et al. 2000; Okunlola and Odeku 2008, 2009). *Dioscorea* are reported to have additional bioactive compounds such as steroidal saponins, steroidal drugs, alkaloids, flavonoids (Martin and Cabanillas 1963; Franklin and Cabanillas 1966; Lkediobi et al. 1988; Poornima and Ravishankar 2007; Avula et al. 2012), and the purine derivative allantoin, Furanoid norditerpenes cyanogens, and Lutein (Martin 1974; Bhandari and Kawabata 2005; Yoon et al. 2008). Many phytochemical studies reveal purine derivatives, saponin, starches, and mucilage are present as the main constituents of *Dioscorea* tubers. Well-known biologically active compounds are present in *Dioscorea* species. Allantoin is present as a nitrogen storage form in plants or as a product in the detoxification process of ammonia in plant tissues. It has been demonstrated that *Dioscorea* species contain higher levels of allantoin than any other plants (Fu et al. 2006). Allantoin could be a good 'standard' for quality control of *Dioscorea* tubers due to pharmacological activities and abundance in *Diosorea* species. To date, there have been many reports of allantoin in biofluid with HPLC, LC-MS/MS, and GC-MS (Misra et al. 2013). The technique capillary electrophoresis has been published for measuring allantoin content in *Dioscorea* tubers. *D. bulbifera* has greater allantoin content than other species. *Dioscorea* species have been reported to have anti-oxidative, anti-fungal, anti-mutagenic, hypoglycemic, and immunomodulary effect. *D. bulbifera* is also used in the pharmaceutical and cosmetic industries. *D. bulbifera* inhibits the α-amylase and α-glucosidase activity, responsible for its anti-diabetic action. It has anti-oxidant, anti-hyperglycemic, and anti-dyslipidemic activity.

Research complements the wide tribal claims of use of *Dioscorea* species against microbial infection. Tubers and other parts of *Dioscorea* possess different types of phenolic compounds, which may be responsible for antimicrobial and anti-oxidant activity (Adetoun and Ikotun 1989; Berthemy et al. 1999; Aderiye et al. 1996; Czauderna and Kowaleczyk 2000; Kaladhar et al. 2010; Prakash and Hosetti 2010; Roy et al. 2012; Seetharam et al. 2003; Xu et al. 2008; Sonibare and Abegunde 2012a, b; Ghosh et al. 2012). Some reports confirmed the folkloric uses of the *Dioscorea* species of Nigeria, providing evidence that tuber extracts could be potential sources of natural antimicrobial agents (Araghiniknam et al. 1996; Hou et al. 2001, 2002; Gao et al. 2002; Dong et al. 2004; Yu et al. 2004; Chang et al. 2004; Liu et al. 2006; Sonibare and Abegunde 2012a; Chandra et al. 2013; Begum and Anbazhakan 2013).

9.7   Metabolic Pathways of Active Compounds: Biosynthesis, Precursor Molecules of Active Compounds, and Elicitation

Metabolism is essentially a linked series of chemical reactions that begins with a specific molecule converting into other particular molecules (Ariens 1996; Aich and Yarema 2008; Landecker 2011). There are many such defined pathways in the cell that constitute a series of enzymatic reactions called metabolic pathways (Horton et al. 2005). These pathways are interdependent and their activity is coordinated by exquisitely sensitive means of communication in which allosteric enzymes are predominant (Berg et al. 2002). Metabolic pathways play a dual role: (1) they provide precursors for cellular components and (2) provide energy for additional processes (Markus et al. 2015).

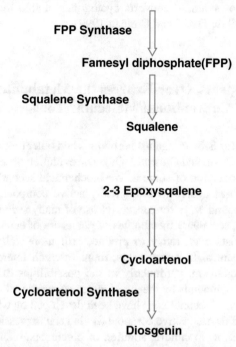

Dimethylallyl diphosphate(DMAPP)+Isopentenyl diphosphate(IPP)

FPP Synthase

Farnesyl diphosphate(FPP)

Squalene Synthase

Squalene

2-3 Epoxysqalene

Cycloartenol

Cycloartenol Synthase

Diosgenin

Fig. 9.7 Schematic representation of enzymes and metabolites of diosgenin biosynthesis pathway (Tal et al. 1984)

The anabolic and catabolic pathways may operate independently. As such, the metabolic pathways occur in localized cellular compartments. Pathways are either anabolic or catabolic, depending on the energy conditions in the cell. Those with characteristics of both are referred to as amphibolic pathways. Biosynthesis of the major bioactive compound diosgenin (Fig. 9.7), found in *Dioscorea* species, is amphibolic.

Synthesis starts in the cytosol with the Gibberellic acid precursor (Hedden and Sponsel 2015) MVA (Mevalonic Acid). Elicitation by the catalyzing activity of cycloartenol synthase on cycloartenol results in Diosgenin. Cholesterol plays an important role in diosgenin synthesis. Cholesterol is of interest as a precursor of many signal molecules, including the steroids progesterone, testosterone, estrogen, and cortisol. Biosynthesis of cholesterol exemplifies fundamental mechanisms for extended carbon skeletons assembly. Cholesterol is a precursor of diosgenin and isopentyl pyrophosphate, which is an activated precursor of cholesterol (Yun et al. 2014) (Fig. 9.5). The combination of isopentenyl pyrophosphate (C_5) units forming squalene (C_{30}) is a fundamental mechanism for the assembly of carbon skeletons of biomolecules. Squalene is the precursor of isoprenoids. Cholesterol is one of phytosterols which is synthesized from isoprenoids. So the processes are interlinked. Squalene is formed by the catalyzing activity of squalene synthase on farnesyl diphosphate (FPP). The squalene is converted into 2–3 Epoxysqualane and cycloartenol. Cycloartenol synthase converts cycloartenotol into diosgenin (Tal et al. 1984; Rojas et al. 1999; David and Croteau 1999).

9.8   Strategy to Express, Over-Express the Metabolites: Application of Conventional/Molecular Tools

Plants develop strategies for defence against herbivores and other biotic and abiotic factors during evolution (David and Croteau 1999). The evolution of self-defence is directly related to the production of some active biochemical secondary metabolites. Secondary metabolites evolve as biologically active compounds. It is not surprising that man has found ways for successful use of many secondary metabolites. As a consequence, secondary metabolites of plants are of economic importance. The challenge exists even today to produce still more valuable natural products from phytoresources. In recent years, more research interest has been shown in secondary metabolism, particularly in the possibilities of altering the production of bioactive compounds by means of either conventional or molecular biological tools. A number of technologies have been developed by which one can regulate the production of desired active compounds via gene expression and over-expression, tissue culture, organ culture, addition of precursors, elicitation, hairy root cultures, bioreactor scaling, and metabolic engineering (Hussain et al. 2012; Prelich 2012). Gene expression relates to production of specific components in response to the functioning of specific genes. Among the biotechnological tools, over-expression comes under genetic engineering for production of desired characters/secondary metabolites (Beggs 1978). It began as a screening tool in molecular genetics shortly after the development of transformation techniques (Herskowitz 1987; Measday et al. 2005).

In *Dioscorea* species, there is a need to upgrade the biochemical pathways involving the production of secondary metabolites as well as to target the gene responsible for production of bitter compounds in the tubers of the species. This entails the alteration of endogenous metabolic pathways to intensify flux towards particular molecules (diosgenin) of interest. Over the past few decades, considerable advances have been made in this context through metabolic engineering via applications of gene expression/over-expression. All molecular and genetic events associated with metabolite production including primary metabolism, its interaction with secondary metabolism, and post-synthetic actions need to be completely characterized for manipulation or reprogramming of metabolic pathways. In this context, combinatorial biosynthesis is a new tool in the generation of novel natural products/secondary metabolites (Chapell 1995; Neumann et al. 2009). The basic concept is combining metabolic pathways in different organisms at a genetic level. There are several pharmaceuticals available in the market that are highly expensive due to the fact that these compounds are only found in rare plants and in low concentrations. Therefore, to achieve a sustainable production of bioactive compounds, there is need for using such approaches under genomics for the alternative production as well as for higher production. The most importance need for *Dioscorea* in this context is the reduction of bitter/toxic compounds from *Dioscorea* species to make them important nutraceutical foods. The plant cell culture is also an effective way of producing desired secondary metabolites (Ramachandra and Ravishankar 2002). Plant cell culture has unique benefits such as independence from geographical variables and ecological factors. It provides quality production with a uniform yield of metabolites. It can yield novel compounds which are not found in the mother plant and above all quick and efficient production at mass level (Larkin and Scowcroft 1981; Larkin 1998).

The strategy of somaclonal variation is a generation of genetic modification in in vitro culture. It is manifested as an inheritable mutation in regenerated plantlets. Plant cell growth in vitro is an asexual process that involves only mitotic cell division and theoretically should not cause any mutation (Karp 1994). However somaclonal variation has a use in improving crops through creation of novel enhanced varieties. Such crops may have increased commercial value, improved biomass yield and increased production of secondary metabolites (Karp 1994; Naqvi et al. 2010).

Once transformed cells are obtained, cell lines are selected expressing the integrated DNA. A number of marker genes have proven effective in plant transformation. Reporter genes are used to demonstrate the transient or stable transformation of plant material and encode enzymes whose substrate is not normally found in plants. Transformation with multiple genes simultaneously allows researchers to study and manipulate an entire metabolic pathway producing secondary metabolites and proteins of interest. However, there are several barriers to transformation with multiple genes simultaneously since the first plant transformation methods were developed for insertion of code for one or two genes. Consequently, as more genes are introduced, the probability of integration into host genomes reduces. To solve this problem, the transfer of multiple genes may be achieved using conventional methods, crossing transgenic lines or sequential processing of the same transgenic line, though such methods are time- and labour-intensive. The problem has been partially addressed with co-transformation and design vectors, which enable multiple gene insertion in a single transformation event (Pickens et al. 2011).

The most important application for transfer of multiple genes is that they can create or modify metabolic pathways, which may include one or multiple genes producing specific metabolites. Among the most ambitious studies in this area is the production of β-carotene in rice. Three carotenoid genes and a marker gene in a single transformation event produce what is now known as the "Golden rice" (Ye et al. 2000). Production of polyunsaturated fatty acids into long chain mustard uses nine genes encoding desaturases and elongases via co-transformation with *A. tumefaciens* (Wu et al. 2005). These reports encourage the vast possibilities with *Dioscorea* species for increased production of diosgenin. Alternatively, metabolic engineering, defined as one or more redirection of enzymatic processes that produce new compounds in an organism, may improve the production or prevent the degradation of existing compounds. Its aim is overproduction of specific compounds, but often, the interconnection of metabolic pathways is such that the number of possible ways to attach a substrate to a product is enormous. These applications allow designing, reconstruction, and visualization of metabolic networks by 'omics' sciences. Through the above-mentioned biological tools, research can lead to the production of desired bioactive compounds as well as suppression of the specific compounds from *Dioscorea* species. Yun et al. (2014) documented that squalane synthase (SQS) catalyzes the condensation of two molecules of farnesyl diphosphate to form squalene, the first committed step for biosynthesis of plant sterols including cholesterol. This is thought to play an important role in diosgenin biosynthesis.

9.9   Findings and Future Prospects

Natural resources are consistently beguiling for researchers who work on food and medicines. The population of the world is growing at an alarming rate, emphasizing the urgent need for self-sufficiency in food and medicine production. Biodiversity provides sources for screening of new food and medicines. Present interpretation concluded that among the different indigenous medicinal foods, *Dioscorea* species play a paramount role as a source of nutraceutical food plants. They provide a safety net as food and conventional medicine during famine/critical periods. The antioxidant activities and rich content of tannin, saponin, and total phenols in selected *Dioscorea* species make them strong nutraceutical vines. The presence of Diosgenin in *Dioscorea* species is the key for the formulation of many more derivatives of future drugs. Natural saponin is of prime importance for the pharmaceutical industry as starting material for the partial synthesis of sex hormones. Despite having much importance, *Dioscorea* species are in the list of neglected tuber crops. This chapter proposes formulation of policies, adopting strategies, and executing research in exploration of the multiple uses of diosgenin and *Dioscorea*, creating awareness for domestication and conservation of the genus. The need for scientific research on removing bitter components to make them a domesticated food crop is also a prime necessity. Priority can be given for attaining food self-sufficiency through proper utilization of tuber crops such as *Dioscorea* species. Necessary measures should be taken to ensure an effective collaboration at institutional, regional, national, and international levels for scientific research works on *Dioscorea* species to establish *Dioscorea* as a true nutraceutical. Conservation and domestication of *Dioscerea* remain one of the challenges of plant metabolism, which can be facilitated by the use of mathematical methods.

References

Aderiye BI, Ogundana SK, Adesanya SA et al (1996) Antifungal properties of Yam (*Dioscorea alata*) peel extract. Folia Microbiol 41(5):407–412

Adetoun A, Ikotun T (1989) Antifungal activity of dihydrodioscorine extracted from a wild variety of *Dioscorea bulbifera* L. J Basic Microbiol 29(5):265–267

Aich U, Yarema KJ (2008) Non-natural sugar analogues: chemical probes for metabolic oligo-saccharide engineering. Glycoscience. doi:10.1007/978-3-540-30429-6_55

Albert S (2012) Food security for Africa: an urgent global challenge. Agric Food Secur. doi:10.1186/2048-7010-1-2

Alexander J, Coursey DG (1969) The origins of yam cultivation. In: Ucko PJ, Dimbleby GH (eds) The domestication and exploitation of plants and animals. Proceedings of a meeting of the research seminar in archaeology and related subjects held at the Institute of Archaeology, London University. Gerald Duckworth, London, pp 405–425

Araghiniknam M, Chung S, Nelson-White T et al (1996) Antioxidant activity of *Dioscorea* and dehydroepiandrosterone (DHEA) in older humans. Life Sci 59:147–157

Ariens EJ (1996) Molecular pharmacology, a basis for drug design. Prog Drug Design 10:429–529

Arinathan V, Mohan VR, Maruthupandian A (2009) Nutritional and antinutritional attributes of some under-utilized tubers. Trop Sub Agric 10:273–278

Asha KI, Nair GM (2005) Screening of *Dioscorea* species for diosgenin from southern Western Ghats of India. Indian J Plant Genet Resour 18(2):227–230

Avula B, Wang YH, Wang M et al (2012) Stuctural characterization of steroidal saponins from *Dioscorea* species using UHPLC-QTOF-MS. Planta Med. doi:10.1055/s-0032-1321072

Beggs JD (1978) Transformation of yeast by a replicating hybrid plasmid. Nature 275:104–109

Begum AT, Anbazhakan S (2013) Evaluation of antibacterial activity of the mucilage of *Dioscorea esculenta* (Lour.) Burkill. Int J Mod Biol Med 4(3):140–146

Behera KK (2006) Ethnomedicinal plants used by the tribals of Similipal Bioreserve Orissa, India: a pilot study. Ethanobot Leaflets 10:149–173

Behera KK, Sahoo S, Prusti A (2010) Productivity co-efficient of tuber yield and dry matter percentage in the tuber of different collections of greater yam (*D. alata* L.) found in Orissa. Libyan Agric Res Center J Int 1(2):108–114

Berg JM, Tymoczko JL, Stryer L (2002) Biochemistry. WH Freeman, New York

Berthemy A, Newton J, Wu D, Buhrman D (1999) Quantitative determination of an extremely polar compound allantoin in human urine by LC-MS/MS based on the separation on a polymeric amino column. J Pharm Biomed Anal 19:429–434

Bhandari MR, Kawabata J (2005) Bitterness and toxicity in wild Yam (*Dioscorea* spp.) tubers of Nepal. Plant Foods Hum Nutr 60:129–135

Bhandari MR, Kasai T, Kawabata J (2003) Nutritional evaluation of wild yam (*Dioscorea* spp.) tubers of Nepal. Food Chem 82:619–623

Bohin L, Goransson U, Alsmark C et al (2010) Natural products in modern life sciences. Phytochem Rev 9(2):279–301

Capell T, Christou P (2004) Progress in plant metabolic engineering. Curr Opin Biotechnol 15 (2):148–154

Carlson SR, Rudgers GW, Zieler H et al (2007) Meiotic transmission of an *in vitro* assembled autonomous maize minichromosome. PLoS Genet 3(10), e179

Chandra S, Saklani S, Mishra AP (2013) *In vitro* antimicrobial activity of Garhwal Himalaya Medicinal Plant *Dioscorea deltoidea* Tuber. Int J Herb Med 1(4):67–70

Chang SJ, Lee YC, Liu SY et al (2004) Chinese yam (*Dioscorea alata* cv. Tainung No. 2) feeding exhibited antioxidant effects in hyperhomocysteinemia rats. J Agric Food Chem 52:1720–1725

Chapell J (1995) Biochemistry and molecular biology of the isoprenoid biosynthetic pathway in plants. Annu Rev Plant Physiol Plant Mol Biol 46:521–547

Choudhury K, Singh M, Pillai U (2008) Ethno botanical survey of Rajasthan—an update. Am Eur J Bot 1(2):38–45

Constabel CP, Bergey DR, Ryan CA (1996) Polyphenol oxidase as a component of the inducible defence response in tomato against herbivores. Recent Adv Phytochem 30:231–252

Cordain L, Eaton SB, Sebastian A et al (2005) Origins and evolution of the western diet: health implications for the 21st century. Am J Clin Nutr 81(2):341–354

Cosa D, Moar B, Lee SB et al (2001) Overexpression of the Bt cry2Aa2 operon in chloroplast leads to formation of insecticidal crystals. Nat Biotechnol 19(1):71–74

Coursey DG (1967) Yams: an account of the nature, origins, cultivation and utilization of the useful members of Dioscoreaceae. Longmans, Greens, London

Czauderna M, Kowaleczyk J (2000) Quantification of allantoin, uric acid, xanthine and hypoxanthine in ovine urine by high-performance liquid chromatography and photodiode array detection. J Chromatogr B 744:129–138

David M, Croteau R (1999) Strategies for bioengineering the development and metabolism of glandular tissues in plants. Nat Biotechnol 17:31–36

Dong M, Feng XZ, Wang BX et al (2004) Steroidal saponins from *Dioscorea panthaica* and their cytotoxic activity. Die Pharm 59(4):294–296

Edison S, Unnikrishnan M, Vimala B et al (2006) Biodiversity of tropical tuber crops in India. National Biodiversity Authority, Chennai, pp 3–60

Endicott K, Bellwood P (1991) The possibility of independent foraging in the rain forest of Peninsular Malaysia. Hum Ecol 19:151–185

FAO (Food and Agriculture Organization of the United Nations) (1994) The state of the Food and Agriculture. Annual report, Rome

FAO (Food and Agriculture Organization of the United Nations) (1998) The state of the world's plant genetic resources for Food and Agriculture. Annual report, Rome

Faostat (2009) http://www.fao.org. Accessed 7 July 2015

Fellmann C, Lowe SW (2014) Stable RNAi rules for silencing. Nat Cell Biol 16:10–18

Fire A, Xu S, Montgomery MK et al (1998) Potent and specific genetic interference by double-stranded RNA in caenorhabditis elegans. Nature 391(6669):806–811

Franklin WM, Cabanillas E (1966) The F1 hybrids of some sapogenin-bearing *Dioscorea* species. Am J Bot 53(4):350–358

Fu YC, Ferng LH, Huang PY (2006) Quantitative analysis of allantoin and allantoic acid in yam tuber, mucilage, skin and bulbil of *Dioscorea* species. Food Chem 94:541–549

Fullagar R, Field J, Denham T, Lentfer C (2006) Early and mid Holocene tool-use and processing of taro (*Colocasia esculenta*), Yam (*Dioscorea* sp.) and other plants at Kuk Swamp in the highlands of Papua New Guinea. J Arch Sci 33:595–614

Fuller DQ, Denham T, Kalin MA, Lucas L et al (2014) Convergent evolution and parallelism in plant domestication revealed by an expanding archaeological record. Proc Natl Acad Sci 111 (17):6147–6152

Gamble JS (1928) Flora of Presidency of Madras, vol 3. Adlard & Son, London, pp 1507–1513

Gao H, Kuroyanagli M, Wu L et al (2002) Antitumor-promoting constituents from *Dioscorea bulbifera* L. in JB6 mouse epidermal cells. Biol Pharm Bull 25(9):1241–1243

Ghosh S, Patil S, Ahire M et al (2012) Synthesis of silver nanoparticles using *Dioscorea bulbifera* tuber extract and evaluation of its synergistic potential in combination with antimicrobial agents. Int J Nanomed 7:483–496

Gouda S, Rosalin R, Das G, Patra JK (2013) Free radical scavenging potential of extracts of *Gracilaria verrucosa* (L) (Harvey): an economically important seaweed from Chilika lake. Indian J Pharm Pharm Sci 6(1):707–710

Gustavo AR (1998) *Agrobacterium tumefaciens*: a natural tool for plant transformation. Elect J Biot 1(3). doi:10.2225/vol1-issue3-fulltext-1

Hahn SK (1995) Yams. In: Smartt J, Simmonds NW (eds) Evolution of crop plants, 2nd edn. Longman, London, pp 112–120

Haines HH (1921–1925) The botany of Bihar and Orissa, vol 5–6. Adlard & Son and West Newman, London, pp 1115–1124

Harijono ET, Sunarharum WB et al (2013) Hypoglycemic effect of biscuits containing water-soluble polysaccharides from wild yam (*Dioscorea hispida* Dennts) or lesser yam (*Dioscorea esculenta*) tubers and alginate. Int Food Res J 20(5):2279–2285

Hart TB, Hart JA (1986) The ecological basis of huntergatherer subsistence in African Rain Forests: the Mbuti of Eastern Zaire. Hum Ecol 14:29–55

Hedden P, Sponsel V (2015) A Century of Gibberellin Research. J. Plant Growth Regul. 34:740–760. doi:10.1007/s00344-015-9546-1

Herskowitz I (1987) Functional inactivation of genes by dominant negative mutations. Nature 329:219–222

Herwig OG, Jutta LW (2014) Plant natural products: synthesis, biological functions and practical applications. Wiley, Weinheim

Horton R, Moran AL, Scrimgeour G et al (2005) Principles of biochemistry, 4th edn. Pearson Education, Upper Saddle River

Hou WC, Chen HJ, Lin YH (2000) Dioscorins from different *Dioscorea* species all exhibit both carbonic anhydrase and trypsin inhibitor activities. Bot Bull Acad Sini 41:191–196

Hou WC, Hsu FL, Lee MH (2002) Yam (*Dioscorea batatas*) tuber mucilage exhibited antioxidant activities in vitro. Planta Med 68(12):1072–1076

Hou WC, Lee MH, Chen HJ et al (2001) Antioxidant activities of dioscorin, the storage protein of yam (*Dioscorea batatas* Decne) tuber. J Agric Food Chem 49(10):4956–4960

Hussain MS, Fareed S, Ansari S et al (2012) Current approaches toward production of secondary plant metabolites. J Pharm Bioallied Sci 4(1):10–20

Jadhav VD, Mahadkar SD, Valvi SR (2011) Documentation and ethnobotanical survey of wild edible plants from Kolhapur District. Recent Res Sci Technol 3(12):58–63

Jain N, Ramawat KG (2013) Nutraceuticals and antioxidants in prevention of diseases. Nat Prod. doi:10.1007/978-3-642-22144-6_70

Kaladhar DSVGK, Rao VN, Barla S et al (2010) Comparative antimicrobial studies of *Dioscorea Hamiltonii* hook.f.tubers with *Azadirachta Indica* Stem. J Pharm Sci Technol 2(8):284–287

Kamble SY, Patil SR, Sawant PS et al (2010) Studies on plants used in traditional medicines by Bhilla tribe of Maharashtra. Indian J Tradit Knowl 9(3):591–598

Karp A (1994) Origins, causes and uses of variation in plant tissue culture. In: Vassil IK, Thorpe TA (eds) Plant cell and tissue culture. Kluwer Academic, Dordrecht, pp 139–152

Kesavan PC, Swaminathan MS (2008) Strategies and models for agricultural sustainability in developing Asian countries. Philos Trans R Soc Lond B Biol Sci 363(1492):877–891

Khare CP (2007) Indian medicinal plants: an illustrated dictionary. Springer, Berlin, p 215

Kuete V, Teponno RB, Mbaveng AT, Tapondjou LA, Meyer JM, Barboni L, Lall N (2012) Antibacterial activities of the extracts, fractions and compounds from *Dioscorea bulbifera*. BMC Complement Altern Med 12(228):1472–6882

Kumar S, Satpathy MK (2011) Medicinal plants in an urban environment; plants in an urban environment; herbaceous medicinal flora from the campus of Regional Institute of Education, Bhubaneswar, Odisha. Int J Pharm Life Sci 2(11):1206–1210

Kumar S, Behera SP, Jena PK (2013a) Validation of tribal claims on *Dioscorea pentaphylla* L. through phytochemical screening and evaluation of antibacterial activity. Plant Sci Res 35 (1–2):55–61

Kumar S, Jena PK, Tripathy PK (2012) Study of wild edible plants among tribal group of Similipal Biosphere reserve forest, Odisha, India: with special reference to Dioscorea species. Int J Biol Technol 3(1):11–19

Kumar S, Parida AK, Jena PK (2013b) Ethno-medico-biology of ban aalu (*Dioscorea* species): a neglected tuber crops of Odisha, India. Int J Pharm Life Sci 4(12):3143–3150

Landecker H (2011) Food as exposure: nutritional epigenetics and the new metabolism. Biosocieties 6:167–194

Larkin P, Scowcroft W (1981) Somaclonal variation—a novel source of variability from cell cultures for plant improvement. Theor Appl Genet 60:197–214

Larkin PJ (1998) Introduction. In: Jain SM, Brar DS, Ahloowalia BS (eds) Somaclonal variation and induced mutations in crop improvement. Kluwer Academic, Dordrecht, pp 3–13

Leja M, Kaminska I, Kramer M et al (2013) The content of phenolic compounds and radical scavenging activity varies with carrot origin and root color. Plant Food Hum Nutr 68 (2):163–170

Lev LS, Shriver AL (1998) A trend analysis of yam production, area, yield, and trade (1961–1996). In: Berthaud J, Bricas N, Marchand JL (eds) L'igname, plante séculaire et culture d'avenir. Actes du séminaire international CIRAD-INRA-ORSTOMCORAF, 3–6 1997. Centre de coopération internationale en recherché agronomique pour le d'eveloppement (CIRAD), Montpellier

Lila SN, Bastakoti R (2009) Ethnobotany of Dioscorea L. with emphasis on food value in Chepang communities in Dhading district, central Nepal. Bot Orient J Plant Sci 6:12–17

Liu YG, Shirano Y, Fukaki H et al (1999) Complementation of plant mutants with large genomic DNA fragments by a transformation-competent artificial chromosome vector accelerates positional cloning. Proc Natl Acad Sci U S A 96(11):6535–6540

Liu YH, Liang HJ, Liu YW et al (2006) Comparisons of antioxidant activities of two species of yam tuber storage proteins in vitro. Bot Stud 47:231–237

Lkediobi CO, Egwim IC, Ikoku O (1988) Acid phosphatase in the tubers of Dioscorea species and its purification from the white yam (D. rotundata Poir). J Sci Food Agric 43(1):27–36

Ma JKC, Hiatt A, Hein M et al (1995) Generation and assembly of secretory antibodies in plants. Science, New York, pp 716–719

Maneenoon K, Sirirugsa P, Sridith K (2008) Ethnobotany of Dioscorea L. (Dioscoreaceae), a major food plant of the Sakai tribe at Banthad Range, Peninsular, Thailand. Ethnoboy Res Appl 6:385–394

Markus AK, Gabriel P, Markus R (2015) The widespread role of non-enzymatic reactions in cellular metabolism. Curr Opin Biotechnol 34:153–154

Martin FW (1969) The species of Dioscorea containing sapogenin. Econ Bot 23(4):373–379

Martin FW (1974) Tropical yams and their potential. Part-3. Dioscorea alata, In: Agriculture handbook no. 495. USAD, Washington, DC

Martin FW, Cabanillas E (1963) A wild hybrid of sapogenin-bearing Dioscorea species. Bull Torrey Bot Club 90(4):232–237

McRae J, Yang Q, Crawford R et al (2007) Review of the methods used for isolating pharmaceutical lead compounds from traditional medicinal plants. Environment 27(1):165–174

Measday V, Baetz K, Guzzo J et al (2005) Systematic yeast synthetic lethal and synthetic dosage lethal screens identify genes required for chromosome segregation. Proc Natl Acad Sci U S A 102:13956–13961

Milton K (1985) Ecological foundations for subsistence strategies among the Mbuti Pygmies. Hum Ecol 13:71–78

Mishra BK (2010) Conservation and management effectiveness of Similipal Biosphere Reserve, Orissa, India. Indian Forester 136(10):1310–1326

Mishra N, Rout SD, Panda T (2011) Ethno-zoological studies and medicinal values of Similipal Biosphere reserve, Orissa, India. Afr J Pharm Pharmacol 5(1):6–11

Mishra RC (1997) The status of rare, endangered and endemic flora of Similipal forest. In: Tripathy PC, Patro RN (eds) Similipal: a natural habitat of unique biodiversity. Orissa Environmental Society, Bhubaneswar, pp 60–72

Mishra RK, Upadhyay VP, Mohany RC (2008a) Vegetation ecology of the Similipal Biosphere Reserve, Orissa, India. Appl Ecol Environ Res 6(2):89–99

Mishra S (1989) An enumeration of Orchids of Similipal hills in Orissa, India. Plant Sci Res 11 (2):73–85

Mishra S, Swain S, Chaudhary SS et al (2008b) Wild edible tubers (Dioscorea spp.) and their contribution to the food security of tribes of Jaypore tract, Orissa, India. Plant Genet Resour 156:63–67

Misra RC, Kumar S, Pani DR, Bhandari DC (2012) Empirical tribal claims and correlation with bioactive compounds: a study on *Celastrus paniculata* Willd., a vulnerable medicinal plant of Odisha. Indian J Tradit Knowl 11(4):615–622

Misra RC, Sahoo HK, Mohapatra AK et al (2011) Additions to the flora of Similipal Biosphere Reserve, Orissa, India. J Bombay Nat Hist Soc 108(1):69–76

Misra RC, Sahoo HK, Pani DR, Bhandari DC (2013) Genetic resources of wild tuberous food plants traditionally used in Similipal Biosphere Reserve, Odisha, India. Genet Res Crop Evol. doi:10.1007/s10722-013-9971-6

Misra RK, Upadhyay VP, Mohanty RC (2003) Vegetation diversity of Similipal Biosphere Reserve. E-Planet 1(1):4–9

Mittal M (2013) To assess the nutritional status and morbidity patterns among non-pregnant non-lactating rural women of reproductive age group (18-40 years). Int J Sci Res Publ 3 (9):1–47

Mohanta RK, Rout SD, Sahu HK (2006) Ethnomedicinal plant resources of Similipal biosphere reserve, Orissa, India. Zoos Print J 21(8):2372–2374

N'danikou S, Achigan-Dako E, Wong JLG (2011) Eliciting local values of wild edible plants in Southern Bénin to identify priority species for conservation. Econ Bot 65(4):381–395

Napoli C, Lemieux C, Jorgensen R (1990) Introduction of a chimeric chalcone synthase gene into petunia results in reversible co-suppression of homologous genes in trans. Plant Cell 2 (4):279–289

Naqvi S, Farré G, Sanahuja G et al (2010) When more is better? Trends Plant Sci 15(1):48–56

Nashriyah M, Athiqah MYN, Amin HS et al (2011) Ethnobotany and distribution of wild edible tubers in Pulau Redang and nearby islands of Tereengganu, Malaysia. Int J Biol Vert Agric food Eng 5(12):110–113

Nayak S, Behera SK, Misra MK (2004) Ethno-medico botanical survey of Kalahandi district of Odisha. Ind J Tradit Knowl 3(1):72–79

Norman JT, Denis PB (1993) Towards a new system of health: the challenge of western diseases. J Community Health 18(1):37–47

Obadoni BO, Ochuko PO (2001) Phytochemical studies and comparative efficacy of the extracts of some haemostatic plants in Edo and Delta states of Nigeria. Global J Pure Appl Sci 8:203–208

Odingo RS (1981) New perspective on natural resources development in developing countries. GeoJournal 5(6):521–530

Okunlola A, Odeku OA (2008) Comparative evaluation of starches obtained from *Dioscorea* species as intragranular tablet disintegrant. J Drug Delivery Sci Technol 18(6):445–447

Okunlola A, Odeku OA (2009) Compressional characteristics and tableting properties of starches obtained from four *Dioscorea* species. Farmacia 57(6):756–770

Onyilagha IC, Lowe J (1986) Studies on the relationship of *Dioscorea cayenesis* and *Dioscorea rotundata* cultivars. Euphytica 35:633–739

Ozo ON, Caygill JC, Coursey DG (1984) Phenolics of five yam (*Dioscorea*) species. Phytochemistry 23(2):329–331

Penna D (2001) Plant metabolic engineering. Plant Physiol 125(1):160–163

Pickens LB, Tang Y, Chooi YH (2011) Metabolic engineering for the production of natural products. Annu Rev Chem Biomol Eng 2:211–236

Poornima GN, Ravishankar RV (2007) In vitro propagation of wild yams, *Dioscorea oppositifolia* (Linn.) and *Dioscorea pentaphylla* (Linn.). Afr J Biotechnol 6(20):2348–2352

Prakash GH, Hosetti BB (2010) Investigation of antimicrobial properties of *Dioscorea pentaphylla* from Western Ghats, India. Sci World 8(8):91–96

Prelich G (2012) Gene over expression: uses, mechanisms, and interpretation. Genetics 190:841–854

Qi B, Fraser T, Mugford S et al (2004) Production of very long chain polyunsaturated omega-3 and omega-6 fatty acids in plants. Nat Biotechnol 22(6):739–745

Rajyalakshmi P, Geervani P (1994) Nutritive value of the foods cultivated and consumed by the tribals of South India. Plant Food Hum Nutr 46:53–61

Ramachandra RS, Ravishankar GA (2002) Plant cell cultures: chemical factories of secondary metabolites. Biotechnol Adv 20:101–153

Reddy KN, Pattanaik C, Reddy CS, Raju VS (2007) Traditional knowledge on wild food plants in Andhra Pradesh. Ind J Tradit Knowl 6(1):223–229

Rojas R, Alba J, Plaza MI et al (1999) Stimulated production of diosgenin in *Dioscorea galeottiana* cell suspension cultures by abiotic and biotic factors. Biotechnol Lett 21:907–911

Roy A, Geetha RV, Lakshmi T (2012) Valuation of the Antibacterial Activity of ethanolic extract of *Dioscorea villosa* tubers—an *in vitro* study. Int J Pharm Pharm Sci 4(1):314–316

Sadasivam S, Manickam A (2010) Biochemical methods. New Age International Publishers, New Delhi, pp 5–262

Sahu SC, Dhal NK, Mohanty RC (2010) Potential medicinal plants used by the tribal of Deogarh district, Orissa, India. Ethno Med 4(1):53–61

Samanta AK, Biswas KK (2009) Climbing plants with special reference to their medicinal importance from Midnapore Town and its adjoining areas. J Econ Taxon Bot 33:180–188

Santhilkumar MSS, Vaidyanathan D, Sivakumar D, Basha MG (2014) Diversity of ethnomedicinal plants used by Malayali tribal in Yelagiri hills of Eastern Ghats, Tamilnadu, India. Asian J Plant Sci Res 4(1):69–80

Sato H (2001) The potential of edible wild yams and yam-like plants as a staple food resource in the African Tropical Rain Forest. Afr Stud Monogr 26:123–134

Saxena HO, Brahmam M (1995) The flora of Orissa, vol 3. Orissa Forest Development Corporation Ltd. and Regional Research Laboratory, Bhubaneswar, pp 1940–1956

Seetharam YN, Jyothishwaran G, Sujeeth H et al (2003) Antimicrobial activity of *Dioscorea bulbifera* bulbils. Ind J Pharm Sci 65(2):195–196

Sheikh N, Kumar Y, Misra AK et al (2013) Phytochemical screening to validate the ethnobotanical importance of root tubers of *Dioscorea* species of Meghalaya, North East India. J Med Plant Stud 1(6):62–69

Singh N, Pangtey YPS, Khatoon S et al (2009) Some ethnobotanical plants of Ranikhet region, Uttaranchal. J Econ Taxon Bot 33:198–204

Sonibare MA, Abegunde RB (2012a) In vitro antimicrobial and antioxidant analysis of *Dioscorea dumetorum* (Kunth) Pax and *Dioscorea hirtiflora* (Linn.) and their bioactive metabolites from Nigeria. J Appl Biosci 51:3583–3590

Sonibare NA, Abegunde RB (2012b) Ethnobotanical study of medicinal plants used by the Laniba village people in South Western Nigeria. Afr J Pharm Pharmacol 6(24):1726–1732

Sosa AJ, Byarugaba DK, Cuevas CFA et al (2010) Antimicrobial resistance in developing countries. Springer, New York

Sudha G, Sangeetha PM, Shree RI et al (2011) In vitro free radical scavenging activity of raw Pepino fruit (*Solanum muricatumaiton*). Int J Curr Pharm Res 3(2):137–140

Swarnkar S, Katewa SS (2008) Ethnobotanical observation on tuberous plants from tribal area of Rajasthan (India). Ethnobot Leaflets 12:647–666

Sylla VM, Ferroud C (2014) New activation methods used in green chemistry for the synthesis of high added value molecules. Int J Energy Environ Eng. doi:10.1007/s40095-014-0148-7

Tal B, Tamir I, Rokem JS (1984) Isolation and characterization of an intermediate steroid metabolite in diosgenin biosynthesis in suspension culture of *Dioscorea deltoidea* cells. Biochem J 219:619–624

Tang G, Galili G, Zhuang X (2007) RNAi and micro RNA: breakthrough technologies for the improvement of plant nutritional value and metabolic engineering. Metabolomics 3 (3):357–369

Tao FJ, Zhang ZY, Zhou J et al (2007) Contamination and browning in tissue culture of Platanus occidentalis L. For Stud China 9(4):279–282

Twyman RM, Christou P, Stoger E (2002) Genetic transformation of plants and their cells. Plant Biot Trans Plants. Marcel Dekker, New York, pp 111–141

Verpoorte R, Memelink J (2002) Engineering secondary metabolite production in plants. Curr Opin Biotechnol 13(2):181–187

Wack AL, Baricas N (2002) Ethical issues related to food sector evolution in developing countries: about sustainability and equity. J Agric Environ Ethics 15(3):323–334

Wada N, Kajiyama SI, Akiyama Y et al (2009) Bioactive beads-mediated transformation of rice with large DNA fragments containing *Aegilops tauschii* genes. Plant Cell Rep 28(5):759–768

Webster J, Beck W, Ternai B (1984) Toxicity and bitterness in Australian *Dioscorea bulbifera* L. and *Dioscorea hispida* Dennst from Thailand. J Agric Food Chem 32:1087–1090

Wu G et al (2005) Stepwise engineering to produce high yields of long-chain polyunsaturated acids in plants. Nat Biotechnol 23:1013–1017

Xu L, Zhou L, Zhao J et al (2008) Fungal endophytes from *Dioscorea zingiberensis* rhizomes and their antimicrobial activity. Lett Appl Microbiol 46(2):68–72

Yadav M, Dugaya D (2013) Non-timber forest products certification in India: opportunities and challenges. Environ Dev Sustain 15:567–586

Ye X et al (2000) Engineering the provitamin A (beta-carotene) biosynthetic pathway into (carotenoide-free) rice endosperm. Science 287:303–305

Yoon DK et al (2008) Determination of allantoin in *Dioscorea rhizoma* by High Performance Liquid Chromotography using cyano columns. Nat Prod Sci 14(4):254–259

Yu ZL, Liu XR, Mc M, Gao J (2004) Anticancer effects of various fractions extracted from *Dioscorea bulbifera* on mice bearing HepA. China J Chin Mater Med 29(6):563–567

Yun Y, Wang R, Liang J et al (2014) Molecular cloning and differential expression analysis of a squalance synthase gene from *Dioscorea zingiberensis*, an important pharmaceutical plant. Mol Biol Rep. doi:10.1007/s11033-014-3487-9

Chapter 10
The Effect of Climate Change on Watershed Water Balance

M.J. Zareian, S. Eslamian, A. Gohari, and J.F. Adamowski

Abstract Climate change refers to a statistically significant long-term shift in the pattern (mean state and variability) of regional or global climate. This phenomenon is attributed to human activities, which have resulted in an increased concentration of greenhouse gases in the global atmosphere. Climate change is already having major effects on the physical environment and biota. The highlighted study investigated the effects of climate change on the sustainability of water resources at a watershed scale, by defining different patterns of climate change (i.e. ideal, medium, and critical) and weighing the output of different general circulation models (GCM). Based on given climate change patterns, the meteorological data (particularly near and far future temperature and precipitation data) were downscaled to the local or regional scale using stochastic weather generators (WGs). The link between climate change and surface runoff was then developed by modelling unit hydrographs within the conceptual framework of rainfall-runoff models. This allowed a reasonable estimate of the future state of surface water resources in the face of climate change to be made; as well as an assertion of the impact of climate change on intra-watershed water consumption (e.g. agricultural, industrial, domestic). Water resource sustainability indices were used to assess the impact of climate change on water resource dynamics within watersheds.

Keywords Climate change • GCM models • Runoff modelling • Water Resources Sustainability Index • Zayandeh-Rud River Basin • Iran

M.J. Zareian (✉) • S. Eslamian • A. Gohari
Department of Water Engineering, Isfahan University of Technology,
8415683111 Isfahan, Iran
e-mail: mjzareian@gmail.com

J.F. Adamowski
Department of Bioresource Engineering, Faculty of Agricultural and Environmental Sciences,
McGill University, 21111 Lakeshore Road, Ste Anne de Bellevue, QC, Canada H9X 3V9

© Springer International Publishing Switzerland 2017
J.N. Furze et al. (eds.), *Mathematical Advances Towards Sustainable Environmental Systems*, DOI 10.1007/978-3-319-43901-3_10

10.1 Introduction

Climate change refers to long-term changes in the Earth's climate dynamics, which in recent years have been increasingly attributed to the expansion of human industrial activity and the attendant increase in the atmospheric concentration of greenhouse gases such as carbon dioxide $[CO_2]_{atm}$ and methane (CH_4), as well as aerosols (IPCC 2007). The increase in the concentration of CO_2 (some 40 % since the previous century) is the major driving force behind climate change (National Research Council 2011). Global climate change has resulted in shifts in various meteorological parameters, particularly temperature, which has risen globally by 0.74 °C per year between 1906 and 2005 (Ganachaud et al. 2011), and precipitation, which has altered in quantity and distribution (Eslamian et al. 2011). Combined with changes in evapotranspiration, these climatic alterations can lead to changes in crop water requirements, agricultural water consumption, and, by extension, watershed runoff (You et al. 2009). Indeed, in a number of regions, climate change has led to a reduction in crop production and an increase in water use by the agricultural sector (Roudier et al. 2011; Hadian et al. 2012; Mokhtari et al. 2012; Gohari et al. 2013a). If current climate change projections prove correct, alterations in depth and timing of stream flow will likely occur (Stewart et al. 2004); therefore, these shifting conditions may significantly alter the water balance in affected watersheds. Facing these possible consequences, policy makers have had to make difficult decisions to prevent further increases in greenhouse gas concentrations (Jaffe et al. 2003; Stern 2006; Brunner et al. 2012; Hadian et al. 2013; Gohari et al. 2014b).

In recent years, the Intergovernmental Panel on Climate Change (IPCC), the main organization active in the field of climate change, has produced a number of general circulation models (GCMs) to quantify the potential effects of climate change on meteorological parameters (IPCC 2007). These models consider the changes in $[CO_2]_{atm}$ and other components that influence climate change and estimate future meteorological parameters (Taylor et al. 2012). Among the several climate change studies which have used these models, some have recommended using a single GCM (Thomson et al. 2006; Liu et al. 2008). However, this approach leads to uncertainty in predictions (Hawkins and Sutton 2009); the use of combined models has become increasingly popular. Their use should, however, be weighted based on their ability to accurately predict meteorological parameters (Dufresne et al. 2013; Gohari et al. 2013a, 2014c; Lee and Wang 2014; Zareian et al. 2014b).

For the purpose of predicting metrological data over large-scale networks and for high atmosphere strata, GCMs pose problems when addressing smaller scale surface events (IPCC 2007). Thus, downscaling methods must be employed, whereby GCMs' outputs are converted to the local scale using data from local meteorological stations (Semenov 2007).

Shifts in climatic parameters can cause changes in watershed characteristics. Besides temperature and precipitation, evapotranspiration can also be altered by climate change. As a result, future crop water requirements may be affected, and consequently, agricultural water consumption (Gohari et al. 2013a). While river flow rates may decline as a result of climate change (Gohari et al. 2014a; Zareian et al. 2014a), domestic and industrial water demands will continue to increase, leading to a water imbalance at the watershed level. Water managers should re-evaluate the operation of existing water resources systems to consider the full range of potential changes in climatic variables (Gohari et al. 2014c).

The present study seeks to determine the effects of climate change on temperature, precipitation, and evapotranspiration in the Zayandeh-Rud River Basin (as a case study), which is one of Iran's most complicated basins, in terms of water resources management. Additionally, the balance between the Zayandeh-Rud River Basin's future water resources and water demands will be investigated.

10.2 Case Study of Zayandeh-Rud River Basin

Located in the Isfahan Province of central Iran, the Zayandeh-Rud River Basin ($26,917 \text{ km}^2$) receives most of its surface water from the Zayandeh-Rud River, at an average rate of $1.4 \times 10^9 \text{ m}^3 \text{ year}^{-1}$, and is characterized by a cool semiarid climate based on the Koppen climate classification (Eslamian et al. 2012). In recent years, the river's annual flow has increased due to natural and transferred flow from nearby basins (Gohari et al. 2013b) and eventually flows into the Gav-Khouni Marsh at the eastern boundary of the basin (see Fig. 10.1).

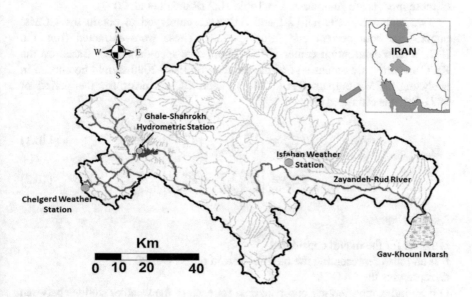

Fig. 10.1 The Zayandeh-Rud River Basin in central Iran

Table 10.1 Characteristics
of the selected weather and
hydrometric stations

Station	Parameter	Longitude	Latitude
Isfahan	Temperature	51° 39′	32° 36′
Chelgerd	Precipitation	50° 07′	32° 27′
Ghale-Shahrokh	Runoff	50° 27′	32° 40′

Since the majority of the basin's natural flow arises from precipitation in its western section, the Chelgerd weather station, situated in this region, was selected as the primary source for precipitation data. The Ghale-Shahrokh hydrometric station, located upstream of the Zayandeh-Rud Reservoir, was selected for estimating the runoff. A number of different agricultural, urban, and industrial sites are highly dependent on the river's water (Salemi and Murray-Rust 2004). Since the considerable development of cities, farms, and industrial towns along this river has mainly occurred in the eastern portion of the basin, data from the Isfahan weather station was selected as most representative of this region. Table 10.1 summarizes the main characteristics of the selected weather and hydrometric stations.

10.3 Methodology

10.3.1 Weighting of the GCM Models

To assess the effects of climate change on temperature and precipitation, the outputs of 15 GCMs were developed for the 2015–2044 period. The characteristics of these models are summarized in Table 10.2 (Randall et al. 2007).

Two emission scenarios (A2 and B1) were employed to obtain the GCMs' outputs for both current and future periods. These were extracted from the IPCC's data distribution center (DDC: http://www.ipcc-data.org/). Based on the IPCC's fourth assessment report, the baseline of 1971–2000 should be chosen in analyzing GCMs outputs (IPCC 2007). Each GCM's error for the period of 1971–2000 is estimated as:

$$\mathrm{TE}_m^{G_i} = \left| \left(\overline{T}_m^B \right)_{G_i} - \left(\overline{T}_m^B \right)_{\mathrm{O}} \right| \tag{10.1}$$

$$\mathrm{PE}_m^{G_i} = \left| \left(\overline{P}_m^B \right)_{G_i} - \left(\overline{P}_m^B \right)_{\mathrm{O}} \right| \tag{10.2}$$

where:

m stands for the month considered
B is an index representing the baseline period of 1971–2000
G_i represents the ith GCM
O is an index representing observed data for each of the weather stations between 1971 and 2000

Table 10.2 Description of the 15 GCMs of IPCC's Fourth Assessment Report (AR4)

Model	Developer	Resolution
HadCm3	UKMO (UK)	$2.5° \times 3.75°$
ECHAM5-OM	MPI-M (Germany)	$1.9° \times 1.9°$
CSIRO-MK3.0	ABM (Australia)	$1.9° \times 1.9°$
GFDL-CM2.1	NOAA/GFDL (USA)	$2.0° \times 2.5°$
MRI-CGCM2.3.2	MRI (Japan)	$2.8° \times 2.8°$
CCSM3	NCAR (USA)	$1.4° \times 1.4°$
CNRM-CM3	CNRM (France)	$1.9° \times 1.9°$
MIROC3.2	NIES (Japan)	$2.81° \times 2.81°$
IPSL-CM4	IPSL (France)	$2.5° \times 3.75°$
GISS-E-R	NASA/GISS (USA)	$4° \times 5°$
ECHO-G	MIUB/M&D (Germany)	$3.9° \times 3.9°$
INM-CM3.0	INM (Russia)	$4° \times 5°$
CGCM3-T63	CCCMA (Canada)	$1.9° \times 1.9°$
NCAR-PCM	NCAR (USA)	$2.8° \times 2.8°$
BCM2.0	BCCR (Norway)	$1.9° \times 1.9°$

$\overline{P}$ is the mean value of precipitation for 30 year records

$PE_m^{G_i}$ is the absolute error of a given GCM in estimating precipitation

$\overline{T}$ is the mean value of temperature for 30 year records

$TE_m^{G_i}$ is the absolute error of a given GCM in estimating temperature

The weight for each of the GCMs used in estimating temperature and precipitation in each month, and for each weather station is given as:

$$PE_m^{G_i} = \left| \left(\overline{P}_m^B \right)_{G_i} - \left(\overline{P}_m^B \right)_O \right| \tag{10.3}$$

$$WP_m^{G_i} = \frac{1/PE_m^{G_i}}{\left(\sum_{i=1}^{15} 1/PE_m^{G_i} \right)} \tag{10.4}$$

where,

$WT_m^{G_i}$ and $WP_m^{G_i}$ show the weight of each of the GCMs in the prediction of temperature and precipitation in each month, respectively. These weights vary with different GCMs, weather stations, and months.

10.3.2 Definition of Climate Change Patterns

Three climate change patterns were identified based on ΔT and ΔP values. The final change of temperature at the Isfahan weather station (ΔT) and change in precipitation at the Chelgerd weather station (ΔP) were calculated as:

$$\Delta T = \frac{\sum_{i=1}^{15} \left(\mathrm{WT}_m^{G_i} \times \Delta T_i \right)}{\sum_{i=1}^{15} \mathrm{WT}_m^{G_i}} \tag{10.5}$$

$$\Delta P = \frac{\sum_{i=1}^{15} \left(\mathrm{WP}_m^{G_i} \times \Delta P i_i \right)}{\sum_{i=1}^{15} \mathrm{WP}_m^{G_i}} \tag{10.6}$$

The ΔT and ΔP values derived from (10.5) and (10.6) were considered as median patterns (ΔT_m and ΔP_m). The GCM models where $\Delta T > \Delta T_m$ were used with (10.5) to calculate a pattern for critical conditions (ΔT_c) and the lower value of ΔT was considered to calculate ideal patterns ($\Delta T < \Delta T_m$). A similar process was employed using (10.6) to develop ΔP_i and ΔP_c, with the difference being that the latter critical value was for reducing precipitation, not increasing it.

10.3.3   Downscaling of the Large-Scale GCM Outputs

The Long Ashton Research Station Weather Generator (LARS-WG), one of the best known stochastic weather generators, can produce daily time series of mete-orological data. Drawing on observed meteorological data in the baseline period and climate change patterns as inputs, this generator can predict a future daily time series of meteorological data. This is based on semi-empirical distribution functions, which can predict when dry and humid periods will occur in the future. To ensure the accuracy of the meteorological data obtained, this model makes some comparisons for goodness of fit, using χ^2, t, and F tests (Semenov 2008).

In the present study, daily time series data for the weather stations and climate change patterns are provided as the input to the LARS-WG model. Daily time series data are then generated for the temperature at the Isfahan weather station and for the precipitation at the Chelgerd weather station.

10.3.4   Rainfall-Runoff Modelling

IHACRES (Identification of unit Hydrographs and Component flows from Rainfall, Evapotranspiration, and Stream flow), a simple model designed to describe the dynamic response characteristics of catchments, was used to predict natural inflow. IHACRES is a catchment-scale rainfall-streamflow modelling methodology, used to characterise the dynamic relationship between rainfall and streamflow using rainfall and temperature data and to predict streamflow. To represent physical features, it incorporates the conceptualization of the relevant large-scale catchment process and is comprised of two modules (non-linear and linear) in series (Fig. 10.2). First in the series is a non-linear loss module, which links rainfall and

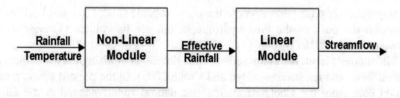

Fig. 10.2 Two modules of the IHACRES

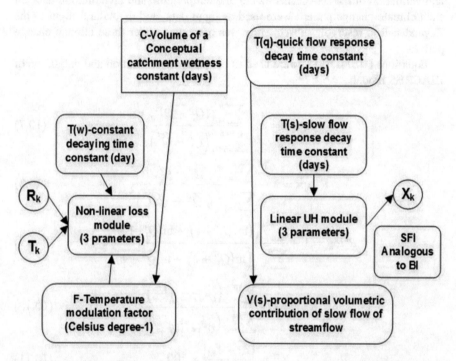

Fig. 10.3 IHACRES model structure and dynamic response characteristics (DRCs). *Source*: Littlewoods (2002)

air temperature (R_k and T_k) to effective rainfall (U_k) with parameters C, $T(w)$, and F (Fig. 10.3). It uses temperature and rainfall data to estimate the relative catchment moisture index; this index indicates the proportion of rainfall that becomes effective rainfall. The second module is a linear unit hydrograph (UH) module, which links effective rainfall (U_k) to stream flow (X_k) with parameters $T(q)$, $T(s)$, and $V(s)$ (Fig. 10.3). It routes the effective rainfall through any configuration of stores in parallel and/or in series, which is identified from the time series of rainfall and stream data, but is typically either one, representing ephemeral streams or two, and represents both slow and quick flows (Croke et al. 2005). In the IHACRES application, it was shown that the parameters in the non-linear module (C, $T(w)$, and F) had significant direct effects on the volume and the peak of the flow hydrograph, while the parameters in the linear module ($T(s)$, $T(q)$, and $V(s)$) had an effect on the peak of the flow hydrograph, but not its volume (Taesombat and Sriwongsitanon 2010).

The model's input data requirements are simple, comprising only precipitation, stream flow, and temperature (Dye and Croke 2003). In the present study, precipitation data from the Chelgerd station and natural runoff gauged at the Ghale-Shahrokh station (1971–2000) were used as IHACRES input data. After calibration and validation of the IHACRES model, the temperature and precipitation data for each climate change pattern were used as input data, and the natural input to the Zayandeh-Rud reservoir subsequently was computed under these climate change patterns.

Equations (10.7–(10.13) were used to evaluate the calibration and validation of IHACRES model:

$$R\,\text{sqrt} = 1 - \frac{\sum_{i=1}^{n}\left(Q_i^o - Q_i^m\right)^2}{\sum_{i=1}^{n}\left(Q_i^o - \overline{Q^o}\right)^2} \tag{10.7}$$

$$R^2\text{sqrt} = 1 - \frac{\sum_{i=1}^{n}\left(\sqrt{Q_i^o} - \sqrt{Q_i^m}\right)^2}{\sum_{i=1}^{n}\left(\sqrt{Q_i^o} - \overline{\sqrt{Q^o}}\right)^2} \tag{10.8}$$

$$R^2\text{log} = 1 - \frac{\sum_{i=1}^{n}\left(\ln\left(Q_i^o + \varepsilon\right) - \ln\left(Q_i^m + \varepsilon\right)\right)^2}{\sum_{i=1}^{n}\left(\ln\left(Q_i^o + \varepsilon\right) - \overline{\ln\left(Q^o + \varepsilon\right)}\right)^2} \tag{10.9}$$

$$R_{\text{inv}}^2 = 1 - \frac{\sum_{i=1}^{n}\left(\frac{1}{Q_i^o + \varepsilon} - \frac{1}{Q_i^m + \varepsilon}\right)^2}{\sum_{i=1}^{n}\left(\frac{1}{Q_i^o + \varepsilon} - \overline{\frac{1}{Q^o + \varepsilon}}\right)^2} \tag{10.10}$$

$$\text{RE} = \left|\frac{Y_o - Y_m}{Y_o}\right| \times 100 \tag{10.11}$$

$$\text{RMSE} = \sqrt{\frac{\frac{1}{n}\sum_{i=1}^{n}\left(Q_i^o - Q_i^m\right)^2}{n}} \tag{10.12}$$

$$C_E = \frac{\frac{1}{n}\sum_{i=1}^{n}\left(Q_i^o - \overline{Q^o}\right)^2 - \frac{1}{n}\sum_{i=1}^{n}\left(Q_i^o - \overline{Q_i^m}\right)^2}{\frac{1}{n}\sum_{i=1}^{n}\left(Q_i^o - \overline{Q^o}\right)^2} \tag{10.13}$$

where:

Q_o is the observed runoff
$\overline{Q_o}$ is the average observed runoff
Q_m is the simulated runoff and
ε is a constant value that is considered for zero observations

10.3.5 Effect of Climate Change on Water Consumption

The effects of climate change on water consumption were evaluated for the agricultural, domestic, and industrial sectors present in the watershed reference evapotranspiration (ET_0) for the Isfahan weather station, which was considered as an index of water use for the agricultural sector, calculated using the Hargreaves and Samani (1982) equation. This was selected due to its low number of parameters, simplicity of calculation, and relatively high accuracy. The equation is given as:

$$ET_0 = 0.0135 \times (K_T) \times (R_a) \times \sqrt{T_{max} - T_{min}} \times (\overline{T} + 17.8) \tag{10.14}$$

where K_T is an empirical coefficient, calculated as:

$$K_T = \left[0.00185 \times (T_{max} - T_{min})^2 - 0.0433 + 0.4023 \times (T_{max} - T_{min}) \right] \tag{10.15}$$

R_a is solar radiation (mm)
$\overline{T}$ is the mean daily temperature (°C)
T_{max} and T_{min} are the maximum and minimum daily temperatures (°C)

Agricultural water consumption was then determined by multiplying the reference evapotranspiration by the area under cultivation.

The predicted population and industrial growth rates were used to estimate the water demand in the Zayandeh-Rud River Basin from the domestic and industrial sectors, respectively.

10.3.6 Water Resources Sustainability Index

The maximum water resource deficit, if such a deficits occur, is considered the worst-case annual deficit, denoted as $\max(D_{Annual})$, for the water user (Moy et al. 1986). A dimensionless maximum deficit can subsequently be calculated by dividing the maximum annual deficit by the annual water demand:

$$Max_{DF} = \frac{\max(D_{Annual})}{Water\ demand} \tag{10.16}$$

10.4 Results

10.4.1 GCM Models Weighting

An assessment of the weighted GCM models' capacity to accurately predict temperatures for the Isfahan weather station and precipitation for the Chelgard weather station is presented in Table 10.3. Different GCM models displayed different accuracies in predicting temperature: The CGCM2.3.2 model, with weight of 0.23, showed the greatest accuracy, while the NCARPCM model, with weight of 0.10, showed the least accuracy in estimating temperature. In general, GCM models with a higher weight had lower frequency than the other GCM models. Four models, CGCM2.3.2, IPSLCM4, ECHAM5OM, and ECHOG, represented 50 % of the GCM models' total weight.

In terms of predicting precipitation at the Chelgerd station, the GISS-ER model showed the greatest accuracy of all the GCMs, with a value of 0.35 (Table 10.3). In contrast, the CSIROMK3.0 model, with the weight of 0.013, displayed the poorest accuracy. Overall, the GISS-ER and CGCM2.3.2 models had the most accurate precipitation forecasts and represented 50 % of the total weight of the GCM models considered (Table 10.3).

Table 10.3 The weight of the different GCM models for prediction of temperature at the Isfahan weather station and precipitation at the Chelgerd weather station

GCM models	Weight of the GCM models	
	Isfahan weather station	Chelgerd weather station
BCM2.0	0.019	0.019
CGCM3-T63	0.016	0.020
CNRMCM3	0.023	0.036
CSIROMK3.0	0.013	0.014
ECHAM5OM	0.053	0.139
ECHOG	0.075	0.136
GFDLCM2.1	0.019	0.017
GISS-ER	0.351	0.109
HADCM3	0.022	0.017
INMCM3.0	0.057	0.026
IPSLCM4	0.051	0.145
MIROC3.2	0.044	0.056
CGCM2.3.2	0.212	0.222
NCARCCSM3	0.025	0.025
NCARPCM	0.013	0.013

10.4.2   Downscaling of the Temperature and Precipitation

To downscale the temperature and precipitation data, observed temperature data from the Isfahan station and observed precipitation data from the Chelgerd station for the period of 1971–2000 served as input to the LARS-WG software. Through the software's site analysis tool, the best probability distribution was fitted to the data and the goodness of fit determined using the Q test section. In this section, goodness of fit tests (e.g. Student's t and χ^2) was conducted and the parameters of both tests computed. Lower values of t and χ^2 indicated a greater conformity to the observed data. Each of these distributions yields a statistical parameter p (p-value), which is compared to a standard significance level of 0.05. In all months, for all tests, all p values for daily time series of temperature and precipitation exceeded 0.05 (Table 10.4), indicating that no significant difference existed between measured and model-predicted data.

10.4.3   Effects of Climate Change on Temperature

Figures 10.4 and 10.5 show the temperature changes at the Isfahan weather station based on the A2 and B1 emission scenarios. Using the A2 scenario, the temperature at the Isfahan weather station increased by between 1.0 and 1.1 °C for the ideal pattern (ΔT_i), 0.4–1.4 °C for the medium pattern (ΔT_m), and 0.7–2.1 °C for the critical pattern (ΔT_c) (Figs. 10.4 and 10.5). Similarly, the values for the B1 scenario under ideal, medium, and critical patterns showed temperature changes of 1.0–1.3 °C, 0.4–1.6 °C, and 0.7–1.8 °C, respectively.

Table 10.4 The statistical details of LARS-WG verification for temperature data at Isfahan weather station and for precipitation data at Chelgerd weather station

Month	Temperature				Precipitation			
	χ^2	p-Value	T	p-Value	χ^2	p-Value	t	p-Value
Jan	0.11	0.96	−0.08	0.69	0.09	0.97	−0.17	0.69
Feb	0.08	1.00	0.59	0.96	0.07	0.99	−0.52	0.85
Mar	0.06	0.88	0.88	0.19	0.06	0.94	1.07	0.56
Apr	0.09	0.71	−0.34	0.53	0.09	0.82	0.19	0.56
May	0.14	0.86	1.11	0.62	0.13	0.86	1.02	0.49
Jun	0.33	0.44	0.42	0.88	0.22	0.66	0.66	0.90
Jul	0.36	0.59	−0.67	0.64	0.24	0.76	−0.26	0.58
Aug	0.09	1.00	0.19	0.52	0.1	0.98	0.16	0.63
Sep	0.28	0.52	0.88	0.71	0.18	0.74	0.30	0.65
Oct	0.13	0.92	0.16	0.25	0.09	0.96	0.56	0.34
Nov	0.07	1.00	0.95	0.36	0.08	0.95	0.05	0.39
Dec	0.31	0.49	−0.29	0.63	0.21	0.72	0.41	0.63

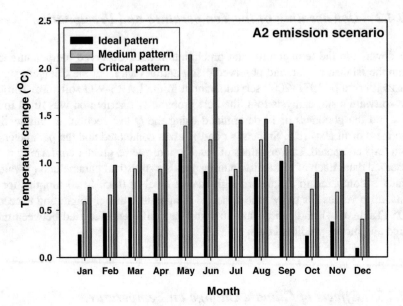

Fig. 10.4 Change in temperature in the Isfahan weather station for A2 emission scenario

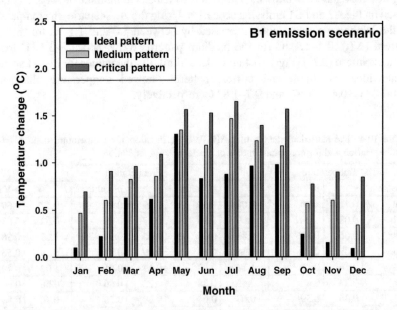

Fig. 10.5 Change in temperature in the Isfahan weather station for B1 emission scenario

Analysis of seasonal and annual changes in temperature showed that summer had the maximum and autumn the minimum temperature increase in both emission scenarios, and for all three climate change patterns. The annual temperature change ranged between 0.63 and 1.31 °C under the A2 emission scenario with a change of 0.95 °C observed under the medium pattern. Under the B1 emission scenario, the temperature changes varied by 0.56–1.15 °C between the ideal and critical patterns. In general, the A2 emission scenario showed a greater increase in the temperature than the B1 scenario.

10.4.4   Effects of Climate Change on Precipitation

Figures 10.6 and 10.7 show the effect of climate change on precipitation. For the A2 emission scenario, the maximum precipitation decrease at the Chelgard weather station occurred in March and varied by 28 and 38 % between the ΔP_i and ΔP_c. The minimum precipitation decrease occurred in September and varied between 4 and 12 % under the different climate change patterns. Under the B1 scenario, the maximum precipitation decrease was observed in May and varied between 19 and 38 %.

The maximum precipitation decrease was observed in winter for both emission scenarios, while the minimum precipitation decrease occurred in summer for the A2 s scenario and in spring for the B1 scenario. In general, the A2 emission scenario showed a greater precipitation decrease than the B1 scenario.

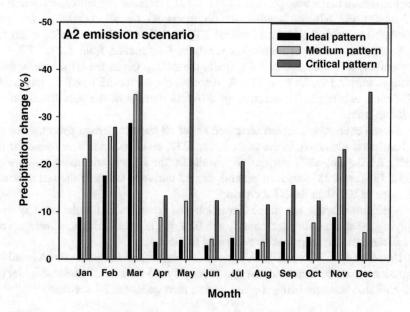

Fig. 10.6 Change in precipitation in the Chelgerd weather station for A2 emission scenario

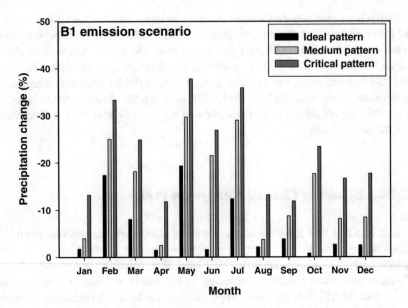

Fig. 10.7 Change in precipitation in the Chelgerd weather station for B1 emission scenario

10.4.5 Effects of Climate Change on Agriculture Water Demand

In comparison to the base period of 1971–2000, reference evapotranspiration (ET_0) values for the Isfahan weather station increased for all months of the year (Table 10.5), with the highest level of ET_0 in both emission scenarios occurring in September. In the A2 emission scenario, ET_0 moved from 6.8 to 7.7 % in September, with a change of 7.5 % under the ΔET_m. Under the B1 scenario, these changes were 7.3 %, 7.8 %, and 8.2 %, respectively, for the ΔET_i, ΔET_m, and ΔET_c. Minimum values of ET_0 occurred in different months of the year based on the scenario used.

Annual evapotranspiration increased under all climate change patterns. For the A2 emission scenarios, increases of 1361, 1372, and 1382 mm were found for the ΔET_i, ΔET_m, and ΔET_c, respectively, while for the B1 scenario, these values were 1359, 1367, and 1376 mm. In general, the A2 emission scenario showed a greater increase in ET_0 than the B1 scenario.

Agricultural water use in the Zayandeh-Rud River Basin is calculated by multiplying the mean cultivated area by the ET_0, while considering the average crop coefficient for the crop species cultivated in the basin.

Based on Figs 10.8 and 10.9, agricultural water demand under the A2 and B1 emission scenarios increased at a rate of roughly 0.5 %, with the increase under the A2 emission scenario being slightly greater than under the B1 scenario.

Table 10.5 Changes in reference evapotranspiration for A2 and B1 emission scenarios for the Isfahan weather station

Month	1971–2000	Evapotranspiration (mm)					
		2015–2044					
		A2 emission scenario			B1 emission scenario		
		Ideal	Medium	Critical	Ideal	Medium	Critical
Jan	33	33	34	34	33	33	34
Feb	43	44	44	45	43	44	45
Mar	71	72	73	73	72	72	73
Apr	106	107	107	109	106	107	108
May	146	149	150	153	149	150	151
Jun	188	192	193	196	192	193	194
Jul	199	203	204	207	205	206	206
Aug	192	202	203	205	203	203	205
Sep	154	164	165	166	165	166	166
Oct	98	99	101	101	99	100	101
Nov	53	55	55	56	54	55	56
Dec	36	36	37	38	36	36	38
Annual	1319	1361	1372	1383	1359	1366	1374

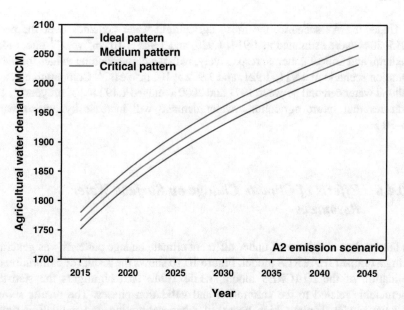

Fig. 10.8 Effect of climate change on agriculture water demand based on the A2 emission scenario

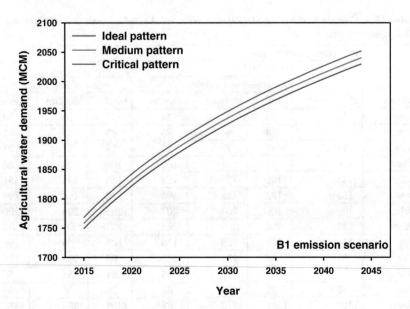

Fig. 10.9 Effect of climate change on agriculture water demand based on the B1 emission scenario

Under the A2 scenario, the mean agricultural water demands over the period 2015–2044 were estimated at 1.914, 1.929, and 1.945×10^9 m^3 year^{-1} under ideal, medium, and critical patterns, respectively, with the corresponding values for the B1 emission scenario at 1.911, 1.921, and 1.932×10^9 m^3 year^{-1}. Comparatively, agricultural water demand between 1971 and 2000 averaged 1.492×10^9 m^3 year^{-1}. This indicates that future agricultural water demand will increase by approximately 25–30 %.

10.4.6 Effects of Climate Change on Surface Water Resources

In this study, annual runoff under different climate change patterns was generated using a lumped IHACRES model. Figure 10.10 shows the results of calibration and validation of the IHACRES model, while Table 10.6 highlights the statistical coefficients related to the calibration and validation phases. The results showed that the model displays high accuracy when calibrating and validating runoff. The annual changes in natural flow of the Zayandeh-Rud River are shown in Figs 10.11 and 10.12. Under the A2 scenario, mean natural runoff for the 1971 to 2000 base period (1.044×10^9 m^3 year^{-1}) declined to 0.727, 0.477, and 0.357×10^9 m^3 year^{-1} for ideal, medium, and critical patterns, respectively, while under the B1 scenario, these values declined by slightly smaller amounts to 0.758, 0.515,

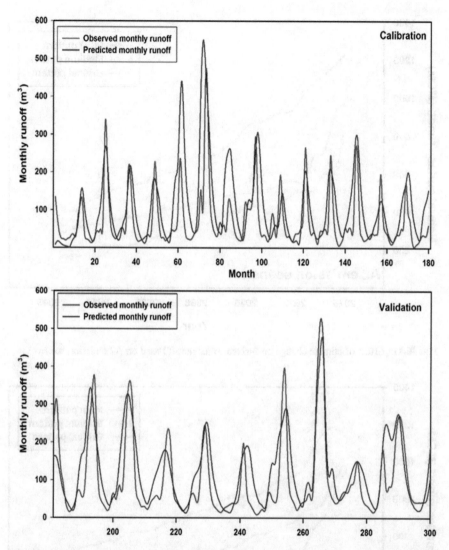

Fig. 10.10 The results of calibration and validation of the IHACRES model for runoff predication

Statistical parameter	Calibration	Validation
R sqrt	0.75	0.71
R^2 sqrt	0.68	0.63
R^2 log	0.71	0.66
R^2 Inv	0.73	0.71
RE	23.7	26.3
RMSE	10.1	11.6
CE	0.81	0.79

Table 10.6 Evaluation values for calibration and validation of flow simulation

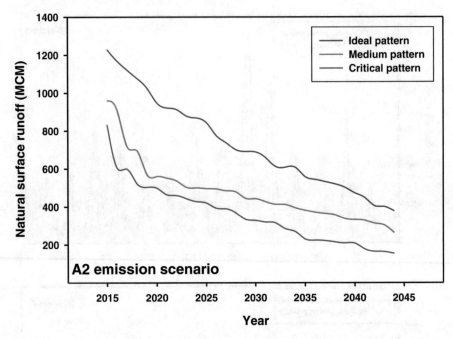

Fig. 10.11 Effect of climate change on surface water runoff based on A2 emission scenario

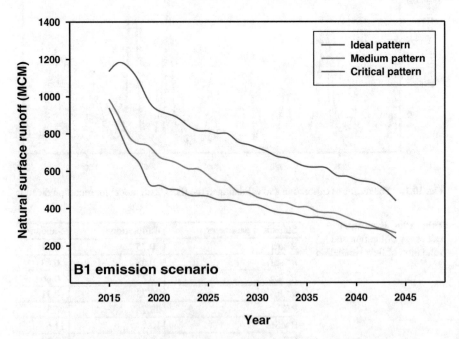

Fig 10.12 Effect of climate change on surface water runoff based on B1 emission scenario

and $0.442 \times 10^9 \, \mathrm{m}^3 \, \mathrm{year}^{-1}$. Overall, it is predicted that mean surface runoff in the Zayandeh-Rud River Basin will decline by roughly 50 %.

10.4.7   Changes in Domestic and Industrial Water Demand

The population of the Zayandeh-Rud River Basin increased at an average rate of $1.5 \% \, \mathrm{year}^{-1}$ during the period 1992–2010, increasing domestic water consumption from 0.330×10^9 to $0.450 \times 10^9 \, \mathrm{m}^3 \, \mathrm{year}^{-1}$. Due to the relative uniformity of the increase in population, a linear regression model was fitted to predict future water consumption in the domestic sector (Fig. 10.13). By extrapolation, it was determined that domestic water consumption will reach $0.630 \times 10^9 \, \mathrm{m}^3 \, \mathrm{year}^{-1}$ by the end of 2044, and that the mean domestic water consumption during the period of 2015–2044 will be roughly $0.546 \times 10^9 \, \mathrm{m}^3 \, \mathrm{year}^{-1}$ (Fig. 10.13).

A similar method was employed to estimate industrial water demand. Review of historical data shows that the industrial water demand increased by $1.3 \% \, \mathrm{year}^{-1}$ between 1992 ($0.125 \times 10^9 \, \mathrm{m}^3 \, \mathrm{year}^{-1}$) and 2000 ($0.202 \times 10^9 \, \mathrm{m}^3 \, \mathrm{year}^{-1}$). A linear regression fit to historical data was extrapolated to predict future industrial water demand (Fig. 10.14). Based on this, industrial water consumption in 2044 was predicted to be $0.318 \times 10^9 \, \mathrm{m}^3 \, \mathrm{year}^{-1}$, and average $0.269 \times 10^9 \, \mathrm{m}^3 \, \mathrm{year}^{-1}$ between the years 2015 and 2044 (Fig. 10.14).

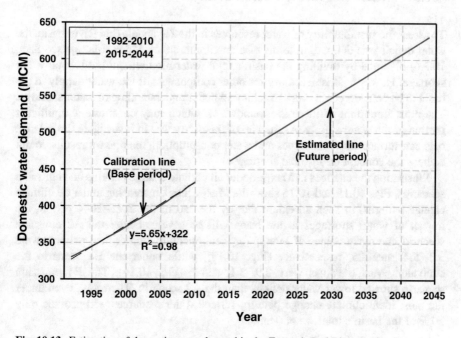

Fig. 10.13 Estimation of domestic water demand in the Zayeneh-Rud River Basin

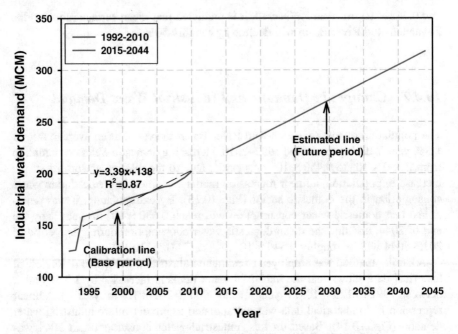

Fig. 10.14 Estimation of industrial water demand in the Zayeneh-Rud River Basin

10.4.8 Water Resources Sustainability

To assess the sustainability of water resources in the Zayandeh-Rud River basin, the water deficit was defined, in terms of a maximum deficit index (Max_{DF}, see Sect. 10.3.6) obtained by dividing the basin's total water requirement by its total water shortage. However, it is necessary for other components of the water supply in the basin to be considered, in order to achieve a better understanding of water shortage. The most important of these is groundwater, which may constitute a significant portion of the water supply. As the aim of this study was to investigate the balance between surface water resources and water consumption, the present results do not address the concept of absolute shortage.

Illustrating the changes in Max_{DF} under all climate change patterns and emission scenarios, Figs 10.15 and 10.16 show the Max_{DF} index increasing under all climate change patterns, to such an extent that by the end of the 2015–2044 period, the impact of water shortages in the basin will be severe. Under the A2 emission scenario, the mean values of Max_{DF} for ideal, medium, and critical patterns were 3.3, 5.4, and 8.5, respectively (Fig. 10.15), while under the B1 scenario the equivalent Max_{DF} values were 2.9, 5.1, and 6 (Fig. 10.16). The B1 emission scenario showed lower water deficits than the A2 scenario. Therefore, even under the most ideal climate change pattern, surface water resources will provide only 30 % of the basin's total water demand.

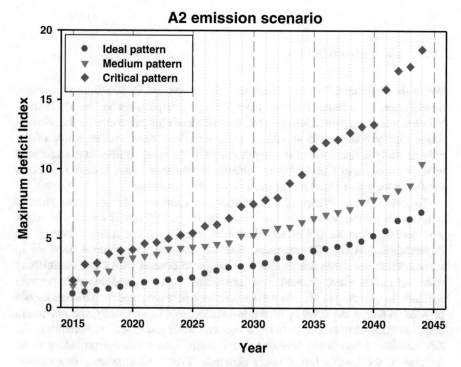

Fig. 10.15 Water sustainability index in the Zayeneh-Rud River Basin for A2 emission scenario

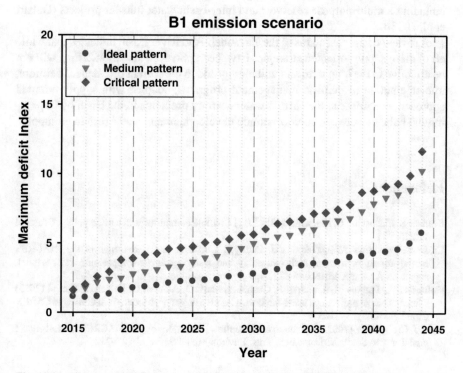

Fig. 10.16 Water sustainability index in the Zayeneh-Rud River Basin for B1 emission scenario

10.5 Conclusion

While the detrimental effects of climate change on the Earth's future are clear, investigators in different domains have different perspectives on climate change. While water management is important, the effects of climate change on the relevant water supply components should be measured. The most notable result of the present study is that each of the highlighted GCM models differs in the accuracy of their estimates of climate change effects. If greater accuracy is desired, models should be weighted based on the accuracy of their forecasts.

To investigate the effects of climate change on surface water resources, simulation models should be used. In this study, the IHACRES runoff simulation model was used to estimate the effect of climate change on the natural flow of the Zayandeh-Rud River. The resulting forecasts showed a continuous decrease in surface water resources due to climate change. In concert with these diminishing water resources, water demand from agriculture, industry, and domestic pursuits will increase, leading to water imbalance. In such cases, water deficit indices should be used to assess the severity of the imbalance, and in this study, the maximum deficit index used to assess the balance between water resources and demands in the Zayeneh-Rud River Basin revealed that surface water resources provide a small fraction of the basin's future water demands. Trends also indicate that climate change effects would lead to severe water shortages. To address this problem in the basin, a number of conventional engineering solutions have been implemented, including a multi-purpose reservoir and inter-basin water transfer projects (Gohari et al. 2013b).

A review of the past status of the Zayandeh-Rud River Basin indicates that while all of these supply-oriented strategies have been practiced since 1952, they will not be effective in the long term and will merely add to the current problems. Therefore, implementation of demand management programs, coupled with supply-oriented approach, can mitigate the current water scarcity problems in the basin by improving the balance between socioeconomic development and water resources supply.

References

Brunner S, Flachsland C, Marschinski R (2012) Credible commitment in carbon policy. Clim Pol 12:255–271

Croke BF, Andrews WF, Jakeman AJ, Cuddy S, Luddy A (2005) Redesign of the IHACRES rainfall-runoff model. In: 29th hydrology and water resources symposium, water capital, engineers Australia. ISBN 085 825 8439

Dufresne J-L, Foujols M-A, Denvil S, Caubel A, Marti O, Aumont O, Balkanski Y et al (2013) Climate change projections using the IPSL-CM5 Earth System Model: from CMIP3 to CMIP5. Clim Dyn 40:2123–2165

Dye PJ, Croke BFW (2003) Evaluation of streamflow predictions by the IHACRES rainfall-runoff model in two South African catchments. Environ Model Softw 18:705–712

Eslamian SS, Gilroy KL, McCuen RH (2011) Climate change detection and modeling in hydrology. In: Climate change research and technology for adaptation and mitigation. Tech press,
Rijeka, pp 87–100

Eslamian SS, Gohari SA, Zareian MJ, Firoozfar A (2012) Estimating Penman–Monteith reference
evapotranspiration using artificial neural networks and genetic algorithm: a case study. Arab J
Sci Eng 37:935–944

Ganachaud AS, Sen Gupta A, Orr JC, Wijffels SE, Ridgway KR, Hemer MA, Maes C, Steinberg
CR, Tribollet AD, Qiu B, Kruger JC (2011) Observed and expected changes to the tropical
Pacific Ocean. In: Bell J, Johnson JE, Hobday AJ (eds) Vulnerability of tropical pacific
fisheries and aquaculture to climate change. Secretariat of the Pacific Community, Noumea,
pp 115–202

Gohari A, Eslamian S, Abedi-Koupaei J, Massah Bavani A, Wang D, Madani K (2013a) Climate
change impacts on crop production in Iran's Zayandeh-Rud River Basin. Sci Total Environ
442:405–419

Gohari A, Eslamian S, Mirchi A, Abedi-Koupaei J, Massah Bavani A, Madani K (2013b) Water
transfer as a solution to water shortage: a fix that can Backfire. J Hydrol 491:23–39

Gohari A, Bozorgi A, Madani K, Elledge J, Berndtsson R (2014a) Adaptation of surface water
supply to climate change in Central Iran. J Water Clim Change. doi:10.2166/wcc.2013.189

Gohari A, Zareian MJ, Eslamian S (2014b) A multi-model framework for climate change impact
assessment. Handbook of climate change adaptation. Springer, Dordrecht, pp 17–35

Gohari A, Madani K, Mirchi A, Bavani A (2014b) System-dynamics approach to evaluate climate
change adaptation strategies for Iran's Zayandeh-Rud Water System. World Environmental
and Water Resources Congress 2014, June 2014, pp 1598–1607. doi:10.1061/
9780784413548.158

Hadian S, Madani K, Rowney C, Mokhtari S (2012) Toward more efficient global warming policy
solutions: the necessity for multi-criteria selection of energy sources. World Environmental
and Water Resources Congress, Albuquerque

Hadian S, Madani K, Mokhtari S (2013) A systems approach to energy efficiency assessment.
World Environmental and Water Resources Congress, Cincinnati

Hargreaves GH, Samani ZA (1982) Estimating potential evapotranspiration. J Irrig Drain Eng
ASCE 108:225–230

Hawkins E, Sutton R (2009) The potential to narrow uncertainty in regional climate predictions.
Bull Am Meteorol Soc 90:1095–1107

IPCC (2007) Climate change 2007: the physical science basis. Contribution of Working Group I to
the IPCC fourth assessment report. Cambridge University Press, Cambridge

Jaffe A, Newell R, Stavins R (2003) Technological change and the environment. In: Maler K,
Vincent J (eds) Handbook of environmental economics. North-Holland, Amsterdam, pp
461–516

Lee JY, Wang B (2014) Future change of global monsoon in the CMIP5. Clim Dyn 42:101–119

Littlewood IG (2002) Improved unit hydrograph characterisation of the daily flow regime (including low flows) for the River Teifi, Wales: towords better rainfall-streamflow models for
regionalization. Hydrol Earth Syst Sci 6:899–911

Liu J, Fritz S, van Wesenbeeck CFA, Fuchs M, You L, Obersteiner M, Yang H (2008) A spatially
explicit assessment of current and future hotspots of hunger in Sub-Saharan Africa in the
context of global change. Glob Planet Change 64:222–235

Mokhtari S, Madani K, Chang NB (2012) Multi-criteria decision making under uncertainty:
application to California's Sacramento-San Joaquin Delta Problem. In: World environmental
and water resources congress, Albuquerque, 20–24 May 2012, pp 2339–2348

Moy WS, Cohon JL, Revelle CS (1986) A programming model for analysis of reliability,
resilience and vulnerability of a water supply reservoir. Water Resour Res 22:489–498

National Research Council (2011) Climate stabilization targets: emissions, concentrations and
impacts over decades to millennia. National Academies, Washington, DC

Randall DA, Wood RA, Bony S, Colman R, Fichefet T, Fyfe J, Kattsov V, Pitman A, Shukla J, Srinivasan J, Stouffer RJ, Sumi A, Taylor KE (2007) Climate models and their evaluation. Contribution of Working Group I to the fourth assessment report of IPCC. Cambridge University Press, Cambridge

Roudier P, Sultan B, Quirion P, Berg A (2011) The impact of future climate change on West African crop yields: what does the recent literature say? Glob Environ Change 21:1073–1083

Salemi HR, Murray-Rust H (2004) An overview of the hydrology of the Zayandeh Rud Basin, Iran. Water Wastewater (Isfahan) 1:2–13

Semenov MA (2007) Development of high-resolution UKCIP02-based climate change scenarios in the UK. Agric For Meteorol 144:127–138

Semenov MA (2008) Simulation of extreme weather events by a stochastic weather generator. Clim Res 35:203–212

Stern N (2006) Stern review on the economics of climate change. Her Majesty's Treasury, London

Stewart IT, Cayan DR, Dettinger MD (2004) Changes in snowmelt runoff timing in western North America under a business as usual climate change scenario. Clim Change 62:217–232

Taesombat W, Sriwongsitanon N (2010) Flood investigation in the Upper Ping River Basin using mathematical models. Kasetsart J (Nat Sci) 44:152–166

Taylor KE, Stouffer RJ, Meehl GA (2012) An overview of CMIP5 and the experiment design. Bull Am Meteorol Soc 93:485–498

Thomson AM, Izaurralde RC, Rosenberg NJ, He X (2006) Climate change impacts on agriculture and soil carbon sequestration potential in the Huang-Hai Plain of China. Agr Ecosyst Environ 114:195–209

You L, Rosegrant MW, Wood S, Sun D (2009) Impact of growing season temperature on wheat productivity in China. Agric For Meteorol 149:1009–1014

Zareian MJ, Eslamian S, Hosseinipour EZ (2014a) Climate change impacts on reservoir inflow using various weighting approaches. World Environmental and Water Resources Congress, Portland

Zareian MJ, Eslamian S, Safavi HR (2014b) A modified regionalization weighting approach for climate change impact assessment at watershed scale. Theor Appl Climatol 122:497–516

Chapter 11
Modelling Challenges for Climate and Community Resilient Socioecological Systems

A. Dey, Anil K. Gupta, and Gurdeep Singh

Abstract Variability, complexity, simultaneity and change in environmental parameters affect social groups. Homeostatic advantages due to resource surplus, institutional access technology and social networks alter perception and community response to climate risks. Modelling requires consideration of socioecological and eco-institutional interactions with social, biological and climatic parameters. We model and manage the multilayer interactions among social institutions, climatic fluctuations and the resultant changes in the rules governing these interactions with the objective of increasing resilience of social and ecological systems to climate change.

We enumerate coping strategies adopted by local communities to suggest modelling approaches for climate-resilient socioecological systems. Statistical tools enable discrete and continuous perspectives in different classes, of institutional and heterogenous social communities; in different time frames and with varying degrees of freedom. Interactions among crops, weeds, pest, temperature, fluctuating rainfall, agro-biodiversity, at farmer's plots affected by different flooding levels in eastern India were studied at decadal intervals over 30 years.

Ecological systems under high climatic risks include drought/flood-prone regions and are inhabited by some of the poorest communities. Modelling communities compulsions, preferences, and the consequences of their choices on socioecological systems will enable a sustainable outcome. This study suggests modelling needs for knowledge-rich, economically poor communities in tropical contexts, which may enable future resilience.

A. Dey (✉)
Indian Institute of Management, Vastrapur, Ahmedabad, Gujarat 380 015, India
e-mail: anamikad@iimahd.ernet.in

A.K. Gupta
Indian Institute of Management, Vastrapur, Ahmedabad, Gujarat 380 015, India

Society for Research and Initiatives for Sustainable Technologies and Institutions,
Ahmedabad 380 015, Gujarat, India
e-mail: anilg@iimahd.ernet

G. Singh
Vinoba Bhave University, Hazaribagh, Jharkhand, India

© Springer International Publishing Switzerland 2017
J.N. Furze et al. (eds.), *Mathematical Advances Towards Sustainable Environmental Systems*, DOI 10.1007/978-3-319-43901-3_11

Keywords Coping strategies • Modelling gaps • Climate resilience • Risk-prone regions

11.1 Introduction

Modelling of socioecological systems requires parameterizing the relationships between various endogenous and exogenous variables so that technological, institutional, cultural and social expectations of different community members can be converged with the goals of climate resilience. The challenge arises when different social and institutional conditions prevent convergence of coping strategies or adaptive approaches. The paper is divided into four parts. In Sect. 11.2, we review limitations of current modelling approaches for dealing with climate change and socioecological resilience, consequently in Sect. 11.3 the socioecological paradigm (Gupta 1984) is revisited and modelling imperatives are drawn. In Sect. 11.4, a case study of coping with climate change in a flood-prone region is provided. Section 11.5 concludes the chapter and provides implications for future studies.

11.2 Limitations of Existing Approaches for Modelling Climatic Systems

Despite limitations of climate resilience models, it is generally agreed that developing countries would suffer not only from warmer climates and increased frequency of extreme events, also from much greater vulnerability in agriculture and a further worsening in the state of malnourishment among people (Fischer et al. 2005). The authors acknowledge the uncertainty with regard to, 'the magnitude of climate change and its spatial and temporal distribution' and socioeconomic implications in heterogeneous populations. Extreme events such as drought and flood may further impair the ability of vulnerable developing countries to adapt. There is archaeological evidence of historical devastation of several urban areas because of an inability to cope with climate change effects (Grimm et al. 2000). Examples include: "the salinization of southern Mesopotamia 4000 years ago (Redman 1992), valleywide erosion in ancient Greece (Van Andel et al. 1990), and almost complete depopulation of large tracts of Guatemala (Rice 1996) and highland Mexico (O'Hara et al. 1993) from 1000 AD to 1400 AD". It is concluded that models integrating human decisions, culture, institutions and socioeconomic systems with ecological endowments are required.

11.2.1 Different Scales of Description, Prediction and Prescription

Sen (1980) argues that description can be used for prediction as well as prescription. It is important to pay attention to the underlying human values enabling these transitions from predictions to prescriptions. Springman et al. (2016) concluded that dietary changes and undernutrition will cause 560,000 deaths worldwide by 2050. The International Model for Policy Analysis of Agricultural Commodities and Trade (IMPACT) has been used extensively for projections spanning into different trades and commodities. Evaluating IMPACT software, Ryan (2003) noticed that accuracy was better for global aggregative scenarios rather than at the local or individual commodity level. Given the high variability in crop yield in dry regions, and a large gap between potential and actual on-farm yields, the effect of climate change has not been realistically simulated (Qingxiang et al. 2004, Grassini et al. 2015). Moreover, the simulation studies do not validate the prospective changes at every spatial grid point. Models which use gridded weather data may have nearly complete geo-spatial coverage but are derived (interpolated) rather than observed (FAO and DWFI 2015). Accurate estimations are possible for scenarios which have a linear weather–crop relationship (Bussel et al. 2011). Projected climatic variability and the expected occurrence of extreme events may generate negative anomalies incompatible with regional and global projections (Soussana et al. 2010). The literature referring to the effect of certain unit changes in temperature or precipitation doesn't fully capture the complexity of climate change phenomena at different scales. The scale at which parameters such as runoff, infiltration rate and evapotranspiration are monitored in agro-climatic models may not be compatible with the feedback effects that take place at river basin level. Future modelling exercises should tackle the complexity of downscaling regional climate impact models, improved crop model based on empirical evidence, better anticipation of extreme events and more precise understanding of decision making and management approaches of the community vis-à-vis various biodiversity changes. Systems with many different components (e.g. species, actors or sources of knowledge) are generally more resilient than those with few components. Redundancy provides 'insurance' within a system by allowing some components to compensate for the loss or failure of others. Redundancy is even more valuable if the components react differently to change and disturbance (response diversity). However, it also incurs costs to maintain which need to be accounted as inputs to models used to determine the overall system resilience.

11.2.2 Dynamic Ecosystems

Natural resource-based enterprises are eco-specific and hence, farmers would need local level projections. Incorporating farmers' knowledge of the systems will increase effectiveness and adaptability among the users (Gupta 1984, 1990). Existing models do not provide an adequate basis for incorporation of local knowledge, as well as participation of people distributed over large expanses in dealing with climate change effects (Swetnam et al. 2011). On one hand there are gaps within ecological studies, and on the other socioecological and participatory approaches for understanding and modelling the community preferences are weak (Gupta 1995). Tietjen and Jeltsch (2007) reviewed 41 models published during 1995–2005 involving the simulations of semi-arid and arid livestock grazing systems. Three general model types were distinguished: (a) state and transition, (b) difference and differential equation and (c) rule-based models. Over time there is an increase in models for improvement of management strategies, while there is a decline in those dealing with system dynamics. Given the heterogeneity of spatial conditions, the adaptability of models to different ecological niches is not always explicit. Maraun et al. (2010) highlight the gap between dynamic models and their users due to downscaling effects, through climate change. It is not always easy to downscale macro projections to micro level locations. The complexity increases when monitoring sites are remote, as spatial heterogeneity affecting a differential capacity of communities to cope with the risks.

There is a growing body of evidence impacting social systems and ecological and evolutionary trends in various vulnerable regions. It is argued that both climatic and non-climatic factors need to be taken into account to estimate biodiversity loss under climate change. The design of increasingly complex models incorporating multifarious interactions and processes is not necessarily a good idea (McMahon et al. 2011). Regretfully there is an absence of monitoring sites of climate change impacts in some of the worlds' most vulnerable regions. Consequently, variables such as soil microbial diversity have been ignored while building climate change models, though its effect on the emergence of socioecological systems is well known. Additional gaps in model development and validation include: the effect of local and general human population's choices on species composition, succession not being modelled adequately; the effect of demographic variability on socioeconomic preferences for biodiversity-based products, and consequent impacts on evolutionary biology of ecosystems. McMahon et al. (2011) conclude: 'This "big science" effort requires collaboration between ecologists, physiologists, climatologists, statisticians, computer scientists, and other disciplines. Above all, we must incorporate isolated, place-based study of communities and populations into continental and global frameworks of quantitative biodiversity modelling. Such an effort can incorporate the abundance of insights gained from ecological field studies into a quantitative framework applicable to conservation policy'.

Weitzman (2009) acknowledges the limits of a model in predicting certainty in the future distribution of species under climate change given so many factors influence the outcome. He quotes George Box to stress, 'models are never true, but fortunately it is only necessary that they be useful. For this it is usually needful only that they are not grossly wrong' (Box 1979, p. 2).

The "critical questions" for species–climate envelope models are as follows: How can the realism of model assumptions, algorithms and parameters be improved? And which questions make particular model applications useful?

11.2.3 Community Heterogeneity

Individual communities and sub-communities have varying levels of vulnerability and consequent resilience over different time and space units creating disparity, heterogeneity and complexity (Cutter et al. 2008). Existing models, have not integrated the built environment or natural processes with socio-economic variability over space and time. Cutter et al. 2008 state 'the relationship between vulnerability, resilience and adaptive capacity is still not well articulated...',

There is a need for better models, indicators and metrics that generate implementable policies for empowering local communities in dealing with risks. Socio-economic development in this century may change the patterns of production, trade, distribution and consumption of food products because of demographic, economic and diet changes (Fischer et al. 2005). There is a paucity of adequate models which restore the relationship between climate change and health (McMichael et al. 2006). The direct and indirect pathways through which socio-economic and demographic distractions may be caused by health changes still need to be properly modelled. The sustainability of socioecological systems crucially depends upon the way communities deal with health impacts and mitigation of adverse consequences.

11.2.4 Socioecological Resilience

Together with ecological resilience, socio-economic resilience needs to be factored. The relationship between poverty and climate change is nonlinear and hence not always as expected (Campbell et al. 2016). Sometimes vulnerable communities transform their resource use practices, instead of becoming more vulnerable they gain strength after an extreme event by improving their institutional conditions, land relations and health of Common Property Institutions. Kotzee and Reyers (2016) tested a concept of socioecological index for measuring flood resilience. They realize that variability over space, sector, season and social categories is not random. The coping strategies of the communities can be compared and lessons can be shared across communities. It is imperative that there are collaborations between impacts, adaptation and vulnerability researchers, and that climatic and integrated assessments are modelled (Moss et al. 2010). Socio-economic scenarios should be developed for assessing climate change risks and vulnerabilities, to overcome the limitations of existing models. It is recognized that the self-organizing ability of social systems along with scale dependence modifies the capabilities of systems to adapt to climate change because of various feedback loops, social variability and technological choices (Collins et al. 2010). The consequences for socioecological systems are uncertain and very complex. Social scientists have generally ignored the relationship between socio-economic and political system in a biophysical context. Simonsen et al. (2014) state the requirement for management to acknowledge the value of diversity and redundancy in management of social-ecological systems in resilience building.

There are no coping strategies addressing climatic risk in flood-prone paddy growing regions in isolated areas (Cutter et al. 2008; Dey 2015). Instead farmers work with a portfolio of enterprises (Gupta 1984, 1997) within an overall umbrella of a portfolio of institutions (Gupta 1992). Farmers attempt to be resilient through redundancy, reduction, reallocation, re-purposing and re-appropriating resources. However resilience depends on an ability to manage within available resources, replenish lost resources, and sometimes forgoing present individual losses for future collective profits. Socioecological endowments, economic condition, community relationship, market and other formal institutions seemed to define the limits of resilience for farmers. Solutions of climatic and agro-technology information; greater access to nurseries of the right kind, creating 'nursery commons' for higher cooperation and exchange of resources and thus collective well-being assist farmers. Easier access to credit and inputs, and persuasive insurance packages are additional steps that help small farmers in coping with climate risks.

Sakai and Dessai (2015) have tried to map the debate conceptualizing resilience. It has been used in different knowledge domains; "such as political ecology (Turner 2013), disaster risk reduction (Manyena 2006), development (Béné et al. 2014), business (Linnenluecke et al. 2012), planning (Wilkinson 2012), and adaptation is no exception (Berkes and Jolly 2002; Davoudi et al. 2012)". The engineering resilience framework (ERF) and socioecological systems (SES) are distinguished. Models must provide scope for learning, changing course, reorganizing and redesigning coping strategies. Single-loop, double-loop and triple-loop learning frameworks are useful in explaining processes that take place once any situation comes to the scene. (Argyris and Schön, 1996, 1978; Pelling and High, 2005; Pelling et al. 2008).

The single loop implies instructional models, which are essentially cause–effect or trigger stimuli–response. Double-loop models are recursive in which the response from the first cycle becomes stimuli for the second one. In double loop, feedback effects change the criteria and response. In triple loop, the heuristics underlying the adaptation through feedback go through transformation. ERF and SES could be double loop, though if changes in social and ecological systems are modified through knowledge, experimentation and innovation, which question the underlying principles, the third loop comes into play.

Socioecological resilience in the Arctic community in western Canada suggested cross-scale institutional linkages and feedback effects for improving adaptive capacity under climate change exist (Berkes and Jolly 2001). The ability of an ecosystem return to its original form of productivity and functionality after an extreme shock, including climatic events such as prolonged drought and flood suggests climatic resilience. However, systems have variable performance after climatic anomolies. The enhancement or reduction could be cascading, fluctuating or stable. While Adger (2000) looks at migration as a breakdown of social resilience, it need not be so. Seasonal migration patterns are important as coming and going from regions affects the dynamic conditions of resilence of the individuals and ecosystems alike. For instance, shepherds move from the arid west in India to the southeastern region for penning the sheep and other animals. In the absence of soil fertilization, the resilience of the livestock receiving regions will go down. Technological and institutional innovations can significantly alter the degree of vulnerability and thus, the resilience.

Further to socioecological resilience, Townsend and Masters (2015) suggest a cross species, spaces and scale conceptual model for biodiversity conservation and socioecological resilience. The role of incentives, in converging socioecological preferences in a mountain context has been further illustrated. In the view of watershed effects and various ecological services, synergy between upland and lowland communities becomes vital for conservation and resilience. However, Townsend and Masters do not take into account that although downstream benefits from upstream conservation are obvious, upstream benefits from downstream consumption and conservation are less apparent. To achieve resilience across elevation, the downstream beneficiaries of upstream conservation should either have direct stakes by way of owning plots upstream or may be required to pay for the services obtained. A good example of such incentives is Switzerland's policy of converting a loan into grant if the livestock farmers stay in the alpine-highland zone for at least 3 years, without people landscape conservation would become almost impossible. Such policies will follow when climate change impact models identify the losers and winners in any conservation adaptation scheme and simulate the resilience benefits of cross subsidization. Simonsen et al. (2014) suggest the need for constant experimentation to consider risks so that different kinds of knowledge systems may contribute towards developing resilience in socioecological systems scenarios. It is asserted that wider participation of communities may help in uncovering perspectives that conventional exercises may fail to notice. Polycentric governance systems should be advocated which support collaboration among different actors across institutions and scales to generate higher resilience.

11.2.5 Dealing with Uncertainty

Heikkinen et al. (2006) recount limitations of the existing models:

> Sources of uncertainty in models have received a considerable attention in the statistical and ecological literature (Chatfield 1995; Harrell et al. 1996; Buckland et al. 1997; Guisan and Zimmermann 2000; Vaughan and Ormerod 2003; Johnson and Omland 2004; Rushton et al. 2004). This attention is warranted because there is increasing evidence that the 'best' model developed for a given studied region is only one among many alternative models. . . .
> In fact, developing hybrid models that bring together the best of correlative bioclimatic modelling with the best of mechanistic and theoretical models is one of the most important challenges for modellers (Araújo and Rahbek 2006).

Four important conclusions about the limitations of models may be drawn: First, in order to generate quantitative insights into the consequences of alternative policy decisions, the participatory component of scenario building must be clearly linked to quantitative modelling and these links at least partly envisaged beforehand; second, complexity needs to be managed, otherwise time will be wasted in implementation; third, it is critical to think from the start about how policy and decisions are made in the particular region of study, otherwise a disconnect may arise between carefully constructed and model scenario exercises and the actual needs of the policymakers for whom they are designed; fourth, such tailored scenario-building exercises can provide critical calibration of larger scale scenarios, ensuring the results do mirror local expectations of change.

National and international governing bodies have a responsibility to recognize and incorporate the value of diverse sources of knowledge. Connectivity can both enhance and reduce the resilience of social–ecological systems and the ecosystem services they produce. Well-connected systems can overcome and recover from disturbances more quickly, but overly connected systems may lead to the rapid spread of disturbances across the entire system so that all components of the system are impacted.

11.3 The Socioecological Paradigm Revisited: Modelling Imperatives

Social choices are shaped by the ecological endowments in most rain-fed regions at risk from drought. The scale at which various social groups pursue different economic activities depends upon their access to factor and product markets, besides non-monetized exchange relations (Gupta 1984, 1989). Given the climatic uncertainties and fluctuations, we propose to revisit the socioecological paradigm and identify the new modelling needs for higher community level resilience (Fig. 11.1). Gupta (1990) observes:

> Various groups of rural households diversify portfolios of their economic enterprises within a range defined by the ecological endowments. The access to factor and product markets, kinship networks, intra and inter household risk adjustments, public and private relief systems and finally common property resources or common pool institutions determine the composition and evolution of portfolios of different enterprises.

In any ecological context, the dynamic relationships between biotic and abiotic factors are constantly in flux due to varying anthropogenic pressures. The potential of the existing portfolio of farms, non-farm enterprises, labour employment and migration patterns, together with gender and composition of ecosystems needs to be identified to cope with risks. The majority of climate change models have ignored or underplayed the role of institutional parameters, particularly those influenced by historically shaped cultural biases, due to the lack of quantifiable data. There are four kinds of portfolio outcomes: High Mean–High Variance; High Mean–Low Variance; Low Mean–Low Variance; Low Mean–High Variance. The mean and the variance in different productive activities are influenced by exogenous and endogenous variables influencing the production system. The composition of crops, their

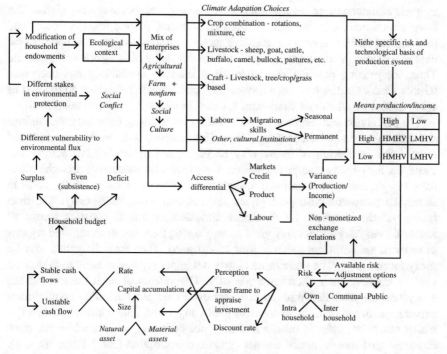

Fig. 11.1 Socioecological paradigm for analysing problems of the poor in dry regions (Gupta 1984)

varieties, so-called weeds (companion plants), associated microbial and insect diversity, the implications of this diversity for nutrient flows and input applications determine the productivity in the wake of accumulated and concurrent climate change impacts. If soil is hard because of successive droughts and has been fallow, often nitrogen fixation is high, carbon may be low resulting in difficult ploughing. Access to draught power determines whether different members of the community can harness the advantage of rains if at all. The solar reflexivity and temperature sinks are influenced by the land use and land cover both naturally and through human interventions, the reflexivity changes through change in land use/land cover contributing to global warming and has been attributed to human influences rather natural processes in recent times. The same climatic change phenomena can affect the choices of different community members divergently, depending upon their access and ability to mobilize intra- and inter-household risk adjustments, community adaptations and state interventions for surviving during stress. Local knowledge and innovations can modify these choices significantly. However, we have not developed indices of knowledge intensity of local institutions to model their resilience potential. One of the ways in which one can measure social capital as a proxy for local exchange relationships is to look at the quantum and direction of inward remittances. The communities with high emigration of labour due to frequent climate change impacts such as floods and droughts may have high or low inward remittances. In the case of high remittances, despite low ecological and institutional resilience, communities can survive through exogenous support. In the case of a pastoral community, the pressure for disposing of livestock will go down. Thus, the grazing pressure may increase if biomass availability has decreased (Gupta 1986). Studies have also shown that opportunistic behaviour on the part of communities can facilitate short-term survival but impair long-term adaptability.

The perception and response to risk is not uniform among different communities and their members. Given past experience (immediate past as well as accumulated effects) the perception of future may be influenced by contingent factors. The strategies for climate change adaptation in regions with successive drought; alternate drought and rain; occasional drought need to be modelled carefully, so as to factor the socioecological perceptions and respond more realistically. The time frame and the discount rate for future choices may not be constant across all resource markets. Farmers may grow slow-growing tree species in rain-fed regions in order to get sufficient income after 15–20 years when their daughters may be ready for marriage. The same farmers may not apply organic manure to the leased land to improve their short-term pay-offs. The effects of such choices on the ecosystem are mediated through different economic outcomes that communities experience. Some manage to have surplus budget and thus accumulate savings, whilst others are able to subsist and may not have any savings while the most disadvantaged have a deficit budget (greater consumption than current income). Farmers may be under debts and also may have reduced productive life due to sustained impairment of nutritional budget. The cases of weak populations in highly stressed environments show such contingencies. When external aid provides short-term relief without investing in long-term institutions it may weaken the community resilience. The predictions of a mathematical model can be as good as the assumptions of the model.

The stakes in environmental conservation are not constant for different classes of community members. The role of social and cultural institutions in modifying ecological resilience is very significant. For instance, certain communities in the Indian arid west do not rear sheep or goat due to the regional cultural background, leaving these for lower status communities. However, when regeneration of overgrazed pastures is less, the cattle cannot survive for long and sheep and goat become viable choices. The cultural resistance gives way and all communities start maintaining sheep and goat. In the absence of any major regeneration effort, after a while pasture gets further degraded changing the ratio of palatable to unpalatable species. For instance, *Tephrosia* is an indicator of overgrazed pastures (Cartwright 1998). The ecological context gets modified and accordingly the cycle of adaptation begins again. In a dynamic socioecological context, the climate change models need to factor the conditions under which estimation of regeneration rates is affected by social, ecological and institutional factors. Modelling of climate change impacts at different scales needs to be informed by the dynamics of local risk adjustments and adaptations.

The ecological landscapes have to be subdivided into institutional landscapes so that simulations can generate different trajectories of adaptations vis-à-vis varying dynamics of institutions. The feedback loops at community level, regional level and supra-regional level will depend upon the available opportunities to communities to blend their informal knowledge with formal science, technology, innovation and institutional knowledge.

Conflicts are inevitable as resources become scarce. The urban communities have relatively faster response time than the remote rural communities. The role of power in shaping the choices and influencing the supply chain for delivering opportunities has to be understood.

11.4 Case Study of Coping with Climate Change in Flood-Prone Regions of India

The study was carried out in three villages: Kharella, Isoulibhari and Shivnathpur in the Milkipur block of Faizabad district of Uttar Pradesh in India. We conducted semi-structured interviews during 1988–1989, 2002–2004 (Gupta and Chandok 2010) and more recently in 2013–2015. The total number of households surveyed was 123 in 2002–2004 and 127 in 2013–2015. A survey of all plots was carried out in all three rounds to understand the coping strategies employed by farmers with different economic and ecological endowments for climate change. Implications on agro-biodiversity changes were explored, individual case studies were pursued to gain deeper insights about farmers' decision-making systems. Located in the watershed of Sarayu River, the region is historically flood prone. While 2013 had floods in the early stage of paddy, 2014 and 2015 had less rainfall to an extent that 2015 was declared droughted in more than 60 % of the district. The district is reported to have undergone significant land use changes accompanied by changes in the cropping pattern with paddy, wheat, potato and sugarcane seeing vast increase in acreage and productivity. Traditional varieties of paddy, minor millets and barnyard millet have been lost. The region is typically divided into three land types based on the inundation level and the elevation of the plots (see Table 11.1).

In upland ecosystems, resourceful farmers could cope effectively as they had prepared nursery stocks pre the onset of floods in 2013 and could manage drainage. Fertilizers were applied at an early stage to manage the nursery and make use of damaged plants. In the event of the early arrival of the monsoon, farmers adjusted to damage in the nursery by using available household resources hence cutting down expenses on fertilizers at the later stages of the crop life cycle. Farmers without resources had to depend on loans to re-sow their nursery. Some farmers who avoided damage had land available, sowed seed in the nursery and sold saplings. Farmers who could manage nursery stocks did not have a requirement for loans to invest in fertilizers.

Table 11.1 Village-wise land types and diversification in Kharif 2013 (Dey 2015)

Village	Land type	Total crop varieties	Total paddy varieties	Total no. of plots	Diversity per unit plot	Paddy diversity per unit plot
Shivnathpur	Lowland	2	3	18	0.111	0.167
	medium land	4	6	67	0.060	0.090
	Upland	4	12	59	0.068	0.203
Isoulibhari	Lowland	3	2	104	0.029	0.019
	medium land	4	7	560	0.007	0.013
	Upland	3	6	61	0.049	0.098
Kharella	Lowland	3	6	42	0.071	0.143
	medium land	3	11	99	0.030	0.111
	Upland	4	8	68	0.059	0.118

Principal Component Analysis (PCA) was employed using a rotation method of Varimax with Kaiser Normalization, four factors were found with Eigen values more than 1. These were as follows:

Factor 1—*Diversified income*: Farmers migrated outstations in search of other income opportunities.
Factor 2—*Constrained input use*: Wherein households managed risk using available intra-household resources and saving on fertilizers like urea.
Factor 3—*Risk ready*: The nurseries which were ready for transplantation when the monsoon arrived could be used even if they were partially damaged.
Factor 4—*Extensive resource use*: Farmers with bigger land holding use less fertilizer at the later stage of the crop.
Factor 5—*Resource readiness*: Farmers who had their nursery ready managed flood water by using pumpsets or breaching bunds.

In medium lands ecosystems, farmers whose fields were not flooded were able to make their nursery ready. They raised a diversified nursery and also invested in early stage fertilizers. However, farmers could not manage within the intra-household resources as they had no additional inflow of cash (source of inward remittance) or cattle and were unable to invest in fertilizers. Farmers with greater landholdings diversified into additional crops and varieties. Farmers who have cattle grow zaid (summer) crop, usually a fodder crop. Some farmers take loans to sow the nursery again. Poor farmers disposed off assets to deal with the stress. They also managed drainage collectively.

Results of the PCA illustrated the following coping strategies:

Factor 1—*Risk ready*: The nursery was ready and the fields were also ready and natural drainage was sufficient.

Factor 2—*Resourceful risk management*, i.e. bigger farmers could manage within their own resources and achieved resilience through crop diversification.

Factor 3—*Risk aversion*: Resourceful farmers used early stage fertilizers and saved on late stage fertilizers.

Factor 4—*Diversified income, specialized crop/variety*: Farmers with diversified income source chose to grow fewer diversity of crops. Those who had to rely on multiple sources of income (could not manage with intra-household resources) had cattle and received income from someone from the family working outside preferred to grow one or few crops/varieties.

In lowland ecosystems, resilience is shown by farmers who diversified into animal husbandry and had resources to sow the nursery in readiness for the early onset of monsoons. They may use their bullocks for puddling/tilling land, keeping fields ready, unlike the farmers who use mechanical means which require surface water levels to drop.

Farmers with smaller landholding and no additional income source could manage within the available intra-household economic resources as nursery stocks were planned to be ready when the monsoons arrived. Early stage fertilizers are used to make the saplings more robust.

Smaller farmers dependent solely on agriculture could manage drainage as they have available work at home. Risk taking farmers grew the zaid (summer) crop and could manage their resources with the extra income. They also have cattle with fodder available from the zaid crop. Some farmers take loans to re-sow nursery stocks and apply urea whereas farmers who managed by modifying their consumption pattern did not. It is apparent that they did not have resources to pay back or invest in fertilizers.

PCA identified the following:

Factor 1—*Labour constraint*: Farmer families with migrant members lacked labour for managing drainage.

Factor 2—Intensive *resource use*: Farms use fertilizers extensively, both early stage and late stage.

Factor 3—*Riskreadiness*: Farmers could manage within their resources as their nursery was ready.

Factor 4—*Labour constrained delayed sowing*: Families with migrant members faced labour constraints in sowing their nurseries in time.

Factor 5—*Diversification through self-reliance*: Farmers who diversified into other crops could manage within their resources.

In communities, social cohesion and kinship relationships play an important role in managing drainage, sharing nursery stocks, pooling labour resources (especially at the time of transplanting and manual harvesting). Farmers exchange, buy or sell saplings amongst themselves and cooperate by sharing manual resources. The wild variety of rice, "Phasarhi" or "tinwa" (*Oryza rufipogon*) is collected and consumed by poor people. Phasarhi has a high shattering index and is very difficult to harvest. Such rice crops require community-level response in order to be consumed as they need to be manually harvested/collected. Phasarhi commands a high price in the market and is consumed by women in a festival called 'teej' irrespective of social or economic class.

Dey (2015) report that women add resilience to households by engaging in non-monetary activities. They exchange labour, food materials (which they grow in the homestead) and knowledge about weeds and nutrition. Women take decisions on homestead utilization, growing different seasonal fruits and vegetables to fulfil nutritional requirements of the family. Gupta (1987) report that women in Tangail district of Bangladesh de-root sweet potato vines, making them rounder in shape with thicker skins enabling safe storage for longer durations, commanding a higher market price. Women also take care of the cattle and hence have developed extensive knowledge in animal husbandry.

Table 11.2 presents a loss of traditional values over time. Decadal variation in paddy varieties was studied by Gupta (1988–1989), Gupta and Chandok (2002–2003) and Gupta and Dey (2014–2015) in three villages of Faizabad district in eastern Uttar Pradesh, namely Isoulibhari, Shivnathpur and Kharella. It was found that none of the traditional varieties reported in 1988–1989 exist in these villages now, with the exception of the Kalanamak which has been reintroduced only this year.

Table 11.2 indicates that the number of people who are experimenting with new varieties has increased since 2003. The traditional varieties have nearly vanished in the three villages. A particular farmer who had sown Kalanamak variety only in 2014, discontinued it in 2015 as he had left his land fallow due to adverse micro-climatic conditions. Traditional varieties were reported in the surveys of 1988–1989 and 2002–2003. Even among the modern varieties, distribution is highly skewed, a few varieties tend to dominate much more than the rest.

Paddy diversity is the maximum in the uplands and lowlands compared to medium uplands, which are more irrigated and thus suitable for hybrids (Table 11.3). In paddy varieties PA6444 is most dominant and is prevalent in the three types of lands indicating its high adaptability. The Sarjoo 52 is the next most popular variety followed by Mansuri and BPT 5204.

11.5 Conclusions and Further Directions

The purpose of climate change models is not only to anticipate the likely impacts but also to identify the possible strategies that can improve resilience. The brief review of literature covered shows that the socioecological paradigm (Gupta 1984) includes several parameters that affect human choices but have not always been factored in designing models. Simplicity is an important attribute of a pragmatic model that has to match conceptual rigour with operational efficiency and ease. One has to make trade-offs in choosing the parameters and also while projecting their performance at different scales, seasons and at various spaces. Social dynamics are affected by space, season, sector and social segmentation (Gupta 1992), all of

Table 11.2 Loss of traditional rice varieties in three villages

SR.	1988–1989	2002–2003	2014–2015
Desi/local/farmer developed varieties under cultivation			
1	Lalmati	Lalmati	Kalanamak (reintroduced in 2014)
2	Muthmuri	Muthmuri	
3	Dehula	Dehula	
4	Bahgari	Baghari	
5	Jarhan		
6	Gajraj		
7	Bashawa		
8	Dhaneshwar		
9	Kalanamak		
10	Dudhiya		
11	Hiramati		
12	Nebui		
13	Vishnu Parag		
14	Samari		
Improved/hybrid varieties			
1	Sarjoo-52	Sarjoo-52	PA 6444
2	Saket-4	NDR-359	Mansuri
3	Pant-4	Pant-10	5204
4	China-4	Pant-12	5644
5	NDR-80	NDR-90	MPH555
6	Kaveri	NDR-118	Raftar
7	Jaya	HY. Rice	Prithvi
8	IR-8	Mansuri	27 P 31
9	IR-36		Narendra shankar
10	Nahar Punjab		Ganga kaveri
11	Usha		Gorakhnath 509
12	NDR-118		Arize 6129
13	Mansuri		5405
14	Sita		6302
15	Madhukar		Amarnath 502
16	Prasad		27P71
17			6304
18			ShahiDawat
19			CSR 36
20			PHB71
21			NDR 118
23			Sugandha
24			NDR 369
25			Ralli gold
26			Sarjoo 52
Total	30	12	26

Table 11.3 Areas under different paddy diversity

Land type	Total crop varieties	Total paddy varieties	Total no. of plots	Diversity per unit plot	Paddy diversity per unit plot
Lowland	8	11	164	0.070	0.082
medium land	11	24	726	0.032	0.053
Upland	11	26	188	0.059	0.105

which are closely intertwined with ecological dynamics, particularly in rain-fed regions. The resilience or homeostasis can be studied in time slices and over broader time periods. Akin to senescence in nature, resilience can be achieved over time by withdrawal, enhancement, slowing down or speeding up the physiological or operational functions. Resilience may be achieved at times, when in biological systems high metabolic rates occur when faced with stress, however such resilience is difficult to sustain because of adverse environmental changes. In social systems, it can take place through agitation, discussion and exchange of ideas and a kind of churning to generate innovative coping strategies.

Biological systems resilience and social resilience are not discretely distinguished. Lots of small changes take place simultaneously leading to stage changes. Simultaneity, diversity, complexity, variability and change are modelled by making assumptions of *ceteris paribus*. In system dynamics models, simultaneous changes are accepted as an essential feature of a model and various optimization goals are achieved through iterative simulations. Systematic biological approaches attempt to achieve the same in biological systems. The challenge is how to model social changes, which may not be linearly predictable and yet may influence the ecological feedback effects iteratively. Challenges lying ahead of us are summarized here only to illustrate the complexity involved. Unless we communicate this complexity in an easily understandable way supplemented by suitable metaphors, we may not get policy makers and institution builders on board. Without their cooperation, social systems can neither supply restraint nor social retribution for violating the socioecological limits for long-term resilience and survival.

Some estimates of precipitation, temperature and in situ conservation indicate that India might reduce its water supply to 20 % of present quantity by 2030. Given such a challenging forecast, one would have expected declaration of emergency, mobilization of popular participation and rigorous transformation of rules for water use, disposal, recycling and regeneration. In the absence of such implementation, while models predict crisis, policy makers don't realize the gravity of the matter to warrant radical changes in resources use practices.

The watershed effects can be simulated at much higher scale including river basins though the precipitation models cannot be restricted to that scale. A cluster of models will need to be developed with various assumptions about the future of society and its interactions with nature. The role of governance, institutions, culture and technology will further unfold how climate change impacts are observed, assimilated, adapted and responded in favour of disadvantaged communities and regions. The stability of forecast and instability of resilience are incompatible. Modellers should not promise too much certainty when it is known that several intervening variables have either not been incorporated in specific models or are very difficult to measure. A persistent mistake in social sciences is a tendency to explain a phenomena by what is measured and leave the rest as not worth measuring, creating in the process, a false rationality.

The study on agro-biodiversity in a few villages of eastern India over the last 30 years shows almost complete loss of diversity of traditional crop varieties, particularly in paddy. Some of the paddy hybrids developed by the private sector have done very well in medium upland and upland conditions. However, hybrids may be losing resistance to pests, diseases and tolerance of flood risks. Unless formal research and development systems continue to provide new crop alternatives every second or third year for every niche, climate change impact can only worsen. Modelling of resilience in the agro-biodiversity sector also requires taking into account the forecast around local institutions of water management, paddy nursery exchanges (due to early flood), access to credit for buying nursery and hiring labour if possible. Policy makers and research institutes can map macro climatic variables and plug them into local decision support systems, which can scale down the implications for niche-specific farm management decisions.

The role of non-farm enterprises including seasonal or permanent migration for coping with risks has to be studied to forecast the nature of resilience that may be available to communities. It is well known that in many high-risk prone regions, male emigration leads to a large number of households being managed by women. The degree of freedom that women have in terms of managing resources (including firewood, farm management including sowing crops, weeding and exchange of nutritious weeds, storage, sale, exchange of crop nurseries, seeds, seedling, healthy or damaged, managing draft power, water management) is not equal with that of men. Equality is one of the most important proxy variables for social vulnerability and gender centrality, in determining coping capacity with climate change impacts at different stages of farming cycles. In the case of livestock-based economies, the recurrence of drought or sometimes even excessive rains may change the succession of species, ratio of palatable or non-palatable species and consequent stress on the pastoral communities (Gupta 1991). This stress becomes worse when male members migrate out and women have to graze young calves, lambs, kids over shorter distances. The cascading effect of overgrazing and subsequenty bloating in the post rainfall period after a drought or warm weather makes communities more vulnerable. Craft activities play an important role in resilience because they are the only products, which can be sold when nothing else is available to dispose of or sell. In such cases, neglect of gender variables can overestimate resilience and underestimate vulnerability unless local institutions and public interventions effectively neutralize the vulnerability.

Ecological landscapes have to be subdivided into institutional landscapes so that simulations can generate different trajectories of adaptations vis-à-vis varying dynamics of institutions. The feedback loops at community level, regional level and supra-regional level will depend upon the available opportunities to communities to blend their informal knowledge with formal science, technology, innovation and institutional knowledge. In this chapter, we have seen how these challenges may be approached in socioecological systems; however, the implementation and formation of policies which lead to resilience requires more research. The use of mathematical methods has great potential when applied for this purpose, with such rigor we can hope to move further towards rationally formed modelling methods which assist with our coping with climate change and sustain ecosystems.

References

Adger WN (2000) Social and ecological resilience: are they related? Prog Hum Geogr 24 (3):347–364

Araújo MB, Rahbek C (2006) How does climate change affect biodiversity? Science-New York Then Washington 313(5792):1396

Argyris C, Schön DA (1978) Organizational learning: a theory of action perspective. Addison-Wesley, Reading

Argyris C, Schön DA (1996) Organizational learning II: theory, method and practice. Addison Wesley, Reading

Béné C, Newsham A, Davies M, Ulrichs M, Godfrey-Wood R (2014) Resilience, poverty and development. J Int Dev 26(5):598–623. doi:10.1002/jid.2992

Berkes F, Jolly D (2002) Adapting to climate change: social-ecological resilience in a Canadian western Arctic community. Conserv Ecol 5(2):18

Box GE (1979) Some problems of statistics and everyday life. J Am Stat Assoc 74(365):1–4

Buckland ST, Burnham KP, Augustin NH (1997) Model selection: an integral part of inference. Biometrics 53:603–618

Campbell BM, Marín MA, Aggarwal PK et al (2016) Climate change and poverty—regional perspectives, a review of literature. Science Forum CGIAR

Cartwright CR (1998) Seasonal aspects of Bronze and Iron Age communities at Ra's al-Hadd, Oman. Environ Archaeol 3(1):97–102

Chatfield C (1995) Model uncertainty, data mining and statistical inference. J R Stat Soc A Stat Soc 158(3):419–466. doi:10.2307/2983440

Collins SL, Carpenter SR, Swinton SM, Orenstein DE, Childers DL, Gragson TL, Grimm NB, Grove JM, Harlan SL, Kaye JP, Knapp AK (2010) An integrated conceptual framework for long-term social-ecological research. Front Ecol Environ 9(6):351–357. doi:10.1890/100068

Cutter SL, Barnes L, Berry M, Burton C, Evans E, Tate E, Webb J (2008) A place-based model for understanding community resilience to natural disasters. Glob Environ Chang 18(4):598–606. doi:10.1016/j.gloenvcha.2008.07.013

Davoudi S, Shaw K, Haider LJ, Quinlan AE, Peterson GD, Wilkinson C, Fünfgeld H, McEvoy D, Porter L, Davoudi S (2012). Resilience: a bridging concept or a dead end? "Reframing" resilience: challenges for planning theory and practice interacting traps: resilience assessment of a pasture management system in northern afghanistan urban resilience: what does it mean in planning practice? Resilience as a useful concept for climate change adaptation? The politics of resilience for planning: a cautionary note. Plann Theory Pract 13(2):299–333. doi:10.1080/14649357.2012.677124

Dey AR (2015) Coping creatively with climate risks: farmers' adjustment with floods in eastern India. Paper presented in the third International Conference on Creativity and Innovation at Grassroots, IIM, Ahmedabad

Fischer G, Shah M, Tubiello FN, Van Velhuizen H (2005) Socio-economic and climate change impacts on agriculture: an integrated assessment, 1990–2080. Philos Trans R Soc Lond B Biol Sci 360(1463):2067–2083. doi:10.1098/rstb.2005.1744

Grassini P, van Bussel JGV, Van Wart J, Wolf J, Claessens L, Yang H, Boogaard H, de Groot H, van Ittersum MK, Cassman KG (2015) How good is good enough? Data requirements for reliable crop yield simulations and yield-gap analysis. Field Crop Res 177:49–63

Grimm NB, Grove JG, Pickett ST, Redman CL (2000) Integrated approaches to long-term studies of urban ecological systems urban ecological systems present multiple challenges to ecologists—pervasive human impact and extreme heterogeneity of cities, and the need to integrate social and ecological approaches, concepts, and theory. BioScience 50(7):571–584. doi:10.1641/0006-3568(2000)050[0571:IATLTO]2.0.CO

Guisan A, Zimmermann NE (2000) Predictive habitat distribution models in ecology. Ecol Model 135(2):147–186

Gupta AK (1984) Agenda for research in dry regions: socio-ecological perspective. Indian Institute of Management, Ahmedabad

Gupta AK(1987) Role of women in risk adjustment in drought-prone regions with special reference to credit problems. IIM working paper no. 704. IIM, Ahmedabad

Gupta AK (1989) Managing ecological diversity, simultaneity, complexity and change: an ecological perspective. WP no. 825. IIM Ahmedabad. Third survey on Public Administration. Indian Council of Social Science Research, New Delhi, p 115

Gupta AK (1990) Survival under stress in South Asia: a socio-ecological perspective on farmer risk adjustment and innovations. Capital Nat Social 1(5):79–94. doi:10.1080/10455759009358417

Gupta AK (1991) Pastoral adaptation to risk in dry regions: a framework for analysis. Stud Hist 7 (2):325–341

Gupta A (1992) Farmers' innovations and agricultural technologies. In: Jodha NS, Banskota M, Partap T (eds) Sustainable mountain development. Oxford and IBH, New Delhi, pp 394–412

Gupta AK (1995) Sustainable institutions for natural resource management: how do we participate in people's plans. In: People's initiatives for sustainable development: lessons of experience. APDC, pp 341–373

Gupta AK (1997) Portfolio theory of technological change: reconceptualising farming systems research. Accessed at https://dlc.dlib.indiana.edu/dlc/handle/10535/4465

Gupta AK, Vikas C (2010) Cradle of creativity: strategies for in-situ conservation of agro biodiversity (No. WP2010-09-03). Research and Publication Department, Indian Institute of Management, Ahmedabad

Gupta SK (1986) Structure and functioning of the natural and modified grassland ecosystems of western Himalaya (Garhwal Himalaya). 3rd Annual report on MAB/DOE, Govt. of India, New Delhi

Harrell FE, Lee KL, Mark DB (1996) Tutorial in biostatistics multivariable prognostic models: issues in developing models, evaluating assumptions and adequacy, and measuring and reducing errors. Stat Med 15:361–387

Heikkinen RK, Luoto M, Araújo MB, Virkkala R, Thuiller W, Sykes MT (2006) Methods and uncertainties in bioclimatic envelope modelling under climate change. Prog Phys Geogr 30 (6):751–777. doi:10.1177/0309133306071957

Johnson JB, Omland KS (2004) Model selection in ecology and evolution. Trends Ecol Evol 19 (2):101–108

Kotzee I, Reyers B (2016) Piloting a social-ecological index for measuring flood resilience: a composite index approach. Ecol Indic 60:45–53

Li K, Yang X, Liu Z, Zhang T, Lu S, Liu Y (2014) Low yield gap of winter wheat in the North China Plain. Eur J Agron 59:1–12

Linnenluecke MK, Griffiths A, Winn M (2012) Extreme weather events and the critical importance of anticipatory adaptation and organizational resilience in responding to impacts. Bus Strateg Environ 21(1):17–32. doi:10.1002/bse.708

Manyena SB (2006) The concept of resilience revisited. Disasters 30(4):434–450. doi:10.1111/j. 0361-3666.2006.00331.x

Maraun D, Wetterhall F, Ireson AM, Chandler RE, Kendon EJ, Widmann M, Brienen S, Rust HW, Sauter T, Themel M, Venema VKC (2010) Precipitation downscaling under climate change: Recent developments to bridge the gap between dynamical models and the end user. Rev Geophys 48(3). doi:10.1029/2009RG000314

McMahon SM, Harrison SP, Armbruster WS, Bartlein PJ, Beale CM, Edwards ME, Kattge J, Midgley G, Morin X, Prentice IC (2011) Improving assessment and modelling of climate change impacts on global terrestrial biodiversity. Trends Ecol Evol 26(5):249–259. doi:10. 1016/j.tree.2011.02.012

McMichael AJ, Woodruff RE, Hales S (2006) Climate change and human health: present and future risks. Lancet 367(9513):859–869. doi:10.1016/S0140-6736(06)68079-3

Moss RH, Edmonds JA, Hibbard KA, Manning MR, Rose SK, Van Vuuren DP, Carter TR, Emori S, Kainuma M, Kram T, Meehl GA (2010) The next generation of scenarios for climate change research and assessment. Nature 463(7282):747–756. doi:10.1038/nature08823

O'hara SL, Street-Perrott FA, Burt TP (1993) Accelerated soil erosion around a Mexican highland lake caused by prehispanic agriculture. Nature 362(6415):48–51

Pelling M, High C (2005) Social learning and adaptation to climate change. Benfield Hazard Research Centre, Disaster Studies Working Paper 11, pp 1–19

Pelling M, High C, Dearing J, Smith D (2008) Shadow spaces for social learning: a relational understanding of adaptive capacity to climate change within organisations. Environ Plan A 40 (4):867–884. doi:10.1068/a39148

Qingxiang L, Xiaoning L, Hongzheng Z, Peterson TC, Easterling DR (2004) Detecting and adjusting temporal inhomogeneity in Chinese mean surface air temperature data. Advances in Atmospheric Sciences 21:260. doi:10.1007/BF02915712

Redman CL (1992) The impact of food production: short-term strategies and long-term consequences. In: Jacobsen JE, Firor J (eds) Human impact on the environment: ancient roots, current challenges. Westview Press, Boulder

Rice DS (1996) Paleolimnological analysis in the central Peten, Guatemala. University of Utah Press, Salt Lake City, pp 193–206

Rushton SP, Ormerod SJ, Kerby G (2004) New paradigms for modelling species distributions? J Appl Ecol 41(2):193–200

Ryan DJ (2003) Two views on security software liability: let the legal system decide. IEEE Security Privacy 1(1):70–72

Sakai P, Dessai S (2015) Can resilience framing enable adaptation to a changing climate? Insights from the UK water sector. Accessed at www.icad.leeds.ac.uk/WorkingPapers/ ICADWorkingPaperNo9_SRIPs-88.pdf

Sen A (1980) Description as choice. Oxf Econ Pap New Ser 32(3):353–369

Soussana JF, Graux AI, Tubiello FN (2010) Improving the use of modelling for projections of climate change impacts on crops and pastures. J Exp Bot 61(8):2217–2228. doi:10.1093/jxb/ erq100

Simonsen SH, Biggs R, Schlüter M, Schoon M, Bohensky E, Cundill G, Dakos V, Daw T, Kotschy K, Leitch A, Quinlan A, Peterson G, Moberg F (2014) Applying resilience thinking: seven principles for building resilience in social-ecological systems. Stockholm University, Stockholm

Springmann M, Godfray HCJ, Rayner M, Scarborough P (2016) Analysis and valuation of the health and climate change co-benefits of dietary change. Proc Natl Acad Sci U S A 113 (15):4146–4151

Swetnam RD, Fisher B, Mbilinyi BP, Munishi PKT, Willcock S, Ricketts T, Mwakalila S, Balmford A, Burgess ND, Marshall AR, Lewis SL (2011) Mapping socio-economic scenarios of land cover change: a GIS method to enable ecosystem service modelling. J Environ Manage 92(3):563–574. doi:10.1016/j.jenvman.2010.09.007

Tietjen B, Jeltsch F (2007) Semi-arid grazing systems and climate change: a survey of present modelling potential and future needs. J Appl Ecol 44(2):425–434. doi:10.1111/j.1365-2664. 2007.01280.x

Townsend PA, Masters KL (2015) Lattice-work corridors for climate change: a conceptual framework for biodiversity conservation and social-ecological resilience in a tropical elevational gradient. Ecol Soc 20(2):1

Turner MD (2013) Political ecology I. An alliance with resilience? Prog Hum Geogr. doi:10.1177/0309132513502770

Van Andel TH, Zangger E, Demitrack A (1990) Land use and soil erosion in prehistoric and historical Greece. J Field Archaeol 17(4):379–396. doi:10.1179/009346990791548628

Van Bussel LGJ, Müller C, Van Keulen H, Ewert F, Leffelaar PA (2011) The effect of temporal aggregation of weather input data on crop growth models' results. Agric For Meteorol 151(5):607–619

Vaughan IP, Ormerod SJ (2003) Improving the quality of distribution models for conservation by addressing shortcomings in the field collection of training data. Conserv Biol 17(6):1601–1611

Weitzman ML (2009) On modelling and interpreting the economics of catastrophic climate change. Rev Econ Stat 91(1):1–19. doi:10.1162/rest.91.1.1

Wilkinson C (2012) Social-ecological resilience: insights and issues for planning theory. Plann Theory 11(2):148–169. doi:10.1177/1473095211426274

Wise RM, Fazey I, Smith MS, Park SE, Eakin HC, Van Garderen EA, Campbell B (2014) Reconceptualising adaptation to climate change as part of pathways of change and response. Glob Environ Chang 28:325–336. doi:10.1016/j.gloenvcha.2013.12.002

Chapter 12
Introduction to Robotics-Mathematical Issues

S.M. Raafat and F.A. Raheem

Abstract Robotic systems play a crucial role in the world and sustainability. Robots presence and our dependencies on them are progressively growing. This chapter brings together mathematical developments in the important fields of robotics, kinematics, dynamics, path planning, control, and vision. Introduction is made on development of robotics in different areas of application (types of robots and applications). The kinematics of a robot manipulator is briefly described. The formulation of dynamics for the manipulator has been obtained based on Lagrange's energy function. Linear Segments with Parabolic Blends and Third-Order Polynomial Trajectory Planning have been described in detail. Different classical control strategies are presented. Finally, basic concepts of Robot Vision are presented.

Keywords Robot kinematic • Robotic dynamics • Path and trajectory planning • Robot control • Robot vision

12.1 Introduction on Robotics; Robot Types and Applications

Through good design practices and thorough consideration of detail, engineers have succeeded in applying robotic systems to a wide variety of industrial, manufacturing, space, domestic or household, social, and medical situations where the environment is structured or predictable. On a practical level, robots are distinguished from other electromechanical motion equipment by their dexterous manipulation capability in that robots can work, position, and move tools and other objects with far greater dexterity than other machines found in the factory. Process robot systems are functional components with grippers, end-effectors, sensors, and process equipment organized to perform a controlled sequence of tasks to execute a process—they require sophisticated control systems. The combined effects of

S.M. Raafat (✉) • F.A. Raheem
Automation and Robotics Research Unit, Control and System Engineering Department,
University of Technology, Baghdad, Iraq
e-mail: 60154@uotechnology.edu.iq

© Springer International Publishing Switzerland 2017
J.N. Furze et al. (eds.), *Mathematical Advances Towards Sustainable Environmental Systems*, DOI 10.1007/978-3-319-43901-3_12

kinematic structure, axis drive mechanism design, and real-time motion control determine the major manipulation performance characteristics such as reach and dexterity, payload, quickness, and precision. For Cartesian robots, the range of motion of the first three axes describes the reachable workspace. Some robots will have unusable spaces such as dead zones, singular poses, and wrist-wrap poses inside of the boundaries of their reach. Usually motion test, simulations, or other analyses are used to verify reach and dexterity for each application (Lewis et al. 1999).

All common commercial industrial robots are serial link manipulators with no more than six kinematically coupled axes of motion. By convention, the axes of motion are numbered in sequence as they are encountered from the base on out to the wrist. The first three axes account for the spatial positioning motion of the robot; their configuration determines the shape of the space through which the robot can be positioned. Any subsequent axes in the kinematic chain provide rotational motions to orient the end of the robot arm and are referred to as wrist axes. There are, in principle, two primary types of motion that a robot axis can produce in its robot arm: either revolute (rotational) or prismatic (translational). It is often useful to classify robots according to the orientation and type of their first three axes. As the robot arm has only three degrees of freedom, there exist a limited number of possible combinations resulting all together in 36 different structures of robot arms. There are very common commercial robot configurations: Articulated robots (robotic arms), spherical, Selective Compliance Assembly Robot Arm (SCARA), cylindrical, Cartesian/gantry (as shown in Fig. 12.1), and parallel robots. Cartesian coordinate robots use orthogonal prismatic axes, usually referred to as X, Y, and Z,

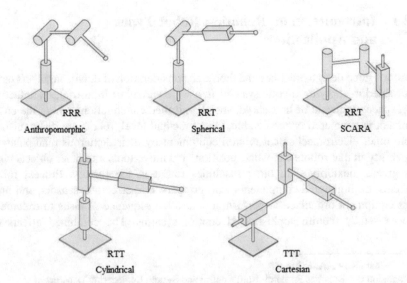

Fig. 12.1 Different types of Robot arms: (RRR) all three joints of the rotational type; (RRT) two joints are rotational and one is translational; (RTT) one rotational and two translational degrees of freedom (Bajd et al. 2010)

to translate their end-effector or payload through their rectangular workspace. One, two, or three revolute wrist axes may be added for orientation. Gantry robots are the most common Cartesian style. Commercial models of spherical and cylindrical robots were originally very common and popular in machine tending and material handling applications (Lewis et al. 1999; Bajd et al. 2010).

Robots are also characterized by the type of actuators employed. Typically manipulators have hydraulic or electric actuation. In some cases, pneumatic actuators are used. A number of successful manipulator designs have emerged, each with a different arrangement of joints and links. Some "elbow" designs, such as the PUMA robots and the SPAR Remote Manipulator System, have a fairly anthropomorphic structure, with revolute joints arranged into "shoulder," "elbow," and "wrist" sections. A mix of revolute and prismatic joints has been adopted in the Stanford Manipulator and the SCARA types of arms. Other arms, such as those produced by IBM, feature prismatic joints for the "shoulder," with a spherical wrist attached. In this case, the prismatic joints are essentially used as positioning devices, with the wrist used for fine motions (Lewis et al. 1999).

The largest number of industrial robot manipulators is found in the car industry (Bajd et al. 2010). They are mainly used for welding. Other important applications of industrial robots are to move objects from point to point. Such examples are found in the process of palletizing. Industrial robots are frequently used in aggressive or dangerous environments, such as spray painting. The request for robot manipulators in the area of industrial assembly of component parts into a functional system is progressively increasing. The interest in robot manipulators in medicine is rapidly increasing as well. They can be found in surgical applications (Boonvisut and Cavusoglu 2013; Keung et al. 2013), drug delivery (Zhou et al. 2013), or in rehabilitation for training of a paralyzed extremity after stroke (Freeman et al. 2012). Exceptional cases of robot manipulators are tele-manipulators. These robots are controlled by a human operator. They are used in dangerous environments or distant places (Bolopion and Régnier 2013).

Wheeled mobile robots can be used on smooth ground. They can effectively be used for vision and other sensors assessing distance or contact with objects in the environment. The biologically inspired legged mobile robots usually have six legs and are used on uneven terrain, as in the forestry robot, which is also capable of cutting trees. Another important class is service robotics where robots are used to help people (predominantly aging populations) in daily activities. The most innovative examples are humanoid robots capable of biped locomotion (Lewis et al. 1999). Tripedal robots, quadrupedal robots, and hexapod robots are other increasingly important legged robots. Flying robots (Nonami et al. 2010) and underwater robots (Javier et al. 2013) are broadly used for observation of distant terrains or for ocean studies.

Finally, humans have sought to establish new dimension of human robot communication, interaction, and collaboration. Sophisticated robotic toys are appreciated by children. Interesting experiments can be found in the art where robots are dancing (Augugliaro et al. 2013), playing musical instruments (Cicconet et al. 2013), and even painting (Lewis et al. 1999). Developing Strategies for Robot soccer competitions has achieved highly advanced stages (Wang et al. 2013).

12.2 Robot Kinematic Modelling

Robot arm kinematics is the science that deals with the analytical study of the motion geometry of the robot manipulator with respect to a fixed and reference coordinate system without regard to the forces or the moments causing the movements. In case of Forward Kinematics (FK) the inputs are the joint angle vectors and the link parameters. The output of the forward kinematics is the orientation and the position of the tool or the gripper. When the joint angles that represent the different robot configuration are computed from the position and orientation of the end-effector, the scheme called as Inverse Kinematics (IK) (Hegde 2008). The representation of FK and IK is shown in Fig. 12.2.

In a serial open loop type of robot manipulators, the links connected to no more than two others via joints at the most. Each pair of a link and a joint gives a single degree of freedom (DOF). Every serial manipulator provides "n" degrees of freedom "n DOF." In general, every link 'k' gets connected at the two ends with the previous link $(k-1)$ and the next link $(k+1)$, forming two joints at the ends of connections (as shown in Fig. 12.3), where O_n is the joint center. However, the kinematic analysis of an n-link manipulator can be solved using Denavit–Hartenberg convention for finding the link parameters (Hegde 2008; Spong et al. 2006):

1. The distance (d_i)
2. The angle (θ_i)
3. The length (a_i)
4. The twist angle (α_k)

After coordinate frames assignment to all robot links, according to Denavit–Hartenberg convention, it is possible to establish the relation between the current frames (i) and the next frame $(i+1)$ by the following transformations in sequence:

- Rotation about Z_i by an angle θ_i
- Translate along Z_i by a distance d_i
- Translate along rotated $X_i = X_{i+1}$ through length a_i
- Rotation about X_i by twist angle α_i

This may result in a product of four homogeneous transformations relating coordinate frames (the current frames (i) and the next frame $(i+1)$) of the serially connected two links. The resulted matrix is known as Arm matrix (A) (Hegde 2008).

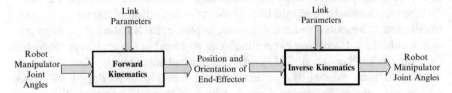

Fig. 12.2 Forward and inverse kinematics representation

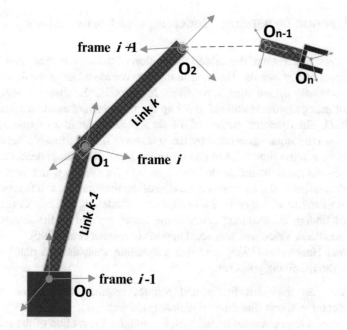

Fig. 12.3 Serial manipulator end-effector frame transformation to base frame

$$A_{i+1}^i = T(z, \theta)T_{\text{trans}}(0, 0, d)T_{\text{trans}}(a, 0, 0)T(x, \alpha) \tag{12.1}$$

$$= T(z, \theta)T_{\text{trans}}(\alpha, 0, d)T(x, \alpha). \tag{12.2}$$

From Equations (12.1) and (12.2) the following matrix may be obtained:

$$A_{i+1}^i = \begin{bmatrix} \cos\theta & -\sin\theta & 0 & 0 \\ \sin\theta & \cos\theta & 0 & 0 \\ 0 & 0 & 1 & 0 \\ 0 & 0 & 0 & 1 \end{bmatrix} \begin{bmatrix} 1 & 0 & 0 & a \\ 0 & 1 & 0 & 0 \\ 0 & 0 & 1 & d \\ 0 & 0 & 0 & 1 \end{bmatrix} \begin{bmatrix} 1 & 0 & 0 & 0 \\ 0 & \cos\alpha & -\sin\alpha & 0 \\ 0 & \sin\alpha & \cos\alpha & 0 \\ 0 & 0 & 0 & 1 \end{bmatrix}$$

$$= \begin{bmatrix} \cos\theta & -\sin\theta\cos\alpha & \sin\theta\sin\alpha & a\cos\theta \\ \sin\theta & \cos\theta\cos\alpha & -\cos\theta\sin\alpha & a\sin\theta \\ 0 & \sin\alpha & \cos\alpha & d \\ 0 & 0 & 0 & 1 \end{bmatrix}.$$

$$\tag{12.3}$$

The coordinate frame at the end-effector of the manipulator is related to the base reference frame by the 'T' matrix in terms of (A) matrices for a six degrees of freedom (6 DOF) robot as example, as follows:

$$T_6^0 = A_1^0 A_2^1 A_3^2 A_4^3 A_5^4 A_6^5. \tag{12.4}$$

12.3   Robotic Dynamics: Modelling and Formulations

Dynamics is the science of describing the motion of massive bodies upon application of forces and moments. The motion can be considered as an evolution of the position, orientation, and time derivatives. In robotics, the dynamic equation of motion for manipulators is utilized to set up the fundamental equations for control (Jazar 2007). The dynamic motion of the manipulator arm in a robotic system is produced by the torques generated by the actuators. The relationship between the input torques and the time rates of change of the robot arm components configurations represents the dynamic modelling of the robotic system which is concerned with the derivation of the equations of motion of the manipulator as a function of the forces and moments acting on it. So, the dynamic modelling of a robot manipulator consists of finding the mapping between the forces exerted on the structures and the joint positions, velocities, and accelerations (Canudas et al. 1996).

Beni and Hackwood (1985) note that a dynamic analysis of a manipulator is useful for the following purposes:

1. It determines the joint forces and torques required to produce specified end-effector motions (the direct dynamic problem).
2. It produces a mathematical model which simulates the motion of the manipulator under various loading conditions (the inverse dynamic problem) and/or control schemes.
3. It provides a dynamic model for use in the control of the actual manipulator.

Equations of motion for the manipulator can be obtained by forming Euler–Lagrange's equation on the basis of Lagrange's energy function. The resulting differential equations describe the motion in terms of the joint variables and parameters of the manipulator.

Let K and V be the total kinetic energy and potential energy stored in the dynamic system. The Lagrangian is defined by (Spong et al. 2006; Min et al. 1992) and (Yamamoto 1992):

$$L(q,\dot{q}) = K(q,\dot{q}) - V(q).\tag{12.5}$$

Using the Lagrangian equations of motion as obtained by Yamamoto (1992):

$$\frac{\mathrm{d}}{\mathrm{d}t}\frac{\partial L}{\partial \dot{q}} - \frac{\partial L}{\partial q} = Q_i, \quad i = 1, \ldots, n,\tag{12.6}$$

where Q_i is the generalized force corresponding to the generalized coordinate q_i. The kinetic energy and potential energy for the link i are given (Spong et al. 2006):

$$K_i = \frac{1}{2}\,\mathrm{trace}\left[\sum_{j=1}^{i}\sum_{k=1}^{i}\frac{\partial T}{\partial q_i}J_i\frac{\partial T}{\partial q_k}\dot{q}_j\dot{q}_k\right],\tag{12.7}$$

$$V_i = -m_i g^{\mathrm{T}} T_i r^{-(i)}. \tag{12.8}$$

The Lagrangian motion equations for the nth link manipulators can be represented as a second-order nonlinear differential equation (Spong et al. 2006):

$$\sum_{j=1}^{n} B_{ij} \ddot{q}_j + \sum_{j=1}^{n} \sum_{k=1}^{n} C_{ijk} \dot{q}_j \dot{q}_j + g_i = Q_i, \quad i = 1, \ldots, n, \tag{12.9}$$

where

$$B_{ij} = \sum_{k=\max(i,j)}^{n} \mathrm{trace} \left[\frac{\partial T_k}{\partial q_i} J_k \frac{\partial T_k^{\mathrm{T}}}{\partial q_i} \right], \tag{12.10}$$

$$C_{ijk} = \sum_{h=\max(i,j,k)}^{n} \mathrm{trace} \left[\frac{\partial T_h}{\partial q_i} J_n \frac{\partial^2 T_h^{\mathrm{T}}}{\partial q_i \partial q_k} \right], \tag{12.11}$$

$$g_i = \sum_{k=i}^{n} m_k g^{\mathrm{T}} \frac{\partial T_k}{\partial q_{ki}} r^{-(i)}. \tag{12.12}$$

Equation (12.9) can be written as a set of second-order vector differential equations:

$$\mathbf{B(q)\ddot{q}} + \mathbf{C(q, \dot{q})} + \mathbf{g(q)} = \mathbf{Q_i}, \tag{12.13}$$

where $\mathbf{B(q)}$ is the symmetric inertia matrix $\mathbf{C(q, \dot{q})}$, the matrix of Coriolis and centrifugal effects, the vector $\mathbf{g(q)}$ denotes the gravity terms, and $\mathbf{Q_i}$ is the generalized force vector.

12.4   Path and Trajectory Planning in Robotics

The definition of the path is the sequence of robot configurations in a particular order without regard to the timing of these configurations. The path planning as a process is the planning of the whole way from the start point to the goal point, including stopping in defined path points. While the trajectory is, the path specified by the time law requirement to move the robot from the starting point to the goal point. In the methodologies of trajectory planning, the task is to attain a specific target from an initial starting point, avoid obstacles, and stay within robot capabilities. Trajectories can be planned either in joint space where the time evolution of the joint angles is specified directly or in Cartesian space specifying the position and orientation of the end-effector frame. In joint space trajectory planning, the joint values that satisfy these robot configurations can be calculated and used by the controller for driving the joints to the desired and new positions. Joint space planning approach is more simple and faster than Cartesian space planning approach because of the inverse kinematics calculations avoidance. Obstacle avoidance is difficult in joint space because of the end-effector pose is not directly controlled, while planning in Cartesian space allows directly satisfying the geometric constraints of the robot work space, but then inverse kinematics must be solved (Niku 2001).

12.4.1 Third-Order Polynomial Trajectory Planning

In this planning method, the problem is to find a trajectory that connects the initial and final configuration. We find the final joint angles for the desired position and orientation using the inverse kinematic equations. Considering one of the joints, which at the beginning of the motion segment at time t_i is at θ_i and has to move to a new value of θ_f at time t_f. Third-order polynomials can be used to plan the trajectory, such that the velocities at the beginning and the end of the motion are zero or other known values (Niku 2001; Spong et al. 2006).

$$\theta(t) = c_0 + c_1 t + c_2 t^2 + c_3 t^3,$$

where the initial and final conditions are as follows:

$$\theta(t_i) = \theta_i; \quad \theta(t_f) = \theta_f; \quad \dot{\theta}(t_i) = 0; \quad \dot{\theta}(t_f) = 0.$$

Taking the first derivative of the polynomial equation:

$$\theta(t_i) = c_0 = \theta_i; \quad \theta(t_f) = c_0 + c_1 t_f + c_2 t_f^2 + c_3 t_f^3; \quad \dot{\theta}(t_i) = c_1 = 0; \quad \dot{\theta}(t_f)$$
$$= c_1 + 2c_2 t_f + 3c_3 t_f^2 = 0.$$

Solving these four equations simultaneously, we get the necessary values for the constants as follows (Niku 2001):

$$\theta(t) = c_0 + c_1 t + c_2 t^2 + c_3 t^3;$$
$$\dot{\theta}(t) = c_1 + 2c_2 t + 3c_3 t^2.$$

$$\text{At } t = 0 \quad (t = t_i = \text{Zero}) \quad \rightarrow \quad \theta(0) = c_0 = \theta_i; \quad \dot{\theta}(0) = c_1$$
$$= 0. \ (\text{initial velocity} = \text{Zero}).$$

$$\text{At } t = t_f: \quad \rightarrow \quad \theta(t_f) = \theta_f = \theta_i + c_2 t_f^2 + c_3 t_f^3 \quad \rightarrow \quad c_2 t_f^2 + c_3 t_f^3$$
$$= \theta_f - \theta_i. \tag{12.14}$$

$$\dot{\theta}(t_f) = 2c_2 t_f + 3c_3 t_f^2 = 0 \quad (\text{Zero velocity at the final time } t_f)$$

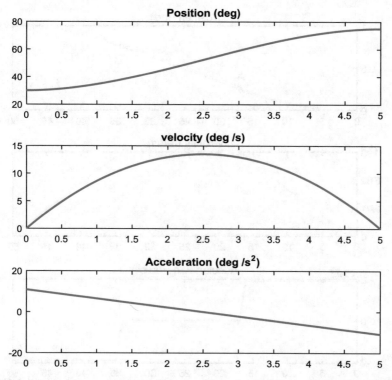

Fig. 12.4 Joint positions, velocities, and accelerations using third-order polynomial equation

$$\rightarrow\ 2c_2 t_f + 3c_3 t_f^2 = 0. \tag{12.15}$$

Solving Eqs. (12.14) and (12.15) gives:

$$c_3 = \frac{-2(\theta_f - \theta_i)}{t_f^3}; \quad c_2 = \frac{3(\theta_f - \theta_i)}{t_f^2}.$$

As an example, Fig. 12.4 shows the joint positions, velocities, and accelerations using third-order polynomial equation.

12.4.2 Linear Segments with Parabolic Blends

This method is a joint space trajectory planning with a trapezoidal velocity profile in which to run the joints at constant speed between the initial and final configurations. Achieving this method in order to create a smooth path, we design the desired trajectory in three parts. Starting with the linear segment and adding a parabolic blend region (quadratic polynomial parts) at the beginning and the end of the

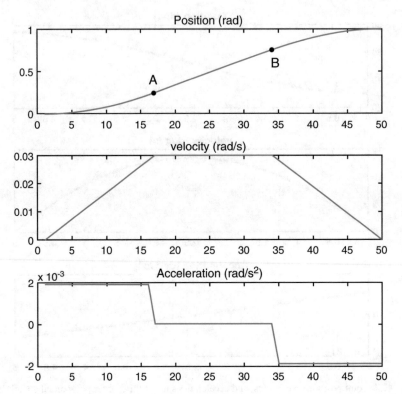

Fig. 12.5 Joint positions, velocities, and accelerations using linear segments with parabolic blends

motion segment for creating continuous position and velocity, as shown in Fig. 12.5. The linear segment is the trajectory segment between points A and B. Assuming that the initial and the final positions are θ_i and θ_f at time $t_i = 0$ and t_f and that the parabolic segments are symmetrically blended with the linear section at blending times t_b and $t_f - t_b$, we can write (Niku 2001; Spong et al. 2006):

$$\theta_i = \theta(t = 0) = \theta(0) = \text{initial position}, \quad \theta_f = \theta(t = t_f) = \theta(t_f) = \text{final position}.$$

$$t_i = 0(\text{starting position}), \quad t_f = \text{final time (ending time)}.$$

For the first parabolic segment:

$$\theta(t) = c_0 + c_1 t + \frac{1}{2} c_2 t^2;$$

$$\dot{\theta}(t) = c_1 + c_2 t;$$

$$\ddot{\theta}(t) = c_2.$$

The acceleration is constant for the parabolic sections, yielding a continuous velocity at the common points (called knot points) A and B. Substituting the boundary conditions into the parabolic equation segment yields (Niku 2001):

$$\text{At } t = 0 \quad \rightarrow \quad \theta(0) = c_0 \quad \rightarrow \quad c_0 = \theta_i;$$
$$\dot{\theta}(0) = c_1 = 0 \text{ (starting velocity = 0)};$$
$$\ddot{\theta}(0) = c_2 \quad \rightarrow \quad c_2 = \ddot{\theta}.$$

Substituting the initial conditions gives parabolic segments in the form:

$$\theta(t) = \theta_i + \frac{1}{2}c_2 t^2;$$
$$\dot{\theta}(t) = c_2 t;$$
$$\ddot{\theta}(t) = c_2.$$

For the linear segment the velocity will be constant and can be chosen based on the physical capabilities of the actuators. Substituting zero initial velocity, a constant known joint velocity ω in the linear portion, and zero final velocity, the joint positions and velocities for points A, B and the final point as follows (Niku 2001):

The general linear equation is,

$$\frac{y}{x} = \frac{y_2 - y_1}{x_2 - x_1} \quad \rightarrow \quad \frac{\theta}{t} = \frac{\theta_B - \theta_A}{t_f - t_b - t_b};$$
$$\frac{\theta}{t} = \omega \quad \rightarrow \quad \omega = \frac{\theta_B - \theta_A}{t_f - 2t_b}$$
$$\rightarrow \theta_B = \theta_A + \omega(t_f - 2t_b).$$
$$\text{At } t = t_b.$$

Because of point A = the end of the first parabolic segment = the start of the linear segment, then the value of θ_A can be found from the end point of the first parabolic segment, so that:

$$\theta_A = \theta_i + \frac{1}{2}c_2 t_b^2;$$

$$\dot{\theta}_A = c_2 t_b = \omega \text{ (constant velocity at the linear segment)}.$$

The necessary blending time t_b can be found as follows:

$$\theta_f = \theta_B + (\theta_A - \theta_i);$$
$$\theta_f = \theta_A + \omega(t_f - 2t_b) + \theta_A - \theta_i; \tag{12.16}$$

$$\theta_A = \theta_i + \frac{1}{2}c_2 t_b^2; \tag{12.17}$$

$$\theta_f = 2\left(\theta_i + \frac{1}{2}c_2 t_b^2\right) - \theta_i + \omega(t_f - 2t_b);$$

$$\theta_f = \theta_i + c_2 t_b^2 + \omega(t_f - 2t_b);$$

$$c_2 = \frac{\omega}{t_b} \tag{12.18}$$

$$\rightarrow \theta_f = \theta_i + \left(\frac{\omega}{t_b}\right)t_b^2 + \omega(t_f - 2t_b);$$

$$\theta_f = \theta_i + \omega t_b + \omega t_f - 2\omega t_b.$$

Then calculating the blending time as:

$$t_b = \frac{\theta_i - \theta_f + \omega t_f}{\omega}.$$

The time t_b cannot be bigger than half of the total time t_f which results in a parabolic speedup and a parabolic slowdown. With no linear segment, a corresponding maximum velocity (Niku 2001):

$$\omega_{max} = 2(\theta_f - \theta_i)/t_f.$$

The final parabolic segment is symmetrical with the initial parabola, but with a negative acceleration, and thus can be expressed as follows (Niku 2001):

$$\theta(t) = \theta_f - \frac{1}{2}c_2(t_f - t)^2, \quad \text{where} \quad c_2 = \frac{\omega}{t_b},$$

$$\rightarrow \begin{cases} \theta(t) = \theta_f - \frac{\omega}{2t_b}(t_f - t)^2, \\ \dot{\theta}(t) = \frac{\omega}{t_b}(t_f - t), \\ \ddot{\theta}(t) = -\frac{\omega}{t_b}. \end{cases}$$

12.5 Classical Control Synthesis and Design

The problem of robot control can be described as a computation of the forces or torques that must be generated by the actuators in order to successfully accomplish the robot task. The robot task can be presented either as the accomplishment of the motions in a free space, where position control is performed, or in contact with the environment, where control of the contact force is required (Bajd et al. 2010).

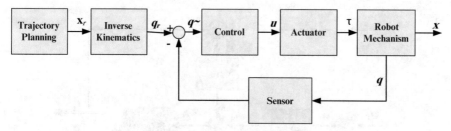

Fig. 12.6 A common robot control system (Bajd et al. 2010)

Control is used to move the robot with respect to the environment as well as to articulate sensor heads, arms, grippers, tools, and implements. Several techniques can be employed for controlling a robot, as in Samad (2001). The choice of the control method depends on the robot task.

A general robot control system consists of the following components: path planning, inverse kinematics, and a closed loop system that contains the controller, actuator, the robot mechanism, and the sensor, as shown in Fig. 12.6. The input to the control system is the desired pose of the robot end-effector, which can be obtained by using trajectory interpolation methods. The variable $\mathbf{x_r}$ represents the reference pose of the robot end-effector. The $\mathbf{x}$ vector, describes the actual pose of the robot end-effector, in general this comprises six variables. Three of them define the position of the robot end point, while the other three determine the orientation of the robot end-effector. Accordingly,

$$\mathbf{x} = \begin{bmatrix} x & y & z & \varphi & \vartheta & \psi \end{bmatrix}^{\mathrm{T}}.$$

The orientation is determined by the angle φ around the z axis (Roll), the angle ϑ around the y axis (Pitch), and the angle ψ around the x axis (Yaw). The internal coordinates q_r represent the desired end-effector position, i.e., the angle ϑ for the rotational joint and the distance d for the translational joint. The desired internal coordinates are compared to the actual internal coordinates in the robot control system. Based on the positional error $\widetilde{q}$, the control system output u is calculated. The actuators ensure the forces or torques necessary for the required robot motion. The robot motion is measured by the sensors.

12.5.1   PD Position Control

For position control of a robot, a Proportional Derivative (PD) is commonly designed. For robot control this closed loop is separate for each particular degree of freedom. The control method is based on calculation of the positional error and determination of control parameters, which enable reduction or suppression of the error. The positional error is reduced for each joint separately, which means that as

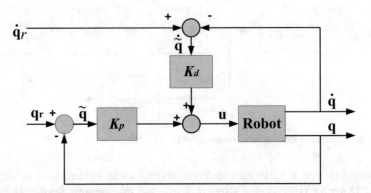

Fig 12.7 PD position control (Bajd et al. 2010)

many controllers are to be developed as there are degrees of freedom. A robot manipulator has several degrees of freedom, therefore the error $\widetilde{q} = q_r - q$ can be stated as a vector, whereas K_p is a diagonal matrix of the gains of all joint controllers. The calculated control input incites robot motion toward reduction of the positional error. The positional error is characterized by the position error $\widetilde{q}$ multiplied by K_p (whereas K_p is a diagonal matrix of the gains of all joint controllers). In addition, to confirm safe and stable robot actions, velocity in closed loop mode is presented with a negative sign. The velocity in closed loop mode brings damping into the system. It is characterized by the actual joint velocities $\dot{q}$ multiplied by a diagonal matrix of velocity gains K_d. The overall control law can be obtained by combining the positional error and the velocity error as given in the following form:

$$u = K_p(q_r - q) + K_d\left(\dot{q}_r - \dot{q}\right),$$ (12.19)

where $\dot{q}_r - \dot{q}$ is the velocity error $\widetilde{q}$. In Eq. (12.19), the reference velocity signal is included in the PD signal in order to avoid unnecessary high damping at fastest part of the trajectory. The synthesis of the PD position controller involves the determination of the matrices K_p and K_d. The K_p gains must be high for faster response. On the other hand by proper choice of the $\mathbf{K}_d$ gains, fast response without overshoot for the robot systems is gained. Figure 12.7 illustrates the PD position control configuration.

12.5.2 PD Control of Position with Gravity Compensation

The robot mechanism is usually known to be under the influence of inertial, Coriolis, centripetal, and gravitational forces. For a simplified model, viscous

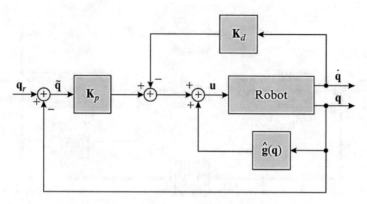

Fig. 12.8 PD control with gravity compensation (Bajd et al. 2010)

friction, which is proportional to the joint velocity, will be considered here. Consequently, robot dynamics of Eq. (12.13) can be rewritten as follows:

$$B(q)\ddot{q} + C(q,\dot{q})\dot{q} + F_v\dot{q} + g(q) = \tau, \qquad (12.20)$$

where τ is the torques in the robot joints. **B, C,** and **g** are defined in Sect. 11.2, $\mathbf{F}_v$ is a diagonal matrix of the joint friction coefficients (Bajd et al. 2010).

In quasi-static conditions, when the robot is moving slowly, it can be assumed that zero accelerations $\ddot{q} \approx 0$ and velocities $\dot{q} \approx 0$. Accordingly, the robot dynamic model is simplified as

$$\tau \approx g(q). \qquad (12.21)$$

The model of gravitational effects $\hat{\mathbf{g}}(\mathbf{q})$ (the circumflex denotes the robot model), which is an acceptable approximation of the actual gravitational forces $\mathbf{g}(\mathbf{q})$. The control algorithm shown in Fig. 12.8 can be written as follows:

$$u = K_p(q_r - q) - K_d\dot{q} + \hat{g}(q). \qquad (12.22)$$

By introducing gravity compensation, the errors in trajectory tracking are significantly reduced. In addition, this control method can be extended to consider the effect of motion of the robot end-effector; starting from the positional error of the robot end-effector which is calculated as:

$$\tilde{x} = x_r - x = x_r - k(q), \qquad (12.23)$$

where $\mathbf{x}_r$ is the reference pose of the robot end-effector and $\mathbf{k}(.)$ represents the equations of direct kinematics.

The velocity of the robot end point is calculated with the help of the Jacobian matrix from the joint velocities. The equation describing the PD controller is:

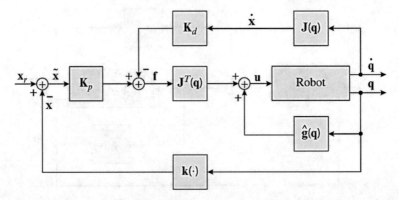

Fig. 12.9 PD control with gravity compensation in external coordinates (Bajd et al. 2010)

$$\mathbf{f} = \mathbf{K_p}\tilde{\mathbf{x}} - \mathbf{K_d}\dot{\mathbf{x}}. \tag{12.24}$$

The negative sign of the velocity error introduces damping into the system. The joint torques are calculated from the force f, acting at the tip of the robot, with the help of the transposed Jacobian matrix and by adding the component compensating gravity. The control algorithm is written as follows:

$$\mathbf{u} = \mathbf{J}^{\mathrm{T}}(\mathbf{q})\mathbf{f} + \widehat{\mathbf{g}}(\mathbf{q}). \tag{12.25}$$

The complete control scheme is shown in Fig. 12.9.

12.5.3 Control of the Robot Based on Inverse Dynamics

This control scheme can be derived from the robot dynamic model described by Eq. (12.20). Assume that the torques τ, generated by the motors, are equal to the control outputs u. Rewrite Eq. (12.20) in order to determine the direct robot dynamic model, which describes robot motions under the influence of the given joint torques. Accordingly, the acceleration $\ddot{q}$ can be expressed in short as follows (Bajd et al. 2010):

$$\ddot{q} = B^{-1}(q)(u - n(q, \dot{q})), \tag{12.26}$$

where $n(q, \dot{q})$ comprising all dynamic components except the inertial component, i.e.

$$\mathbf{n}(\mathbf{q}, \dot{\mathbf{q}}) = \mathbf{C}(\mathbf{q}, \dot{\mathbf{q}})\dot{\mathbf{q}} + \mathbf{F_v}\dot{\mathbf{q}} + \mathbf{g}(\mathbf{q}). \tag{12.27}$$

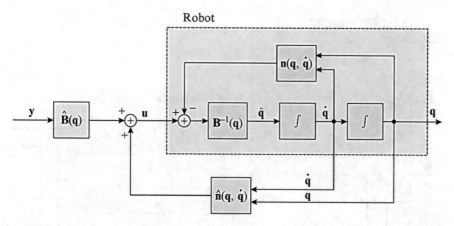

Fig. 12.10 Linearization of the control system by implementing the inverse dynamic model (Bajd et al. 2010)

Assume that the robot dynamic model is known. The inertial matrix $\widehat{\mathbf{B}}(\mathbf{q})$ is an approximation of the real values $\mathbf{B}(\mathbf{q})$, while $\widehat{\mathbf{n}}(\mathbf{q}, \dot{\mathbf{q}})$ represents an approximation of $\mathbf{n}(\mathbf{q}, \dot{\mathbf{q}})$ as follows:

$$\widehat{\mathbf{n}}(\mathbf{q}, \dot{\mathbf{q}}) = \widehat{\mathbf{C}}(\mathbf{q}, \dot{\mathbf{q}})\dot{\mathbf{q}} + \widehat{\mathbf{F}}_v\dot{\mathbf{q}} + \widehat{\mathbf{g}}(\mathbf{q}). \tag{12.28}$$

The controller output $\mathbf{u}$ is based on inverse dynamics as in the following equation:

$$\mathbf{u} = \widehat{\mathbf{B}}(\mathbf{q})\mathbf{y} + \widehat{\mathbf{n}}(\mathbf{q}, \dot{\mathbf{q}}), \tag{12.29}$$

where the approximate inverse dynamic model of the robot was used. The system, combining Eqs. (12.26) and (12.29), is shown in Fig. 12.10.

By simple substitutions we can write the vector $\mathbf{y}$, having the acceleration characteristics:

$$\mathbf{y} = \ddot{\mathbf{q}}_r + \mathbf{K}_p(\mathbf{q}_r - \mathbf{q}) + \mathbf{K}_d(\dot{\mathbf{q}}_r - \dot{\mathbf{q}}). \tag{12.30}$$

It consists of the reference acceleration $\ddot{\mathbf{q}}_r$ and two contributing signals which depend on the errors of position and velocity. These two signals suppress the error arising because of the imperfectly modelled dynamics. The complete control scheme is shown in Fig. 12.11. By considering Eq. (12.30) and the equality $\mathbf{y} = \ddot{\mathbf{q}}$, the differential equation describing the robot dynamics can be written as follows:

$$\ddot{\tilde{q}} + K_d\dot{\tilde{q}} + K_p\tilde{q} = 0, \tag{12.31}$$

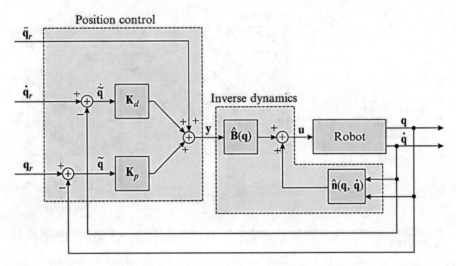

Fig. 12.11 Control of the robot based on inverse dynamics (Bajd et al. 2010)

where the acceleration error $\ddot{\widetilde{q}} = \ddot{q}_r - \ddot{q}$ was introduced. The differential equation (12.31) describes the time dependence of the control error as it approaches zero. The dynamics of the response is determined by the gains K_p and K_d.

Similar to internal coordinates, the derivation of equation that describe the dynamics of the control error, an analogous equation can be written for the error of the end-effector pose. Accordingly, the acceleration $\ddot{x}$ of the robot end-effector can be expressed as follows:

$$\ddot{\widetilde{x}} + K_d\dot{\widetilde{x}} + K_p\widetilde{x} = 0 \quad \Rightarrow \quad \ddot{x} = \ddot{x}_r + K_d\dot{\widetilde{x}} + K_p\widetilde{x}. \tag{12.32}$$

Taking into account the equality $\mathbf{y} = \ddot{q}$

$$\mathbf{y} = \mathbf{J}^{-1}(\mathbf{q})\left(\ddot{x} - \dot{\mathbf{J}}(\mathbf{q}, \dot{\mathbf{q}})\dot{\mathbf{q}}\right). \tag{12.33}$$

Substituting $\ddot{x}$ in Eq. (12.33) with expression (12.32), the control algorithm based on inverse dynamics in the external coordinates is obtained as follows:

$$\mathbf{y} = \mathbf{J}^{-1}(\mathbf{q})\left(\ddot{x}_r + K_d\dot{\widetilde{x}} + K_p\widetilde{x} - \dot{\mathbf{J}}(\mathbf{q}, \dot{\mathbf{q}})\dot{\mathbf{q}}\right). \tag{12.34}$$

The control scheme encompassing the linearization of the system based on inverse dynamics (12.29) and the closed-loop control (12.34) is shown in Fig. 12.12.

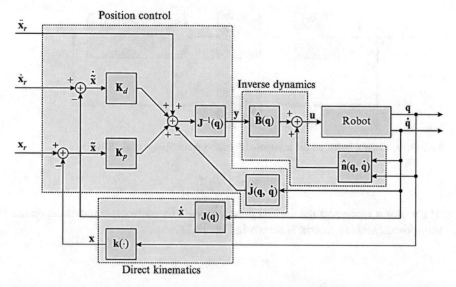

Fig. 12.12 Robot control based on inverse dynamics in external coordinates (Bajd et al. 2010)

12.5.4   *Control Based on the Transposed Jacobian Matrix*

This control method is based on the relation connecting the forces acting at the robot end-effector with the joint torques. The relation is defined by the use of the transposed Jacobian matrix:

$$\boldsymbol{\tau} = \mathbf{J}^{\mathrm{T}}(\mathbf{q})\mathbf{f}, \tag{12.35}$$

where the vector $\boldsymbol{\tau}$ represents the joint torques and f is the force at the robot endpoint. The aim is to control the pose of the robot end-effector, where its desired pose is defined by the vector $\mathbf{x}_r$ and the actual pose is given by the vector x. Robots are usually provided with sensors that measure the joint variables. The pose of the robot end-effector must be therefore determined by using the direct kinematic model $x = k(q)$, where q indicates the vector of joint variables, x indicates the vector of task variables; usually, three position coordinates and three Euler angles The positional error of the robot end-effector ($\tilde{x} = x_r - x$) must be reduced to zero. A simple proportional control system with the gain matrix $\mathbf{K}_p$ (Bajd et al. 2010) is introduced:

$$\mathbf{f} = \mathbf{K}_p \tilde{x}. \tag{12.36}$$

As the robot displacement can only be produced by the motors in the joints, the variables controlling the motors must be calculated from the force f. This calculation is performed using the transposed Jacobian matrix in Eq. (12.35).

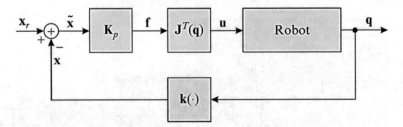

Fig. 12.13 Control based on the transposed Jacobian matrix (Bajd et al. 2010)

$$\mathbf{u} = \mathbf{J}^{\mathrm{T}}(\mathbf{q})\mathbf{f}. \tag{12.37}$$

The vector $\boldsymbol{u}$ represents the desired joint torques. The control method based on the transposed Jacobian matrix is shown in Fig. 12.13.

12.5.5 Control Based on the Inverse Jacobian Matrix

In this method, the control is based on the relation between the joint velocities and the velocities of the robot end point, which is known as the Jacobian matrix (Bajd et al. 2010).

$$\dot{\mathbf{x}} = \mathbf{J}(\mathbf{q})\dot{\mathbf{q}}. \tag{12.38}$$

For small displacements, the relation between changes of the internal coordinates and changes of the pose of the robot end point can be expressed as follows:

$$\mathbf{dx} = \mathbf{J}(\mathbf{q})\mathbf{dq}. \tag{12.39}$$

For small error in the pose, we can calculate the positional error in the internal coordinates by the inverse relation (Eq. 12.39).

$$\tilde{\mathbf{q}} = \mathbf{J}^{-1}(\mathbf{q})\tilde{\mathbf{x}}. \tag{12.40}$$

In this way, the control method is translated to the known method of robot control in the internal coordinates. The control method, based on the inverse Jacobian matrix, is shown in Fig. 12.14.

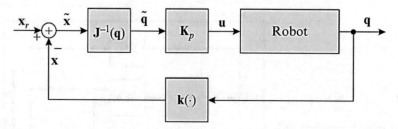

Fig. 12.14 Control based on the inverse Jacobian matrix (Bajd et al. 2010)

12.5.6   Control of the Contact Force

The robot force control method is based on control of the robot using inverse dynamics. A contact force f appears in the inverse dynamic model due to the interaction of the robot with the environment. As the forces acting at the robot end-effector are transformed into the joint torques by the use of the transposed Jacobian matrix, we can write the robot dynamic model in the following form (Bajd et al. 2010)

$$\mathbf{B}(\mathbf{q})\ddot{\mathbf{q}} + \mathbf{C}(\mathbf{q}, \dot{\mathbf{q}})\dot{\mathbf{q}} + \mathbf{F}_v\dot{\mathbf{q}} + \mathbf{g}(\mathbf{q}) = \boldsymbol{\tau} - \mathbf{J}^{\mathrm{T}}(\mathbf{q})\mathbf{f}. \tag{12.41}$$

It can be seen that the force f acts through the transposed Jacobian matrix in a similar way as the joint torques, i.e., it tries to produce robot motion.

12.5.6.1   Linearization of a Robot System Through Inverse Dynamics

Let us denote the control output, representing the desired actuation torques in the robot joints, by the vector $\boldsymbol{u}$. The direct dynamic model was described by Bajd et al. (2010)

$$\ddot{\mathbf{q}} = \mathbf{B}^{-1}(\mathbf{q})\big(\mathbf{u} - \mathbf{n}(\mathbf{q}, \dot{\mathbf{q}}) - \mathbf{J}^{\mathrm{T}}(\mathbf{q})\mathbf{f}\big). \tag{12.42}$$

Equation (12.24) describes the response of the robot system to the control input $\boldsymbol{u}$. Taking into account the initial velocity value, the actual velocity of the robot motion is obtained by integrating the acceleration. While taking into the account the initial position, the actual positions in the robot joints are calculated by integrating the velocity. The described model is represented by the block *Robot* in Fig. 12.15. The system is linearized by including the inverse dynamic model into the closed loop:

$$\mathbf{u} = \widehat{\mathbf{B}}(\mathbf{q})\mathbf{y} + \widehat{\mathbf{n}}(\mathbf{q}, \dot{\mathbf{q}}) + \mathbf{J}^{\mathrm{T}}(\mathbf{q})\mathbf{f}. \tag{12.43}$$

The use of circumflex denotes the estimated parameters of the robot system.

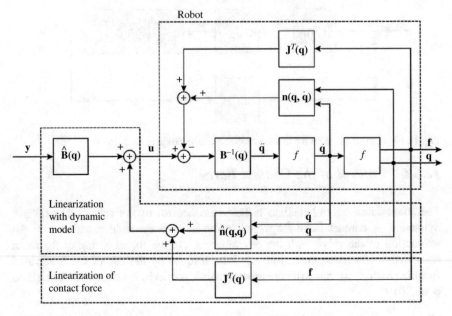

Fig. 12.15 Linearization of the control system by implementing the inverse dynamic model and the measured contact force (Bajd et al. 2010)

12.5.6.2 Force Control

Based on linearized model of the control system, the force control is translated to control the pose of the end-effector. If it is required from the robot to increase the force exerted on the environment, the robot end-effector must be displaced in the direction of the action of the force. The following control system by Bajd et al. (2010) can be used:

$$y = J^{-1}(q)\left(\ddot{x}_r + K_d\tilde{x} + K_P\tilde{x} - \dot{J}(q,\dot{q})\dot{q}\right). \tag{12.44}$$

Accordingly, the control of the robot end-effector (including the linearization) while taking into account the contact force can be determined. This is summarized in Fig. 12.16.

Generally, for appropriate handling of interactions between robot and environment, it is necessary to consider force control strategies, either in an indirect way by means of a suitable use of position control laws or in a direct way by means of true force control laws (Samad 2001).

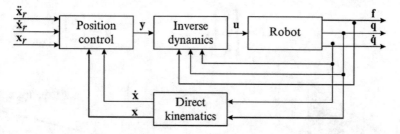

Fig. 12.16 Robot control based on inverse dynamics in external coordinates including the contact force (Bajd et al. 2010)

12.6   Robot Vision and Visual Servoing

Vision technology gives robots intelligent eyes. Using these eyes, robots can recognize the position of objects in space and adjust their working steps accordingly. The benefits of sophisticated vision technology include savings, improved quality, reliability, safety, and productivity. Robot vision is used for part identification and navigation. Vision applications generally deal with finding a part and orienting it for robotic handling or inspection before an application is performed. Sometimes vision-guided robots can replace multiple mechanical tools with a single robot station.

12.6.1   Robot Vision

Recognizing the geometry of the robot workspace from a digital image (Fig. 12.17) is the main task of robot vision which is solved by finding the relation between the coordinates of a point in the two-dimensional (2D) image and the coordinates of the point in the real three-dimensional (3D) robot environment. The basic equations of optics determine the position of a point in the image plane with respect to the corresponding point in 3D space. We will therefore find the geometrical relation between the coordinates of the point $P(x_c, y_c, z_c)$ in space and the coordinates of the point $p(u,v)$ in the image.

Studying the robot geometry and kinematics by attaching the coordinate frame to each rigid robot segments or to objects manipulated by the robot where, the camera itself represents a rigid body and a coordinate frame should be assigned to it. A corresponding coordinate frame will describe the pose of the camera. The z_c axis of the camera frame is directed along the optical axis, while the origin of the frame is positioned at the center of projection. Using a right-handed frame where the x_c axis is parallel to the rows of the imaging sensor and the y_c axis is parallel with its columns. The image plane is in the camera, which is placed behind the center of projection. The focal length is the distance (f_c) between the image and the

$$P = \begin{bmatrix} 0 \\ y_c \\ z_c \end{bmatrix}. \qquad\qquad p = \begin{bmatrix} 0 \\ y \end{bmatrix}.$$

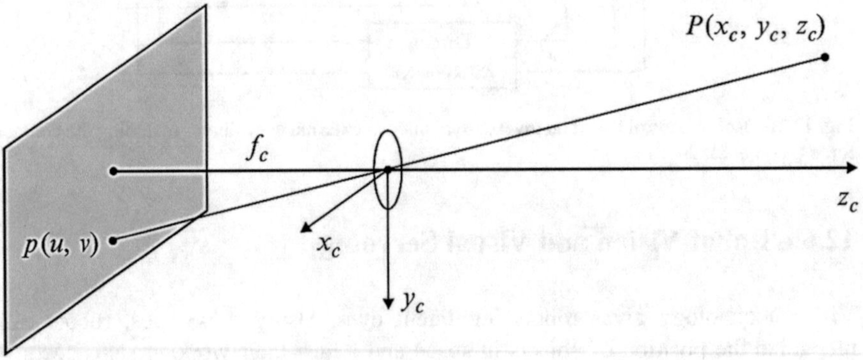

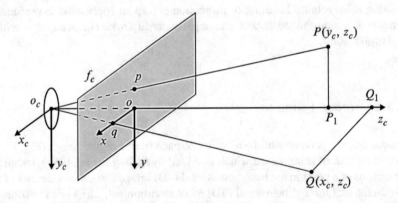

Fig. 12.17 Perspective projection (Bajd et al. 2010)

Fig. 12.18 Equivalent image plane (Bajd et al. 2010)

center of projection and has a negative value, as the image plane intercepts the negative z_c axis. The equivalent image plane placed at a positive z_c value (Fig. 12.18). Both the equivalent image plane and the real image plane are symmetrical with respect to the origin of the camera frame. The origin of this frame is placed in the intersection of the optical axis with the image plane. The x and y axes are parallel to the x_c and y_c axes of the camera frame. The camera has two coordinate frames, the camera frame and the image frame. The point P be expressed in the camera frame, while the point p represents its projection onto the image plane. The point P is located in the y_c, z_c plane of the camera frame (Bajd et al. 2010).

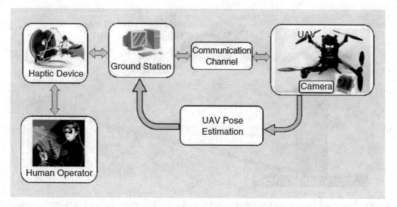

Fig. 12.19 Flying robot system with vision technology (Carloni et al. 2013)

$$P = \begin{bmatrix} 0 \\ y_c \\ z_c \end{bmatrix}, \quad p = \begin{bmatrix} 0 \\ y \end{bmatrix}.$$

A famous case study for using vision technology with robotics is the flying robots such as quadcopters (Carloni et al. 2013). These robots have gained increased interest in research. To navigate safely, flying robots need the ability to localize themselves autonomously using their onboard sensors. Potential applications of such systems include the usage as a flying camera, for example to record sport movies or to inspect bridges from the air, as well as surveillance tasks and applications in agriculture. The main idea of the developed flying robot system is described in Fig. 12.19.

12.6.2   Robot Control Using Visual Servoing Technique

The task is to control the pose of the robot's end-effector using visual information, features, extracted from the image. The Pose is represented by a six element vector encoding position and orientation in 3D space. The camera may be fixed or mounted on the robot's end-effector in which case there exists a constant relationship, between the pose of the camera and the pose of the end-effector. The image of the target is a function of the relative pose between the camera and the target. The relationship between these poses is shown in Fig. 12.17. The distance between the camera and target is frequently referred to as depth or range. The relevant frames required are shown in Fig. 12.20.

The camera contains a lens, which forms a 2D projection of the scene on the image plane where the sensor is located. This projection causes direct depth information to be lost, and each point on the image plane corresponds to a ray in

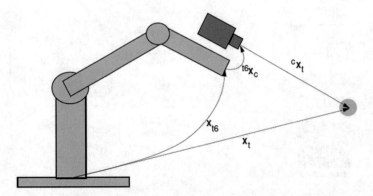

Fig. 12.20 Relevant coordinate frames; world, end-effector, camera, and target

3D space. Some additional information is needed to determine the 3D coordinate corresponding to an image plane point. This information may come from multiple views or knowledge of the geometric relationship between several feature points on the target (Corke 1994).

Robots typically have six degrees of freedom (DOF), allowing the end-effector to achieve, within limits, any pose in 3D space. Visual servoing systems may control six or fewer DOF. Motion so as to keep one point in the scene, the interest point, at the same location in the image plane is referred to as fixation. Animals use fixation to direct the high-resolution fovea of the eye toward regions of interest in the scene. In humans, this low-level, unconscious, fixation motion is controlled by the brain's medulla region using visual feedback from the retina. Fixation may be achieved by controlling the pan/tilt angles of the camera like a human eye or by moving the camera in a plane normal to the optical axis. High performance fixation control is an important component of many active vision strategies. Keeping the target centered in the field of view has a number of advantages that include:

- Eliminating motion blur since the target is not moving with respect to the camera
- Reducing the effect of geometric distortion in the lens by keeping the optical axis pointed at the target

Visual servoing can be classified into position-based visual servoing (Fig. 12.21) and image-based visual servoing. In position-based control (PBVS), the geometric model of the target object is used in conjunction with visual features extracted from the image to estimate the pose with respect to the camera frame, computing the control law by reducing the error in pose space. In this way (Fig. 12.8), the estimation problem involved in computing the object location can be studied separately from the problem of calculate the feedback signal required by the control algorithm (Corke 1994; Miura 2004).

In image-based servoing (IBVS), the last step is omitted, and servoing is done on the basis of image features directly. The structures referred to as dynamic look and move make use of joint feedback, whereas the PBVS and IBVS structures use no joint position information at all. The image-based approach (as shown in

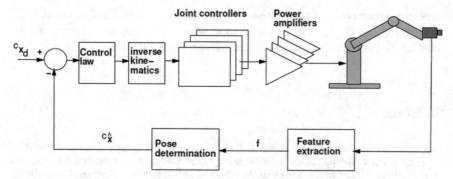

Fig. 12.21 Position-based visual servo (PBVS) structure

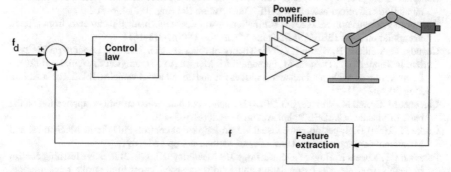

Fig. 12.22 Image-based visual servo (IBVS) structure

Fig. 12.22) may reduce computational delay, eliminate the necessity for image interpretation, and eliminate errors in sensor modelling and camera calibration. However, it does present a significant challenge to controller design since the plant is nonlinear and highly coupled.

12.7 Conclusion

This chapter sheds light on the essentials of a robotic system. First, we exhibited types and applications of robots emphasizing the modelling of the kinematic of a robot manipulator. Then, we formulate the dynamics of a robot manipulator. A third-order polynomial trajectory planning was illustrated as well as linear Segments with Parabolic Blends. Suggestions of some practical control structures have been given; PD position control in different configurations can be realized. Control that based on inverse dynamics is widely used for robots. In order to consider the relation between the joint velocities and the velocities of the robot end point, control based on the (or inverse) Jacobian matrix can be applied. Force control method can find many areas of applications as well. Finally, vision technology for robotic system has been illustrated. The study of robotics and its mathematics has great application in sustainability for the future.

References

Augugliaro F, Schoellig AP, D'Andrea R (2013) Dance of the flying machines. Methods of designing and executing an aerial dance choreography. IEEE Robot Autom Mag 20(4):96–104

Bajd T, Mihelj M, Lenarčič J, Stanovnik A, Munih M (2010) Robotics. Springer, Dordrecht, pp 5–7, 58–60, 77–95

Beni G, Hackwood S (1985) Recent advanced in robotics. Wiley, New York

Bolopion A, Régnier S (2013) A review of Haptic feedback teleoperation systems for micromanipulation and micro-assembly. IEEE Trans Autom Sci Eng 10(3):496–502

Boonvisut P, Cavusoglu MC (2013) Estimation of soft tissue mechanical parameters from robotic manipulation data. IEEE/ASME Trans Mechatron 18(5):1602–1611

Canudas C, Siciliano B, Bastin G (1996) Theory of robot control. Springer, London, pp 21–29

Carloni R, Lippiello V, D'Auria M, Fumagalli M, Mersha AY, Stramigioli S, Siciliano B (2013) Robot vision obstacle-avoidance techniques for unmanned aerial vehicles. IEEE Robot Autom Mag 20(4):22–31

Cicconet M, Bretan M, Weinberg G (2013) Human-robot percussion ensemble, application on the basis of visual cues. IEEE Robot Autom Mag 20(4):105–110

Corke PI (1994) High-performance visual closed-loop robot control. PhD thesis, Mechanical and Manufacturing Engineering, University of Melbourne, pp 186–189

Freeman CT, Rogers E, Huges A-M, Burridge JH, Meadmore KL (2012) Iterative learning control in health care, electrical stimulation and robotic- assisted upper-limb stroke rehabilitation. IEEE Control Syst Mag 32(1):18–43

Hegde GS (2008) A text book on industrial robotics, 2nd edn. University Science Press, New Delhi

Javier J, Prats M, Sanz PJ, Garcia JC, Marin R, Robinson M, Ribas D, Ridao P (2013) Grasping for the seabed. IEEE Robot Autom Mag 20(4):121–130

Jazar RN (2007) Theory of applied robotics: kinematics, dynamics, and control. Springer, New York, pp 507–546

Keung W, Yang B, Liu C, Poignet P (2013) A quasi- spherical triangle- based approach for efficient 3-D soft tissue motion tracking. IEEE/ASME Trans Mechatron 18(5):1472–1484

Lewis FL, Zhou C, Stevens GT, Fitzgeral JJM, Liu K (1999) Robotics. In: Kreith F (ed) Mechanical engineering handbook. CRC Press LLC, Boca Raton

Min Z, Nirwan A, Edwin SH (1992) Mobile manipulator path planning by genetic algorithm. In: Proceedings of the IEEE/RSJ international conference on Intelligent Robot and System, 7–10 July 1992

Miura K (2004) Robot hand positioning and grasping using vision. PhD thesis, Strasbourg I University, Strasbourg

Niku SB (2001) Introduction to robotics: analysis, systems and applications. Prentice Hall, Upper Saddle River

Nonami K, Kendoul F, Suzuki S, Wang W, Nakazawa D (2010) Autonomous flying robots: unmanned aerial vehicles and micro aerial vehicles. Springer, Berlin, pp 2–60

Samad T (2001) Robot control—perspectives in control engineering technologies, applications, and new directions. Wiley-IEEE Press eBook, New York, pp 442–461. doi:10.1109/9780470545485.ch18

Spong MW, Hutchinson S, Vidyasagar M (2006) Robot modeling and control, 1st edn. Wiley, New York, pp 149–185, 215–221

Wang M-L, Wu J-R, Kao L-W, Lin H-Y (2013) Development of a vision system and a strategy simulator for middle size soccer robot. In: International conference on advanced robotics and intelligent systems, May 31–June 2, Tainan, Taiwan, pp 54–58

Yamamoto Y (1992) Coordination locomotion and manipulation of mobile robot manipulator. IEEE Trans Autom Control 39(6):1326–1332

Zhou H, Alici G, Than TD, Li W (2013) Modeling and experimental investigation of rotational resistance of spiral-type robotic capsule inside a real intestine. IEEE/ASME Trans Mechatron 18(5):1555–1562

Chapter 13
Intelligent and Robust Path Planning and Control of Robotic Systems

S.M. Raafat and F.A. Raheem

Abstract Intelligent control has a great influence in improving the performance of robotic systems. Similarly robust control proves to enhance the precision and accuracy of positioning in robotic applications. This chapter describes some powerful intelligent and robust control methods for robots. The development of intelligent methods of path planning and obstacle avoidance is illustrated in detail. An efficient decision-making model for collective search behavior for a swarm of robots in a risky environment has also been demonstrated. In addition, an intelligent variable structure control for robot manipulator has also been shown. Finally, discussion and further directions are given.

Keywords Intelligent control • Iterative learning control • Robust control • Swarm robotics systems • Robot manipulator

13.1 Introduction

The increasingly wide spectrum of complex applications of robotic systems demands that the robots be highly intelligent. In order to guarantee superior control and performance in robotics, new techniques such as intelligent, Iterative Learning Control (ILC), and robust control need to be established. These techniques can deal with task complexity, multiobjective decision making, large amount of perception data, and considerable amounts of heuristic information. Therefore, intelligent, ILC, and robust robotic systems have been in the focus of captivating research recently.

Intelligent control achieves automation via the emulation of biological intelligence. It either seeks to replace a human who performs a control task (e.g., moving a robot manipulator) or it borrows ideas from how biological systems solve problems and applies them to the solution of control problems (e.g., the use of neural networks for control) (Merlet 2014). It can swiftly handle some critical

S.M. Raafat (✉) • F.A. Raheem
Automation and Robotics Research Unit, Control and System Engineering Department,
University of Technology, Baghdad, Iraq
e-mail: safanamr@gmail.com

© Springer International Publishing Switzerland 2017
J.N. Furze et al. (eds.), *Mathematical Advances Towards Sustainable Environmental Systems*, DOI 10.1007/978-3-319-43901-3_13

difficulties that arise in the control of complex technological like the complexity of the system's requirements and goals, the presence of nonlinearities and constraints, and the increased uncertainty in the system modelling and designing. Changes in the environment and performance criteria, immeasurable disturbances, and component failures are some of the characteristics which necessitate intelligent control.

ILC is one of the most effective control methodologies in dealing with frequent tracking control problems. The key design feature of ILC is the effective use of past information to improve tracking performance within a small number of trials, while ensuring robustness of the process against system uncertainty (Zhao et al. 2015).

Robust control techniques can systematically address the control challenges associated with robotic system, e.g., uncertainties due to modelling errors, nonlinearities, and disturbances.

Swarm robotics is a methodology that appeared in the field of artificial swarm intelligence, along with the biological studies of insects (i.e., ants and other fields in nature) which harmonize their actions to achieve tasks that are away from the abilities of a single individual (Martin 2010).

13.2 Intelligent Control

Intelligent control systems may be autonomous, human aided, or computer aided. They have the potential to yield new levels of system performance in terms of reliability and accommodation of significant changes in the system parameters. There are several intelligent paradigms that can effectively solve intelligent control problems in robotics: symbolic knowledge-based systems (AI—expert systems), connectionist theory (NN—neural networks), fuzzy logic (FL), and theory of evolutionary computation (GA—genetic algorithms), which are of great significance in the improvement of intelligent robot control algorithms. Also of great importance in the development of efficient algorithms are hybrid techniques based on integration of particular techniques such as neuro-fuzzy networks, neuro-genetic algorithms, and fuzzy-genetic algorithms (Katic and Vukobratovic 2003).

13.2.1 Neural Network-Based Control for Robotics

The computational capabilities of the systems with neural networks are remarkable and very encouraging; they include many "intelligent functions" such as logical reasoning, learning, pattern recognition, formation of associations, or abstraction from examples. In addition, they have the ability to acquire the most skillful performance in the control of complex dynamic systems. They also calculate a large number of sensors with altered modalities, providing noisy and sometimes unpredictable information. Among the beneficial characteristics of neural networks are as follows: Learning, Generalization, Massive Parallelism, and Fault Tolerance, which make them suitable for system integration. They are also suitable for realization in hardware.

There are some important types of network models that are commonly used in robotic applications. These include multilayer perceptrons (MP), radial basis function (RBF) networks, recurrent version of multilayer perceptron (RMP), Hopfield networks (HN), CMAC networks, ART networks, and Kohonen Self-organizing Feature-Mapping Networks (SOFM) (Katic and Vukobratovic 2003).

Neural networks are inherently plastic structures that call for substantial learning. In fact, the learning occurs under different learning conditions and can be completed under various types of interaction with the environment. Generally, there are three main types of learning, namely: 1—Supervised learning, 2—Reinforcement learning, and 3—Unsupervised learning. It is known that a robot learns from what is happening rather than from what is predicted. Hence, whenever self-adaptive and autonomous capabilities are required, the application of neural networks in robot control has a great chance of success. Neural networks are capable of recognizing the changes in the robots environmental circumstances and reaction to them, or, making decisions based on changing manufacturing events (Katic and Vukobratovic 2003).

There are few different major architectures for supervised robot learning: specialized learning, generalized learning, feedback-error learning, and adaptive learning architectures. The feedback-error learning architecture (Fig. 13.1) is an entirely online architecture for robot control that permits a simultaneous learning and control process. The main concern is the learning of an inverse dynamic model of the robot mechanism for tasks with holonomic constraints, where exact robot dynamics is mostly unknown. The neural network, as part of the feed-forward control, produces the required driving torques at robot joints as a nonlinear mapping of the robot's desired internal coordinates, velocities, and accelerations (Katic and Vukobratovic 2003):

$$P_i = g\left(\omega_{jk}^{ab}, q_d, \dot{q}_d, \ddot{q}_d\right) \quad i = 1, \ldots, n \qquad (13.1)$$

Where $P_i \in R^n$ is the joint driving torque generated by the neural network, ω_{jk}^{ab} are the adaptive weighting factors between the neuron j in the ath layer and the neuron k in the bth layer, and g is the nonlinear mapping.

According to the integral model of robotic systems, the decentralized control algorithm with learning has the form:

$$u_i = u_{ii}^{ff} + u_{ii}^{fb} \quad i = 1, \ldots, n \qquad (13.2)$$

$$u_i = f_i(q_d, \dot{q}_d, \ddot{q}_d, P) - \text{KP}_{ii} e_i - \text{KD}_{ii} \dot{e}_i - KI_{ii} \int e_i dt \quad i = 1, \ldots, n \qquad (13.3)$$

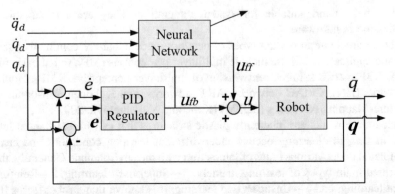

Fig. 13.1 Feedback-error learning design

Where f_i is the nonlinear mapping, which describes the nature of robot actuator model; $KP \in R^{n \times n}$ are the position, velocity, and integral local feedback gains, respectively; $e \in R^n$ is the feedback error.

An example for applying neural network in robotic can be found in Yu and Rosen (2013), where the neural Proportional Integral Derivative (PID) control is designed. An upper limb 7-DOF exoskeleton with this neural PID control has been used for experimental work. The objective of neural PID control is to make the transient performance faster and with less overshoot, such that humans feel comfortable. The robot in joint space has been regarded as in free motion without human constraints. Semiglobal asymptotic stability of the neural PID control and local asymptotic stability of the neural PID control with a velocity observer are proved with standard weight training algorithms.

13.2.2 Fuzzy Logic-Based Control for Robotics

Fuzzy logic has proved to be beneficial in cases where the process is too difficult to be analyzed by conventional quantitative techniques, or where the available information is qualitative, imprecise, or unreliable. Bearing in mind that it is based on precise mathematical theory, fuzzy logic also offers a possibility of integrating heuristic methods with conventional techniques for the analysis and synthesis of automatic control systems, consequently facilitating further refinement of fuzzy control-based systems (Katic and Vukobratovic 2003).

At the heart of the fuzzy set theory is the concept of fuzzy sets, which are used to model statements in natural (or artificial) language. Fuzzy sets are a generalization of the classical (crisp) sets. The classical set concept assumes the possibility of dividing the particles of some universe into two partitions: particles that are members of the given set, and those which are not. This partition process can be described by means of the characteristic membership function. For a given universe of discourse X and the given set A, the membership function μ_A (.) assigns a value to each particle $x \in X$ o that:

$$\mu_A(x)\begin{cases} 1 & \text{if } x \in A \\ 0 & \text{otherwise} \end{cases} \tag{13.4}$$

With fuzzy sets, the boundary between the members and nonmembers of the set is not strict. This softening of the boundary is defined mathematically using the membership degree junction, which assigns to each particle a value that indicates the degree of membership to the given set. Accordingly, the fuzzy set $\tilde{A}$ in the universe of discourse X is defined by its degree of the membership function $\mu_{\tilde{A}}$ (.) of the form:

$$\mu_{\tilde{A}} : X \to [0, 1] \tag{13.5}$$

A choice of basic fuzzy set operations has to be made by considering the context in which these operations shall be carried out. The most often used set of basic standard operations of fuzzy set theory is:

Complement : $\mu_{\tilde{A}}(x) = 1 - \mu_{\tilde{A}}(x)$

Union : $\mu_{\widetilde{A \cup B}}(x) = \max \left[\mu_{\tilde{A}}(x), \ \mu_{\tilde{B}}(x) \right]$ Intersection : $\mu_{\widetilde{A \cap B}}(x)$

$= \min \left[\mu_{\tilde{A}}(x), \ \mu_{\tilde{B}}(x) \right].$

The fuzzy set theory that is based on the operators defined this way is usually referred to as possibility theory. The fuzzy relation is a generalization of the classical concept of relation between elements of two or more sets. Additionally, fuzzy relations allow one to specify different levels of strength of association between individual elements. The levels of association are represented via degrees of membership to the fuzzy relation, in the same manner as the degree of membership to a fuzzy set is represented.

Fuzzy logic is a discipline engaged in formal principles of approximate reasoning. Contrary to classical formal systems, fuzzy logic allows one to evaluate the truth of a proposition as, e.g., a real number in the interval [0, 1]. The basis of fuzzy logic is the theory of fuzzy sets. In order to enable work with imprecise propositions, fuzzy logic allows the use of Fuzzy predicates, Fuzzy truth values, Fuzzy quantifiers, Fuzzy modifiers.

Fuzzy control approaches the control problem in a fundamentally different way compared to the traditional model-based techniques. Instead of precise mathematical models, fuzzy control uses an imprecise and incomplete description of the process and/or the manner in which the system is controlled by human operators, whereby the theory of fuzzy sets is used as a principal tool. The Fuzzy controller consists of four basic components: a condition (fuzzification) interface, a knowledge base, an inference mechanism, and an action (defuzzification) interface.

The control rules frequently used in fuzzy controllers are of the type:

R_k : if $(E$ is A_k and ΔE is B_k) then $U = C_k$.

Where E represents the value of the error e, ΔE represents the error change Δe between successive operation cycles of the controller, and U represents the fuzzy control action that is transferred to the action interface, which in turn generates the control signal u.

Merging fuzzy logic-based control with analytic methodologies exploits the advantages of both approaches in real-time robot control. For force control and hybrid position/force control in robotics, the basic principle of application of fuzzy logic is the same as in the case of position tracking control, except for the input level, where the space of input variables is extended with force tracking errors. The conventional force controllers give acceptable results whenever the parameters of the robot environment are recognized and fixed. Nevertheless, this assumption in most cases is not satisfied, so that it is necessary to adapt the control structure and parameters of robot controllers with respect to the changing environment. One of the techniques for solving these problems is the application of fuzzy logic. In the process of control synthesis, there are many restrictive assumptions connected with the availability of force derivation and the possibility to measure the velocity of the robot end-effector, and these assumptions are effective only when the estimations of environment parameters are close to their real values. However, in practical application, information about force derivation contains measured noise, and the accuracy of position and velocity of the end-effector are not sufficient for satisfactory force control. Also, the range of parameters of the working environment can be very wide because, for example, the stiffness coefficient can vary in a large span, depending on whether the environment is soft or hard. The purpose of the fuzzy scheme of force control (Fig. 13.2) is to achieve exact force regulation when the stiffness of the working environment is an unknown with great variation, using no restrictive assumptions.

13.2.3   Genetic Algorithms in Robotic Systems

These algorithms utilize an iterative approach in solving search problems. GAs represent global search algorithms based on the mechanism of natural selection and natural genetics. This mechanism is based on a Darwinian-type survival-of-the-fittest strategy with reproduction, where stronger individuals in the population have a higher chance of creating an offspring. They are population-based search techniques that rely on the information contained in a broad group of candidate solutions to solve the problem at hand. New populations of candidate solutions are generated by implementing operators inspired by natural genetic variation. The three most popular operators used in almost all GAs, are as follows: (1) selection, (2) recombination (often termed crossover), (3) mutation. These three operators produce an efficient search mechanism that generally converges rapidly to near-optimal solutions. The cycle of evolution is repeated until a desired termination

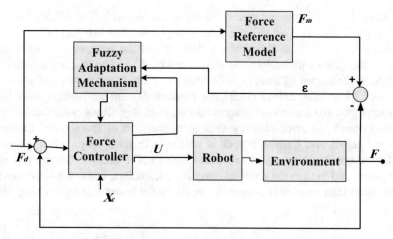

Fig. 13.2 Force control based on fuzzy logic

criterion is reached. This criterion can also be set by the number of evolution cycles, the amount of variation of individuals between different generations, or a predefined value of fitness.

Robot controller parameters are usually determined by trial and error, through simulations, and experimental test. Hence, in such cases, GAs were introduced as an alternative to the hand design of robot controllers, especially for autonomous robots acting in uncertain and noisy domains. The paradigm of genetic algorithms appears to offer an effective way for automatically and efficiently searching for a set of controller parameters, yielding better performance (Katic and Vukobratovic 2003).

13.2.4   Hybrid Intelligent Approach in Robotic Systems

Although fuzzy logic can encode expert knowledge in a direct and easy way using rules with linguistic labels, it often takes a lot of time to design and tune the membership functions which quantitatively define these linguistic labels. Wrong membership functions can lead to poor controller performance and possible system instability. An excellent solution is to apply learning techniques by neural networks, which can be used to design membership functions automatically, simultaneously reducing development time and costs and improving the system performance. These combined neuro-fuzzy networks can learn faster than neural networks. They also provide a connectionist architecture that is easy for very large-scale integration (VLSI) implementation to perform the functions of a conventional fuzzy logic controller with distributed learning abilities. Most of the proposed neuro-fuzzy networks are in fact Takagi–Sugeno controllers, where the consequent parts of linguistic rules are constant values (Katic and Vukobratovic 2003).

One of the most important methods is Adaptive-Network-Based Fuzzy Inference System (ANFIS). The system has a total of five layers. The learning rule is a hybrid method which combines the gradient descent and the least squares estimate to identify the ANFIS parameters, as shown in Fig. 13.3. The training algorithm of ANFIS, as developed by Jang (1993), takes the initial fuzzy model and tunes it by means of a hybrid technique combining gradient descent back-propagation in the backward pass and mean least squares optimization algorithms in the forward pass. At each epoch, an error measure defined as the sum of the squared difference between actual and desired output is reduced. Training stops when either the predefined epoch number or error rate is obtained. The gradient descent algorithm is implemented to tune the nonlinear premise parameters, while the basic function of the mean least squares is to optimize or adjust the linear consequent parameters.

13.3   Iterative Learning Control

The key principle behind the Iterative Learning Control (ILC) is to use current run error to adapt the control signal that will be applied to the system in the next run in order to reduce the tracking error (Mainali et al. 2004). Figure 13.4 shows a typical ILC configuration where the update is based on the tracking error (Ahn et al. 2007)

$$u_{k+1}(t) = u_k(t) + F_{\ln}\left(\frac{d}{dt}\right)e_k(t) \qquad (13.6)$$

Where $u_k(t)$ and $e_k(t)$ are the feed-forward control input and the tracking error of the kth trial, respectively, and $F_{\ln}$ is a linear filter, which performs a filtering operation on the tracking error.

A major application area for ILC comprises robotic operations, such as a gantry robot in a processing application that collects an object, or payload, from a given location, transports it over a finite duration and places it on a conveyor under synchronization, before returning to the same location, and then repeating the procedure as many times as required. Recent improvements in outcome measures have led to use of ILC being recognized as a significant advance by end users, that is, practitioners in healthcare (Freeman et al. 2012).

Another important application of ILC is developed in Zhao et al. (2015) in order to correct a preplanned path of an industrial robot. ILC is used to identify the robot kinematic parameters along the path in a local working zone. In particular, the correction of the path input for an industrial robot controller is derived by learning the actual robot kinematic parameters via ILC. By means of the past path-tracking information, the ILC method will gradually improve the performance of path tracking, thus precise path tracking can be obtained under certain conditions.

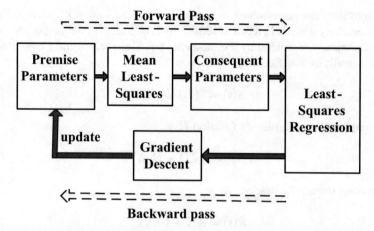

Fig. 13.3 ANFIS learning using hybrid techniques

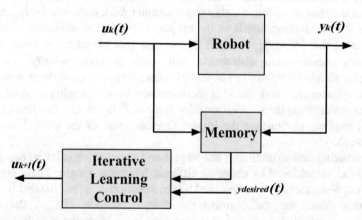

Fig. 13.4 Basic ILC configuration

13.4 Robust Control

The control objectives of high resolution and enhanced robust stability against different uncertainties and large improvement in tracking performance can be achieved by suitable formulation of the robust control design problem.

13.4.1 Robust H_∞ Controller

A typical procedure of designing an H_∞ controller is organized as follows: (1) setting weighting functions, (2) constructing a generalized plant, and (3) calculating an

H_∞ controller. The performance analysis of a standard feedback system is defined in the frequency domain in terms of sensitivity functions at system input and output. These functions are defined by the following equations (Zhou and Doyle 1998).

The sensitivity function $S(s)$:

$$S(s) = 1(1 + G_N K)^{-1} \tag{13.7}$$

The complementary sensitivity function $T(s)$:

$$T(s) = G_N K (1 + G_N K)^{-1} \tag{13.8}$$

And control sensitivity function $R(s)$:

$$R(s) = K(1 + G_N K)^{-1} \tag{13.9}$$

Where G_N is the nominal transfer function of the system. The robust optimal control problem consists of finding a stabilizing controller $K(s)$, such that the H_∞ norm of the sensitivity function matrix at the output, $\|S(j\omega)\|_\infty$ is minimized. However, minimizing the $\|S(j\omega)\|_\infty$ over all frequencies will increase sensitivity to high frequency measurement disturbance and leads to poor stability. Therefore, $\|S(j\omega)\|_\infty$ should be minimized over only where largest magnitude of disturbance occurs. Moreover, good tracking performance to a changing desired output requires maximizing the smallest singular value of $T(s)$. On the other hand, optimal control requires minimizing the largest singular value of the control sensitivity function $R(s)$.

Minimizing and maximizing the largest and smallest singular values (or H_∞ norms) can be achieved by choosing different frequency ranges. These ranges for various optimizations can be specified by the selection of three different frequency weighting functions: the performance weight function $W_e(j\omega)$ the control weighting function $W_u(j\omega)$ and the uncertainty weighting function $W_a(j\omega)$ such that the H_∞ norm of the mixed sensitivity functions is minimized (Tewari 2002). $S(s)$ and $T(s)$ are dependent and must be considered together when choosing weights. The standard H_∞ control problem is to find a $K(s)$ which satisfies:

$$\|F(s)\|_\infty \leq \gamma \tag{13.10}$$

Providing internal stability of the closed-loop system composed of P and K, where ‖.‖∞ represents the H_∞ norm of the transfer matrix $F(s)$, and γ is a positive number. The H_∞ controller can be found by iteratively solving two Riccati equations (Zhou and Doyle 1998).

Many applications of robust control for robotic systems can be found in literature, as in the following two examples; in Siqueira et al. (2011), a combined computed torque plus H_∞ linear control formulation is applied to a robotic manipulator. First, the computed torque method is used to precompensate the dynamics of the nominal system. Then, the H_∞ controller is used to postcompensate the residual error which is not completely removed by the computed torque method. The combined controller is able to perform robust tracking control.

Another application where the control design method for motion control in a global coordinate system of an oscillatory-base manipulator had been developed in Sato and Toda (2015). A sensor error and actuator saturation are explicitly accommodated in the control system design for practical implementation. To cope with practical constraints, an additional weighting function for H_∞ control design and an optimization scheme for PD gains are exploited.

13.4.2 Robust Sliding Mode Control (SMC)

The main focus of sliding mode method is how to adjust certain input control of nonlinear plant that causes variable state system to reach the equilibrium point (Setiawan et al. 2000). To do this we choose a manifold or certain area so that once the state system enters the area or manifold, it will slide to equilibrium point. SMC is a nonlinear control technique which provides an effective and robust means of controlling a nonlinear system in the presence of uncertain parameters and is often used to suppress errors and provide closed-loop robustness. SMC is often the favored basic control approach, especially because of the insensitivity property toward the parametric uncertainties and external disturbances. SMC design consists of defining a sliding or switching surface, which is rendered invariant by the action of switching terms. The invariance of the switching surface implies asymptotic stability of the tracking error in spite of unmodelled dynamics and lack of knowledge of model parameters (Canudas et al. 1996).

The salient advantages of sliding mode control include the property of robustness to structured and unstructured uncertainties once the system enters the sliding mode. Note, however, that system robustness is not assured until the sliding surface is reached. In a variable structure control (VSC) system, control laws are designed to drive the system states toward a specific sliding surface. As the sliding surface is hit, the system response is governed by the surface; consequently, robustness to uncertainty or disturbance is achieved. However, it is known that there is a major drawback in the SMC approach: undesired phenomenon of chattering due to high frequency switching, which often excites undesired dynamics (Hsu et al. 2001). VSCs are robust in the sense that they are insensitive to errors in the estimates of the parameters as long as reliable bounds on the parameters are known. To formulate a VSC law, it is helpful to recast the state equations of a system in terms of the tracking error and its derivative (Schilling 1996).

Finally, it should be mentioned here that sensor-based control approaches have proven to be more efficient than model-based control approaches when accuracy is required in robotized industrial applications. Accordingly, some sensor-based control approaches well adapted for servoing parallel robots involve the presence of virtual robots hidden into the controller, which have assembly modes and singularities different from the real robots (Geoffroy et al. 2014). Nevertheless, the analysis of the robot models hidden into the controllers developed by the sensor-based control community required the use of mathematical tools developed by the mechanical design community.

Every measurement has some random noise. Modelling that noise can result in systems of significantly higher performance. Such performance improvements may lead to improved safety, accessibility, and robustness (Kelly 2013). Optimal estimate can be produced from a continuous sequence of measurements. Maximum Likelihood Estimation, Recursive Optimal Estimation, and Extended Kalman Filter (EKF) are the most known hard working estimators in robotics.

13.5   Swarm Robotics System

Swarm robotics is concentrated on the management of decentralized, self-organized multirobot systems so as to designate such a cooperative behavior as a consequence of local interactions with one another and with their environment (Martin 2010).

General characteristics that should be considered for swarm robotics: robot anatomy, operation of a large number of robot members are that there should be decentralized control, robust and cheap in nature, high reliability is also key. Swarm robotics is fully distributed and self-organized (Al-dulaimy Ahmed Ibraheem 2012).

A swarmanoid (heterogenous robotic swarm) exploits the heterogeneity and complementarity of its constituent robot types to carry out complex tasks in large, three-dimensional (3-D), man-made environments. Humanoid robots are usually assumed to be the most efficient robot type for man-made environments. One of the goals of the swarmanoid project was to refute this assumption (Dorigo et al. 2013).

Several swarm robotics applications have been reported in literature, like in coverage, networking, olfactory-based navigation, and collective behavior. Swarm robotics can be classified according to the algorithms that have been used to implement them; dispersion algorithm has been used on iRobot swarm. It has been divided into two subalgorithms. The first subalgorithm disperses robots uniformly and the second subalgorithm explores boundaries, these distributed localization and mapping algorithms have been used to move the iRobot swarm through an indoor environment and generate a map. Swarm robotics is based on particle systems such as particle swarm optimization (PSO) and ant colony optimization (ACO) (Al-dulaimy Ahmed Ibraheem 2012).

In order to remedy the deficiency of heterogeneity, an interesting swarmanoid had been proposed in Dorigo et al. (2013). The swarm robotics system composed of three different robot types with complementary skills, as shown in Fig 13.5: foot-bots are small autonomous robots dedicated for moving on both even and uneven terrains; hand-bots are autonomous robots that have the ability of climbing vertical surfaces and handling small objects and eye-bots—autonomous flying robots that can attach to an indoor ceiling. The hand-bot has no autonomous mobility on the

Fig. 13.5 The swarmanoid robots: (**a**) three foot-bots are collected around a hand-bot and are ready for cooperative transport; (**b**) an eye-bot attached to the ceiling. It has a bird's-eye view of the environment (Dorigo et al. 2013)

ground and must be carried by foot-bots to the location where it can climb and grip objects of attention; eye-bots are capable of analyzing the environment from an advantaged location to collectively gather information unreachable to foot-bots and hand-bots, they retrieve significant information about the environment and transfer it to robots on the ground.

13.6 Case Studies

13.6.1 Robot Manipulator Online Path Planning and Obstacle Avoidance Using Fuzzy Logic

13.6.1.1 Joint Space Path Planning

The fuzzy joint space planning (Raheem and Gorial 2013) consists of two blocks for two-link robot planar as example. Each fuzzy block (FB) plans the motion of each robot link separately. The first block for first link produces $\Delta\theta_1(i+1)$ depending on the first input $\Delta\theta_{1g}(i+1)$ the error between the goal value and the current value of θ_1 and on the current value $\theta_1(i)$ which represents the second input to the first fuzzy block. Similarly, the output of the second fuzzy block is $\Delta\theta_2(i+1)$ depending on the first input $\Delta\theta_{2g}(i+1)$, the error between the goal value and the current value of θ_2 and on the current value $\theta_2(i)$ which represents the second input to the second fuzzy block (Raheem and Gorial 2013). Figure 13.6 shows structure of the proposed motion planning for a two-link robot.

Simulations have been carried out to test this planning system as follows: The lengths of the robot links are $L_1 = L_2 = 0.45$ m. The robot has to move from the start configuration $\theta_1 = 170^0, \theta_2 = 30^0$) to the goal configuration $\theta_1 = 10^0, \theta_2 =$

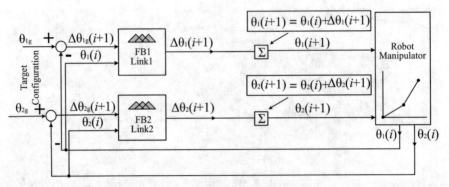

Fig. 13.6 Two-link robot joint space path planning using fuzzy logic

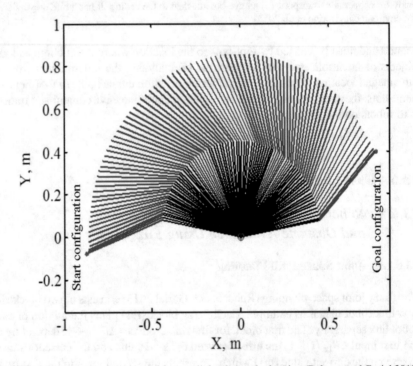

Fig. 13.7 Joint space fuzzy logic used for two-link robot path planning (Raheem and Gorial 2013)

35^0). The path planning is shown in Fig. 13.7. The error in reaching the goal after 363 program iterations was as follows: $0.014087\,^\circ$ for θ_1 and $-0.040168\,^\circ$ for θ_2 (Raheem and Gorial 2013).

13.6.1.2 Cartesian Space Path Planning

The block diagram structure of the motion planning system for a two-link robot (FB means Fuzzy Block) is shown in Fig. 13.8. The suggested fuzzy structure consists of two blocks for two-link planar robot (Raheem and Gorial 2013). The first block for the first robot link produces $(\Delta x(i+1))$ which represents the required change in x-axis for planning the robot motion to the next point $(i+1)$. This is based on the new x-axis error $e_x(i+1)$ and the current value of the robot gripper position $x(i)$Similarly the output of the second fuzzy block $(\Delta y(i+1))$, which represents the required change in y-axis for planning the robot motion to the next point at $(i+1)$, and in turn depends on the new y-axis error $(e_y(i+1))$ and the current value $(y(i))$ of the robot gripper position. A block of the inverse kinematics (IK) is needed to generate the new robot joint variables $(\theta_1(i+1), (\theta_2(i+1))$ from the new Cartesian point $(x(i+1), y(i+1))$ using the following set of equations:

$$D = \frac{x^2 + y^2 - L_1^2 - L_1^2}{2L_1 L_2}; \ \theta_2 = a \ \tan 2\left(D, \pm \sqrt{1 - D^2}\right);$$
$$\theta_1 = a\tan 2 \ (L_1 + (L_2 \ \cos(\theta_2)), \ L_2 \ \sin(\theta_2))$$

The block of the forward kinematics (FK) is needed to convert the current joint variable $\theta_1(i)$ and $\theta_2(i)$ to the desired current position of the robot gripper which represents the main feedback in the suggested planning system:

$$x = L_1 \cos(\theta_1) + L_2 \cos(\theta_1 + \theta_2), \ y = L_1 \sin(\theta_1) + L_2 \sin(\theta_1 + \theta_2).$$

Computer simulation has been done to test this system, as shown in Fig. 13.9. The robot has to move from the start point $(x = -0.87$ m, $y = -0.07$ m) the goal point $x = 0.76$ m, $y = 0.39$ m).The length of the robot links are $L_1 = L_2 = 0.45$ m. Figure 13.14 shows the path planning. The error in reaching the goal point after

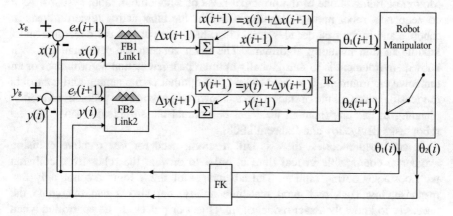

Fig. 13.8 Fuzzy logic block diagram structure for online Cartesian space path planning of a two-link robot manipulator (Raheem and Gorial 2013)

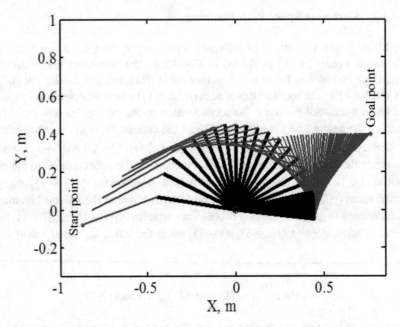

Fig. 13.9 Two-link robot manipulator Cartesian space path planning using fuzzy logic (Raheem and Gorial 2013)

(280) program iterations was as follows: $e_x = 0.00041775$ m and $e_y = -2.8386 \times 10^{-5}$ m. Moreover, no oscillatory motion of robot end-effector has been observed or recorded near the goal point (Raheem and Gorial 2013).

13.6.1.3 Obstacle Avoidance Using Fuzzy Logic

Motion planning is one of the principal tasks of autonomous robot systems. As a consequence, robot motion planning is one of the most active research areas in robotics and in the past decades great effort has been put into the development of flexible motion planning algorithms. The main concern for path planning in a known environment is to find globally optimal path navigation algorithms. For an unknown environment, it is good to said that optimal path planning and control is not feasible since all the obstacles are unknown and solving the problem of motion planning in this environment need not be optimal but sufficiently satisfying the robot tasks (Bulgakov and Raheem 2008).

In advanced robotics, there is still a strong need for fast reactive collision avoidance computable in real time in order to prevent the robot from collision with obstacles during motion. The advantages of fuzzy logic are not only fast response, low cost, and good real-time ability, but also it can overcome the necessity to know the exact model of the object or process to be controlled when the fuzzy logic control is applied. In addition, it can meet the real-time requirements for robot motion planning. A separate fuzzy control unit is usually built for online planning and control of each robot manipulator link in unknown environment (Bulgakov and Raheem 2008).

In this case study, a structure of fuzzy control for solving the problem of an online obstacle avoidance of robot manipulator in unknown environment was proposed. The suggested fuzzy controller consists of two stages and four control blocks for a two-link robot planar as example. Each fuzzy stage controls one robot link separately. Figure 13.10 shows structure of the proposed controller for a two-link robot. The fuzzy blocks FB1 for the first link and the second link are fully identical since their task to generate a step of a motion while a different blocks of FB2 for the first and second links to generate the required change in joint angle. Where $(\Delta\theta g_1(i+1))$ $(\Delta\theta g_2(i+1))$; the difference between the current joint angle and the goal joint angle $\Delta\theta_1(i)$; the previous value of $(\Delta\theta)$ for every link. At time zero the initial conditions $(\Delta\theta_1(0))$ $\Delta\theta_2(0) = 0)$. $S\theta_1(i+1)$, $S\theta_2(i+1)$., the output of the first stage equals a suggested step of motion for every link. d_1, d_2; the minimum distance between the first and the second link with the nearest obstacle. $\Delta\theta_1$ $(i+1)$, $\Delta\theta_2$ $(i+1)$; the output of the second stage and equal to the required change of joint angles $(\Delta\theta_1, \Delta\theta_2)$ (Bulgakov and Raheem 2008).

For a three links and higher, we can apply this controller by increasing the number of internal blocks. The internal feedback gives the ability of producing a step of motion proportional to the goal-joint error and the last value of $((\Delta\theta_j))$. It means when the robot link is near the goal point or when the robot faces an obstacle the produced step will take into account the current configuration of the robot and the situation of environment (Bulgakov and Raheem 2008).

The results of the computer simulation are shown in Fig. 13.11, where the robot has to move from the start configuration $(\theta_1 = 60°,\ \theta_2 = 25°)$ to the goal configuration $(\theta_1 = 180°, \theta_2 = 60°)$. The error in reaching the goal after (554) program iterations was $0.013934°$ for θ_1 $0.014194°$ for θ_2 The four obstacles were successfully avoided and the first link was successfully stopped moving before collision with obstacle number four. Computer Simulation points out that there is an oscillation in motion of the second link. It occurs when this particular link begins to move from two adjacent obstacles till it succeed to pass this situation toward the target. Figure 13.12 shows the graphs of d_1, d_2, $S\theta_1$, $S\theta_2$, $\Delta\theta_1$, $\Delta\theta_2$, θ_1, and θ_2. The results show that the structure of fuzzy logic blocks was suitable for giving a fast reactive action to solve the problem of robot motion planning and control in unknown environment. The fast fuzzy reactive output can be computed in real time. Fuzzy logic structure for online control consists of two stages and four fuzzy blocks. The simulation results and the practical results in Bulgakov and Raheem (2008) show that the robot successfully avoided all the unknown obstacles. Choosing the type of the distance sensors for finishing the experimental results was very important since it affects the real-time computation, accuracy of measurement, and the ability of better avoiding any unknown obstacles (Bulgakov and Raheem 2008).

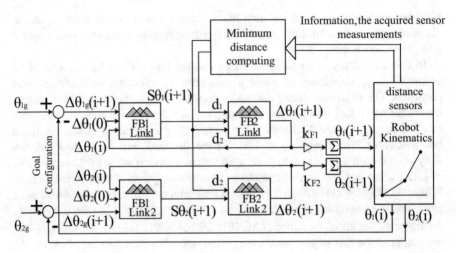

Fig. 13.10 The structure of the fuzzy controller for a two-link robot (Bulgakov and Raheem 2008)

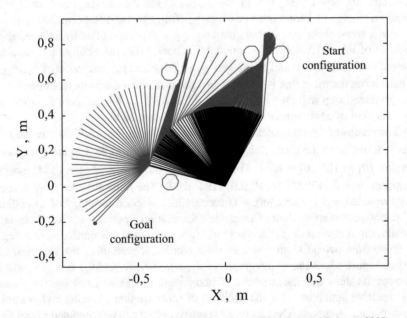

Fig. 13.11 Two-link robot manipulator Obstacle avoidance (Bulgakov and Raheem 2008)

13.6.2 Optimal Collective Search Behavior for the Swarm of Robots Under Risk

Decision-making models are based on rational behavior and risk aversion; they lack capabilities for implementing actual human decision-making behavior under risk. In Al-dulaimy Ahmed Ibraheem (2012), an efficient decision-making model had

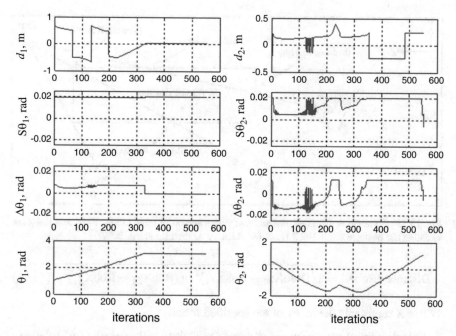

Fig. 13.12 Simulation results (Bulgakov and Raheem 2008)

been proposed to achieve the collective search behavior for the swarm of robots in a risky environment. The framework captures common human decision-making attitude toward risk aversion and risk seeking which are vital for handling the risk of violating environment constraints. The path planning of the swarm robots from the start position to the destination goal in a constrained environment (i.e., with landmines, fire, chemicals, nuclear waste, etc.) implies the risk of colliding with the environmental constraints, being delayed when maneuvering around them or ending up in a deadlock. In the proposed framework, the constraints of the path-planning problem have been used to generate the probability of the risk of violating system constraints during the evolutionary process. Moreover, those constraints are presented as obstacles in the environment and they are stationary with different sizes and locations. In the framework of Al-dulaimy Ahmed Ibraheem (2012), to show how to generate the probability of violating system constraints, with three swarm robots, one polygon obstacle, and initial and goal locations as shown in Fig. 13.13. For simplicity's sake, we will assume that robot1, robot2, and robot3 are referred to as Rob1, Rob2, and Rob3, respectively. Figure 13.13 shows that Rob1 and Rob2 are violating the polygon obstacle in (Rob1x1, Rob1y1), (Rob1x2, Rob1y2) and (Rob2x1, Rob2y1), (Rob2x2, Rob2y2), respectively, while Rob3 is not. The probability of violating system constraints is obtained as follows:

1. Find the distance value between (RobXx1, RobXy1), (RobXx2, RobXy2) for each robot in the swarm that has violated the obstacle using, for example, an Euclidean distance function:

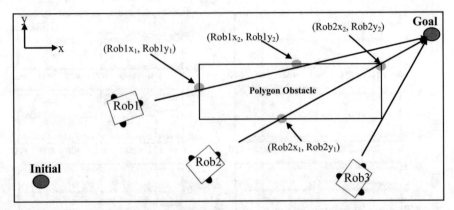

Fig. 13.13 A simple constrained environment with a swarm of three robots: Rob1 and Rob2 are violating the obstacle while Rob3 is not (Al-dulaimy Ahmed Ibraheem 2012)

$$\text{Distance (RobX)} = \sqrt{(\text{RobXx}_2 - \text{RobXx}_1)^2 + (\text{RobXy}_2 - \text{RobXy}_1)^2} \quad (13.11)$$

Where X represents the index of the specified robot.

2. Calculate the algebraic sum of all robots' violations in the swarm (TotViolation):

$$\text{TotViolation} \sum_{i=1}^{N} \sum_{j=1}^{m} \text{Distance}_j(i) \quad (13.12)$$

Where N and m represent the swarm robots' size and the number of obstacles in the constrained environment, respectively.

3. Find the probability of violating system constraints p for each robot in the swarm using the following equation:

$$p(i)1 - (\text{Current Distance (RobX) /TotViolation}) \quad (13.13)$$

Rob1 and Rob2 violated the system constraints as shown in Fig. 13.13; therefore, infeasible path planning solutions have been obtained from Rob1 and Rob2, as compared to Rob3 which provides a feasible path-planning solution.

In this test, the obstacles' boundaries are presented in the following constraint equations:

$$\text{Obstacle } 1 = \begin{cases} x - 1 \geq 0 \\ x - 2 \leq 0 \\ y - 7 \geq 0 \\ y - 9.5 \leq 0 \end{cases}, \text{Obstacle } 2 = \begin{cases} x - 2 \geq 0 \\ x - 6 \leq 0 \\ y - 5 \geq 0 \\ y - 6 \leq 0 \end{cases}, \text{Obstacle } 3$$

$$= \begin{cases} x - 6 \geq 0 \\ x - 8.5 \leq 0 \\ y - 8 \geq 0 \\ y - 9 \leq 0 \end{cases}. \text{Obstacle } 4 = \begin{cases} x - 6 \geq 0 \\ x - 9.5 \leq 0 \\ y - 0.5 \geq 0 \\ y - 2.5 \leq 0 \end{cases}, \text{Obstacle } 5 = \begin{cases} x - 8 \geq 0 \\ x - 9 \leq 0 \\ y - 4 \geq 0 \\ y - 6 \leq 0 \end{cases},$$

$$\text{Obstacle } 6 = \begin{cases} x - 3 \geq 0 \\ x - 10 \leq 0 \\ y - 3 \geq 0 \\ y - 4 \leq 0 \end{cases}$$

Figure 13.14a, b and c show the path taken by three swarms of robots to converge to the same goal using the proposed framework (Al-dulaimy Ahmed Ibraheem 2012), swarm robots' collective search behavior and the best evolving process of the objective function value for presented case. Figure 13.14a shows three typical global paths planned by the swarm of robots, which are presented in three different colors to distinguish between them. The decision-making model provides a feasible solution for the global path-planning problem from the initial position (0, 0) to the goal position (10, 10) using swarm of 10 robots while maneuvering around the obstacles. The robots cooperated to find an optimal global path for the swarm in the environment where risk is associated.

13.6.3 Intelligent Robust VSC Control for Robot Trajectory Tracking

A common problem in robotics is trajectory tracking, in which a robot is required to follow accurately a continuous pathway. Such a task is mainly preprogrammed such that the arm positions are stored in the controller memory for later recall. Define the state vector as $x^T[e^T, \dot{e}^T]$, where $e = q_d - q$ for each joint, and define a time-varying sliding surface, $s(t)$, in the state space R^n by the vector equation $s(e, \dot{e}) = 0$ where:

$$s := \dot{e} + \lambda e \tag{13.14}$$

In which λ is a strictly positive diagonal constant matrix. As can be seen from Eq. (13.13), maintaining system states on the surface $s(t)$ for all $t > 0$ will satisfy the tracking requirements $q \to q_d$ and $\dot{q} \to \dot{q}_d$. Indeed $s \equiv 0$ represents a linear differential equation and it will force e and $\dot{e}$ to approach zero, given any bounded initial conditions $e(0)$ and $\dot{e}(0)$ Thus, we have transformed a second-order tracking

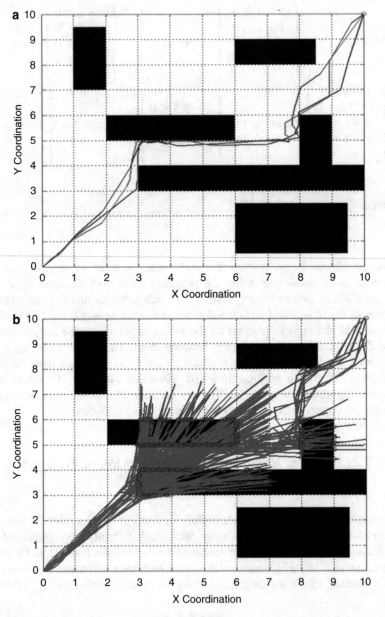

Fig. 13.14 (**a**) Three swarms robots path planning (Al-dulaimy Ahmed Ibraheem 2012), (**b**) Swarm of ten robots collective search behavior (Al-dulaimy Ahmed Ibraheem 2012), (**c**) Best evolving process of the objective function value (Al-dulaimy Ahmed Ibraheem 2012)

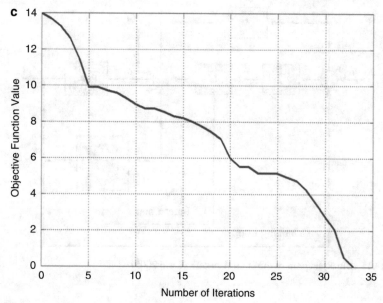

Fig. 13.14 (continued)

problem into a first-order stabilization problem. The control architecture employs Neural Network (NN)s to approximate the unknown functions in the VSC. One hidden layer NNs is considered. Radial-Based Function (RBF) NNs are used to adaptively learn the uncertain bounds. The Gaussian weight vectors of the RBF NNs are computed by an appropriate adaptation algorithm. The control input is designed such that (Raafat et al. 2009):

$$\delta\tau = \widehat{\theta}_D^{\mathrm{T}}(x)\ W_D(x)\ y'' + \widehat{\theta}_d^{\mathrm{T}}(x)\ W_d\ (x) - K_v\mathrm{sgn}\ (s) \qquad (13.15)$$

Where $\widehat{\theta}_D^{\mathrm{T}}(x)$ and $\widehat{\theta}_d^{\mathrm{T}}(x)$ are the weight vectors of the RBF W_D, and W_d are Gaussian type of functions defined as:

$$W_{Di}\ (x) = \exp\left(-\frac{\|x - \mu_{Di}^{\ 2}\|}{\sigma_{Di}^2}\right) i = 1, 2,\ \ldots,\ n_D \qquad (13.16)$$

$$W_{di}(x) = \exp\left(-\frac{\|x - \mu_{di}^{\ 2}\|}{\sigma_{di}^2}\right) i = 1, 2,\ \ldots, n_d \qquad (13.17)$$

Where μ is the center of receptive field and σ is the width of Gaussian function.
The sgn is the sign function:
$\mathrm{sgn}(s) = +1$ if $s > 0$
$\mathrm{sgn}(s) = -1$ if $s < 0$
K_v is the sliding gain given by the next expression

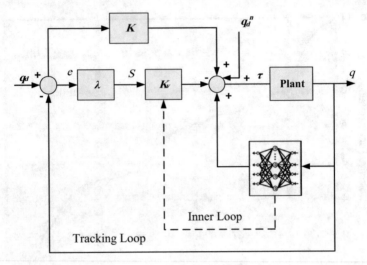

Fig. 13.15 Block diagram of the control system (Tariq 2005)

$$K_v = \varepsilon_D |y^n| + \bar{\varepsilon}_d + D_0^{-1}\zeta \tag{13.18}$$

ε_D and ε_d are the upper bounds of $|\varepsilon_D|$ and ε_d respectively, D_0 is the known dynamic part, ζ is a small positive constant, and the weight vectors are adjusted by using an adaptive mechanism. Accordingly, the output tracking error asymptotically converges to zero.

From Eq. (13.15) it can be also seen that the NNs used are essential to realize the nonlinear adaptive control law because the uncertainty bounds are unknown nonlinear functions. However, if these bounds are constants, the design of the controller and adaptive laws can be greatly simplified without using NNs. The overall structure of the controlled robotic system is shown in Fig. 13.15. Figure 13.16 is a typical result of applying intelligent robust VSC for three-axis SCARA where the load $m_L = 0$ kg and the gain $K = [200\ 25]$, each NN has 20 nodes and 20 weights. It is shown that a good tracking is obtained where the maximum tracking error is 4×10^{-3} rad, 4×10^{-3} rad, and 3 mm for joints 1, 2, and 3, respectively (Tariq 2005).

13.7 Conclusion and Further Directions

The recent mathematical and technological advances in robotics have paved the path for influential achievements, providing new challenges to the fields of robotics systems. Research in robotics continues to be extremely active. Development of control methods for a variety of robotics research areas are in the hot spot of recent research and contributions. It's our hope that this chapter will help to address the

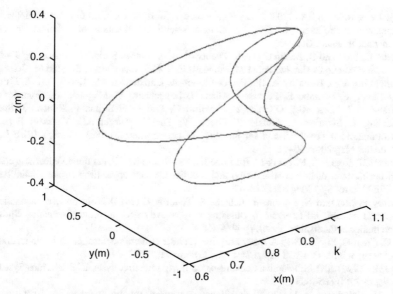

Fig. 13.16 Neural network-based adaptive robust control tracking performance for test 5 with $m_L = 0$ kg, $k_p = 200$, $k_d = 25$, the width of the Gaussian function $\sigma = 0.25$, 20 nodes and initial values equal to 0.5. Actual and desired trajectories (Tariq 2005)

most effective and continuously renewed topics in robotic control systems: intelligent control methods, robust control, Iterative Learning Control, and swarmanoid, and will provide a starting point for further focused discussion on these topics. In this chapter, we have also present few examples in order to study different robotics problems including path planning and obstacle avoidance using fuzzy logic, risk avoidance for swarm robots, and intelligent robust control for robot.

A great amount of research is still required to recognize various fundamental principles, in decision making, control, perception, learning, and to realize an action in open-ended situations. It should be mentioned here that in addition to the importance of assessing the performance of robotic systems, the probability of specifying safety and vital properties such as dependability and flexibility means robotics technology shall feature to be eventually recognized by nonexpert users into innovative application situations applicable across sustainability fields.

References

Ahn H-S, Chen YQ, Moore KL (2007) Iiterative learning control: brief survey and categorization. IEEE Trans Syst Man Cybern Appl Rev 37(6):1099–1021

Al-dulaimy Ahmed Ibraheem Abdulkareem (2012) A new swarm optimal collective searching behaviour framework using decision making under risk. Ph.D. thesis, University of Salford Manchester, pp 65–109

Bulgakov A, Raheem FA (2008) Fuzzy logic structure for on-line control of robot manipulator in unknown environment. In: 53rd international scientific colloquium of technical University Ilmenau, Ilmenau, Germany

Canudas C, Siciliano B, Bastin G (1996) Theory of robot control. Springer, London, pp 93–181

Dorigo M, Floreano D, Gambardella LM, Mondada F, Nolfi S, Baaboura T, Birattari M, Bonani M, Brambilla MM, Brutschy A, Burnier D, Campo A, Christensen AL, Decugnière A, Caro G, Ducatelle F, Ferrante E, Förster A, Guzzi J, Longchamp V, Magnenat S, Gonzales JM, Mathews N, Oca MM, O'Grady R, Pinciroli C, Pini G, Rétornaz P, Roberts J, Sperati Stirling T, Stranieri A, Stützle T, Trianni V, Tuci E, Turgut AE, Vaussard F (2013) Swarmanoid A novel concept for the study of heterogeneous robotic swarms. IEEE Robot Automat Mag 20(4):60–71

Freeman CT, Rogers E, Huges A-M, Burridge JH, Meadmore KL (2012) Iterative learning control in health care, electrical stimulation and robotic- assisted upper-limb stroke rehabilitation. IEEE Control Syst Mag 32(1):18–43

Geoffroy P, Mansard N, Raison M, Achiche S, Todorov E (2014) From inverse kinematics to optimal control. In: Lenarčič J, Oussama K (eds) Advances in robot kinematics. Springer International Publishing, Basel, pp 409–418

Hsu C, Chen G, Li X (2001) A fuzzy adaptive variable structure controller with applications to robot manipulators. IEEE Trans Syst Man Cybern 31(3):331–340

Jang JSR (1993) ANFIS: adaptive-network-based fuzzy inference system. IEEE Trans Syst Man Cybern 23(3):665–685

Katic D, Vukobratovic M (2003) Intelligent control of robotic systems. In: Tzafestas SG (ed) International series on microprocessor-based and intelligent systems engineering, vol 25. Springer Science+Business Media, Dordrecht, pp 21–161

Kelly A (2013) Mobile robotics mathematics, models, and methods. Cambridge University Press, Cambridge, pp 270–369

Mainali K, Panda SK, Xu JX, Senjyu T (2004) Position tracking performance enhancement of linear ultrasonic motor using iterative learning control. In: 35th annual IEEE power electronics specialists conference, Aachen, Germany, pp 4844–4849

Martin EM (2010) Swarm Robotics, from biology to robotics In-Teh Olajnica 19/2, 32000 Vukovar Croatia: V

Merlet J-P (2014) The influence of discrete-time control on the kinematico-static behavior of cable-driven parallel robot with elastic cables. In: Lenarčič J, Khatib O (eds) Advances in robot kinematics. Springer International Publishing, Basel, pp 113–121

Raafat SM, Said WK, Akmeliawati R, Tariq NM (2009) Improving trajectory tracking of a three axis scara robot using neural networks. In: IEEE symposium on industrial electronics and applications (ISIEA 2009), Kuala Lumpur, Malaysia, 4–6 October

Raheem FA, Gorial II (2013) Comparative study between joint space and cartesian space path planning for two-link robot manipulator using fuzzy logic. IRAQI J Comput Commun Control Syst Eng 13(2):1–10

Sato M, Toda M (2015) Robust motion control of an oscillatory-base manipulator in a global coordinate system. IEEE Trans Indus Electron 62(2):1163–1174

Schilling RJ (1996) Fundamentals of robotics analysis and control. Prentice-Hall of India, New Delhi, pp 289–296

Setiawan T, Widodo RJ, Mahayana D, Pranoto I (2000) Design nonlinear system with sliding mode control. In: Proceedings of the sixth AEESEAP triennial conference, Kuta, Bali, Indonesia, 23–25 August.

Siqueira AAG, Bergerman M, Terra MH (2011) Robust control of robots fault tolerant approaches. Springer, London, pp 17–24

Tariq NM (2005) Improving trajectory tracking of SCARA robot using neural networks. Master thesis, Control and System Engineering Department, University of Technology-Baghdad, p 82–124

Tewari A (2002) Modern control design with MATLAB and SIMULINK. Wiley, Hoboken, NJ

Yu W, Rosen J (2013) Neural PID control of robot manipulators with application to an upper limb Exoskeleton. IEEE Trans Cybern 43(2):673–684

Zhao YM, Lin Y, Xi F, Guo S (2015) Calibration-based iterative learning control for path tracking of industrial robots. IEEE Trans Indus Electron 62(5):2921–2929

Zhou K, Doyle JC (1998) Essentials of robust control. Prentice-Hall, Upper Saddle River, NJ 82–87:82–87, 269–282

Chapter 14
Prospects for Sustainability in Human–Environment Patterns: Dynamic Management of Common Resources

M. De Marchi, B. Sengar, and J.N. Furze

Abstract Models envisaged by policy makers and sustainability scientists cooperatively produce tools with desired practical implications. Models presented in the volume are summarized, with existing policy methods for resource management and Global perspective for resource planning methods in context. Models are discussed firstly in the framework of learning environments as an opportunity to test and practice sustainability, combining different tools from models, case studies, and scenarios. Sustainability asks for a multi-scale pragmatics; key elements of the multiple scale approach representing common ground in existing practice and frontier research have been identified. Two case studies are presented as a meta-summary of the issues presented in the book, being Rural South Asia (with special reference to India), and analysis of mega-diverse countries in Latin America, with special focus on Ecuador.

India's case study offers an account of policies implemented by the Ministry of Rural Development (MRD) and the Ministry of Tribal Affairs (MTA) since 1947, creating sustainable models for biodiversity and natural resource management. To comprehend the methods of resource sustainability usage in the present era, while enhancing resource utilization capacity of traditional practices, a tribal (indigenous communities) village—'Ajanta' (Western India)—is elaborated upon. The Latin America case study focuses on declinations of the *buen vivir* (good living) concept

M. De Marchi (✉)
Department of Civil, Environmental and Architectural Engineering, University of Padova, Via Marzolo 9, Padova 35131, Italy
e-mail: massimo.demarchi@dicea.unipd.it

B. Sengar
Department of History and Ancient Indian Culture School of Social Sciences, Dr. Babasaheb Ambedkar Marathwada University, Aurangabad 431004, India
e-mail: binasengar21@gmail.com

J.N. Furze
Faculty of Environment and Technology, University of the West of England, Frenchay Campus, Coldharbour Lane, Bristol BS16 1QY, UK
e-mail: james.n.furze@gmail.com

© Springer International Publishing Switzerland 2017
J.N. Furze et al. (eds.), *Mathematical Advances Towards Sustainable Environmental Systems*, DOI 10.1007/978-3-319-43901-3_14

319

and the debate between putting nature to the service of a nation (extractivism) and Yasunization, the neologism coined in Ecuador by a civil society seeking territorial innovation combining community engagement, biodiversity conservation, and environmental justice.

Through the macro-cosmic to micro-cosmic approach, a 'north-south', 'developed versus developing regions' dichotomy and possible areas of accord for resource management and sustainability in the global perspective is presented.

Keywords Ajanta • Biodiversity • Buen vivir • Heritage • Model • Scale • Yasuní

14.1 Global Trends in Natural Resource Management: "Urban-Rural Cosmos"

The most important discoveries of the twentieth century exist not in the realm of science, medicine, or technology, but rather in the dawning awareness of the Earth's limits and how those limits will affect human evolution. Humanity has reached a crossroad where ecological catastrophes meet what some call sustainable development. A great deal of attention has been given to what governments, corporations, utilities, international agencies, and private citizens can do to help in the transition to sustainability; little thought has been given to what schools, colleges, and universities can do. Ecological literacy queries how the discovery of finiteness affects the content and substance of education. Given the limits of the Earth, what should people know and how should they learn it (Orr 1992)? Changing human–environmental relations and resultant bio-cultural spheres remains a serious challenge in the social policy formation and political–economic policing. Throughout human socio-cultural, civilizations' emergence, and history, there have been conflicts and negotiations to reach out to a consensus, either through political force or peaceful talks to reach a solution which could mitigate human–environment–human (HEH) differences and bring a possible stability. Over approximately the last two centuries during which there was massive production of an industrial and post-industrial society, the human–environmental balance has surpassed its equilibrium; the HEH model is being challenged by the environmental–human–environmental (EHE) challenge; there is increased human pressure on one hand challenged by a resource shortage. Ever growing populations are placing unrealistic demands on land-food-water-vegetative sources. Conversely, we are being consistently threatened by Global pressures of climatic change, which increasingly threaten the cultural landscapes with unpredictable climatic cycles. Threatening environmental challenges are happening in the terrestrial biotopes (Naveh 1998), questioning our economic models of growth based on globalization, causing irregular vulnerabilities on the rural cosmos (Leichenko and O'Brian 2002), and booming urban heat island (UHI). These are outcomes of massive and rapidly growing urban centers/cities/metros (Masson et al. 2014). The challenges of EHE stress are results of insensitive human interventions to their own environments which have caused severe results. Consequences are devastating for nature; the direct victims will be or are humans. With loss of adequate lands, water resources, and vegetation, the human cultures sustenance is null and void. Restoration can only be granted by efforts on all levels from micro-level natural terrain, forests, rural spaces, and large urban settlements. Naveh (1998) raised the aspect of ecological restoration with great emphasis. Thus, the restorative process was meant to mend that which is lost or vulnerable to loss. Restorations, according to the studies by Naveh, are only possible through trans-disciplinary methods; one discipline cannot possibly give practical solution to the vast human miscreants.

Human–community participation in conservation, restoration, and preservation of the ecological and biodiversity spaces is an essential component of environmental sustainability. The dependency of the rural/forest-based communities on natural resources makes it inevitable for them to be part of the precautionary measures to be adopted for the resource sustenance. Inclusive policies for the prevention of biodiversity loss remain one of the integral social phenomena, forest policy formation in South Asia biodiversity, and environmental conservation (Agarwal 2001). The natives of a region carry legacy of co-habitation with a certain ecological surroundings, creating a habitat for themselves that makes them privileged communities. Inclusion of the native or indigenous communities not only creates a balanced growth model, but also helps the planners to bring about result-oriented framework during sustainable environmental planning of regions (Kjosavik and Shanmugaratnam 2004; Moor and Gowda 2014).

Today's challenges are the outcome of collective disasters imposed by humans on their environment with their historically implanted interventions and skills. Cure for many of these disasters are laid in humans' cultural evolutionary methods (Naveh 1998).

Developing a model for human–environmental sustainability in the cultural landscapes requires the restoration of biological, ecological, and cultural diversity, together with structural and functional landscape integrity and heterogeneity—as total landscape eco-diversity. Combined analysis of water relations and cultural diversity represents an effort to examine the complex role water plays as a force in sustaining, maintaining, and also threatening the viability of culturally diverse people. It is argued that water is a fundamental human need, a human right, and a core sustaining element in biodiversity and cultural diversity. The core concepts utilized in this book draw upon a larger trend in sustainability science, a recognition of the synergism, and analytical potential in utilizing a coupled biological and social systems analysis, as the functional viability of nature is both sustained and threatened by humans (Johnston and Hiwasaki 2011).

The present volume is a collective effort to envisage a holistic model for global challenges in EHE studied through all land, water, and atmospheric paradigms. The details of these models will be given in detail in Sects. 14.2 and 14.3.

14.2   Learning Environments for Sustainability: Models and Case Studies

Models presented in previous chapters can be collected as common threads adopting the perspective of the learning environment as an opportunity to test and practice sustainability. The challenge of sustainability requires better use of tools and opportunities. Models (from computer models to theoretical models) are key tools for research and policy design and belong to a wide learning environment including case studies (from local to global) where people experience innovation and scenarios, from narratives to more sophisticated experiments of anticipation science (Boyd et al. 2015; Scolozzi and Poli 2015).

The challenge of sustainability is deeply rooted in the ability to creatively manage the human attitude of modelling and visualization (Di Biase 1990; Andrienko et al. 2013). Models vary in different sciences and different fields of knowledge and practice: from law, to theories, from computer based to mathematic models, from spatial, numerical, and discursive ones.

Modelling is a constitutive human ability from human history to personal life history; a space where generations may meet either privately among parents and children or in social space creating a place for exchanging visions and attitudes.

The social role of models for sustainability is rooted in the synergy among policy development and model management flow; models represent cornerstones in the toolbox of new governance offering an immediate experience of transparency and visibility in decisions and future images of present choices. Models facilitate inclusive decision making processes including actors, exploring interests and conflicts, which shows options related to position of different actors.

From the second chapter, James Furze develops an interesting model exercise of plant distribution and biodiversity combining variables of the plant world (life-history strategy, primary metabolic type, life-form) with location (water-energy topography dynamic). The modelling exercise is not only a crossroad of multidisciplinary paths among botany, ecology, geography, mathematics, and computer science, but is also a combinatorial approach where mathematical analysis is integrated with graphical and spatial representation. In this chapter, the reader cannot avoid to be involved into the theoretical deepness of the work recalling the seminal contribution of Alexander von Humboldt. Humboldt coined scientific communication to citizens (Farinelli 1998), inventing a new geography to avoid which simultaneously narrated fantastic savages and monsters and cold and the aseptic communication of geometrical mapping. The new geography of Humboldt brought intellectual models of the complexity of the world, resulting from the interaction between humans and nature. The *Ansichten der Natur* (1808) is a way to prepare people in understanding diversity, to improve and change the world into a heterogeneous *Kosmos*, not just a uniformed space. Understanding the double task of humans as creators and spectators in this Earth as a theatre is imperative, where humans can both perform and enjoy the performance (Turri 1998). Humboldt's book *Ansichten der Natur* is a manifesto about what a model is and what it could do. The German *Ansichten* means vision, opinion, idea, notion, judgment, and has been translated in different languages with different sensibilities. The Italian scholar of Humboldt, Franco Farinelli, edited the book as *Quadri della natura* (1998), stressing the concept of *Ansichten* as a scene or a sight. Models are painted as something which facilitate figurative and prefigurative issues and visualize the decision making process.

A sort of invisible Ariadne thread connects the reflections of Chap. 4 with Humboldt's *Ansichten* of Chap. 2. The authors highlight the issue of model as "clear box", overcoming the image of the scientist as "owner" of the "black box". This becomes social actors arguing with other social actors about the suitability of specific models to share visions and to avoid increasing error of using the wrong models. Absolute ranking in models should not exist, because each model has "its special capabilities" and the scientist has a social responsibility in offering criteria to choose the right model. The chapter offers an invaluable contribution, summarizing a wide spectrum of models used in water systems from predictive and investigative ones. Watersheds are a further element of climate change. The effect of climate change requires serious consideration, and the use of modelling is justified and elaborated as discussed in Chap. 10 by Mohammed Zareian. In the context of water systems, GIS and Remote sensing add spatial dimension to models: again the issue of visualization recall the consolidation in the past 20 years of 3D participatory mapping, participatory GIS, and community cartography (Craig et al. 2002; Simao et al. 2009).

All model flow steps can exit the specialist rooms and be discussed in the square of *Polis*; the transparent box in modelling requires social inclusive modelling involving people and locale. Over the last 50 years, post-normal sciences exit the cold rooms of government where decision makers and scientists meet to tailor better decisions. Since 1992 and the acclaimed United Nations Agenda 21, sciences share the awareness involved in the new governance of sustainability, visualizing and desiring sustainability before implementing it. The roads opened in Rio 1992 and were confirmed in Rio 2012, challenging society to meet the 2030 sustainable development objectives. The role of a learning environment is in the context of post-civil society characterized by the crisis of political representation and the charismatic drift in governmental affairs where debates are mere political fights between majority and opposition (Jameson and Speaks 1992; Echeverria 1995). Challenges for today and the future are the role of science and learning environments for sustainability, caring for the spaces of public and open debate. The issues of modelling interactions among natural and social system to build community resilience are well discussed in Chap. 11 with respect to the challenge of climate change. The author highlights the role of modelling in building awareness, discovering and improving coping strategies across genders, and linking climatic fluctuations with the rule of governance.

We need to visualize sustainability before, during, and after dealing with sustainability in territories and in different sectors without forgetting relations, as Kelly Swing reminds us in Chap. 6 with the words of John Muir: "When we try to pick out anything by itself, we find it attached to everything else in the universe". Diversity is a key element to understand the world as a complex network of aspects, coherent unity, and as model of organization. From ethno-taxonomy to binomial

Linnean taxonomy, there is a human scientific bridge looking to build coherent knowledge models to grasp the plurality of diversity. Chapters 5 and 6 focus on biodiversity and conservation, offering a reflection of ecological and biological models (Mayr 1997). The importance of modelling is in the field of mathematical symbolization, computer modelling, organism, laboratory or case studies, and before all as a way of thinking. Biodiversity is a combination of variety, variability, and processes (from ecological to evolutionary) and represents a challenge for human knowledge, the unbalanced race between the rate of cataloguing new species and the rate of extinctions. The author stresses the concept of "known" and "unknown" applied to organism diversity and the difference between cataloguing species and understanding the ecological functions and human values: "less than 5 % of the world's flora has been scientifically screened for biochemical activity or potential medical applications". Loss of species is not just about total number, but also about "which particular species are disappearing".

Swing highlights the geographical nature of biodiversity: the majority of species live in the tropics (Novotny 2009), while at the same time, here land cover changes have higher impacts considering the greater concentration of species per hectare. The uneven distribution of biodiversity in terms of species for area unit is one of the aspects of species area relations (more space, more species), which is critically endangered by the processes of habitat fragmentation. The challenge to conservation starting from the point that both "nature and human nature are quite complex" has to do with the spatial organization where normally society looks for separation of the space for humans and the space for ecosystems. Despite its origins in 1969, Environmental Impact Assessment is not unfolding all its potential as a tool to improve environmental performance of decision making, setting developments suitable to nature. Design with Nature (McHarg 1969) still maintains its character of a precursor not yet fully adopted in the quest for sustainability implementation.

Combining biogeochemistry and ecology to adopt a complex approach modelling entire ecosystems instead of individual interactions is the challenging topic of Chap. 7. The author asks for cooperation among ecologists, biochemists, and mathematicians to integrate the different scales: from atoms to ecosystems. The challenge is to develop an idealized model system with a wide array of explanatory power either in the domain of processes or in the domain of landscape system combining macrocosm, mesocosms, and microcosm. Beyond scale, this chapter offers interesting reflections about boundaries in modelling ecological systems.

Biologists use specific organisms as models, for example plant metabolic pathways are explored in Chap. 8 by Hanan Hashem and Dioscorea are elaborated upon in neutraceutical applications in Chap. 9 by Sanjeet Kumar. Ecologists work with communities and environments. The choice of the appropriate ecosystems for modelling is placed in the debate among realism and replication. Replication needs small habitats with clear boundaries (a natural microcosm) where it is possible to replicate experiments adopting solid statistic analysis. In contrast, realism works on unique case studies (learning environments), with long durations, qualitative approaches, flexible boundaries, and looks to grasp the dynamics of complex systems. At the kernel of complexity science, there is the observer and the observation process. The observer frames the area of analysis and defines boundaries among issues to be considered and outside of research interests. In complex systems, criteria of analysis should be clearly stated at the beginning of the modelling exercise (Maturana and Varela 1979; Nir 1990; Prigogine and Stengers 1997). The issue of boundaries is a common challenge in different categories of ecosystems well-represented; for example, in industrial ecology looking for the area of influence in material flow of firms, cities, or regions (Newell and Cousins 2015). Contrastingly, uniqueness, reality, and lack of replication are common elements in case study approaches used in social sciences; observation of specific case and extension to wide systems are not something new and a wide array of tools have been consolidated (Hesse-Biber 2010; Tashakkori and Teddlie 2010; Torrance 2012).

Case studies help in managing the challenge of scale in modelling, but we are developing tools to bridge the scale from operations to territory facing the challenge of sustainability: so we open the next thread—scale.

14.3   Multi-scale Pragmatics Challenges and Solutions

To weave the previous articulated threads of models and the thread of scale, it is useful to recall two interesting models ('Digital Earth' and 'Building Information Model') which combine complexity, spatiality, visualization, and multi-scale approaches. Firstly, the 'Digital Earth' commonly available in our computer from different categories of virtual globes, which are simultaneously so common and as unknown in potentiality for public debate, education, research, and professional sustainability practice. Diffusion as a common tool is encouraged by the International Society of digital Earth. Virtual globes allow people to pan, display, and exit the rigid limitation of cartographic scale. To analyze changes in time, we choose the cosmos to focus from crossroads to the globe; we share position, itinerary areas, share pictures, share changes, share future visions (Goodchild 2000, 2012; Grossner et al. 2008). Each event as a place combining information is referred by a system of four-dimensional coordinates. The other declinations of digital Earth are digital video globes available in schools, museums, or expositions. Digital video globes are powerful tools to visualize the common world and to view events, past, and future changes shared in a community setting (Butler 2006). Digital Earth is not new, launched on 31/01/1998 by Al Gore at California Science Centre of Los Angeles; it was a challenge combining the technology opportunities and Earth as a real learning environment for global knowledge (Gore 1992, 1998). Computer development in hardware and software evolution helped the creation of tools like 'Keyhole Earth viewer' (2001), NASA World Wind (2003), Google Earth (2005), and many other virtual globes (Atzmanstorfer and Blaschke 2013). Concurrently, there has been evolution of hardware with digital video globes (Grossner and Clarke 2007), additionally the consolidation of the conceptual change before and during the construction of digital Earth and the foundation of International society of Digital Earth (2006). The challenge of the collaborative initiative between Earth sciences, Geographic Information Science, human and life sciences offers an opportunity of a social collaborative platform connected by Internet facilitating exchange of data and information, implementing a multi-dimensional, multi-scale, multi-temporal and multi-layered system (Craglia et al. 2012). The digital Earth allows the construction of a global virtual representation of the planet to model Earth systems, including cultural and social aspects represented by the human societies living on the planet (De By and Georgiadou 2014; Mahdavi-Amiri et al. 2015).

Building Information Modelling (BIM) is a process developed starting in the 2000s by which "a digital representation of a building's physical ad functional characteristics is created, maintained and shared as a knowledge resource" (Briscoe 2016). Principally a tool in the domain of architecture and engineering, it is the continuing evolution of modelling knowledge including the life cycle of a building (Miettinen and Paavola 2014; Rokooei 2015). It represents an 'Information Modelling and Management' developed to improve interoperability, coordination, quality, dialogue, and visualization; it is not just software (Wu and Kaushik 2015). BIM practitioners speak of 3D (classical three-dimensional representation derived from Computer Aided Design), 4D (inclusion of time in building operations from construction to maintaining), 5D (costs, economic and financial management of a building), 6D (the sustainability dimension of a building, at the moment limited to energy standards), and 7D (life cycle, from planning to decommissioning) (Jalaei and Jrade 2015; Liu et al. 2015). However, BIM is evolving and integrating with Infrastructure Information Management (IIM) for big operations (railway, water sewage, and others) and Landscape Information Management (LIM) (Blanco and Chen 2014; Briscoe 2016). We are living the transition and interoperability between BIM/IIM/LIM and GIS, developing frameworks and standardization. Clearly we enter a long trip; at the moment, preoccupations are about software, modelling authoring, standardization, definitions of interoperability. However, there is a potentiality not completely explored link with GIS, integrating human built environments with biological and ecological dimension of territories and secondly an opportunity to integrate visualisation and social inclusion (Gill et al. 2010; Zajíčková and Achten 2013). These two models open an interesting focus on scale: this conceptual tool is used in many social and natural sciences, sometimes used in different ways, but represents a possible common ground in shaping sustainability.

The concept of scale, apparently clear and unquestionable, can be split into three areas of complexity: the size, the level, and the relationships (Howitt 1998; Sayre 2005). Size is the most immediate issue that directly links territory and cartographical scale or world and models.

The issue of size is the first immediate contact people have with scale and models at the same time. As pointed by Mohammad Hadi Bazrkar in Chap. 4, scale models, or reduced representations of real world, are very common from children's toys to scientific and technical operations. Scale models of building, infrastructures, machines, transportation, and raised relief maps were very important tools before computer development. Now, the world of 3D printers gives new life to many scale modelling operations, allowing the physical manipulation of models after elaboration and visualization on the computer screen.

Size still remains a central point in scale approach in space and time, so distances appropriate for walking are commonly elements used in architecture and planning to design sustainable spaces, allowing people to perform the majority of human activities through proximity without using individual cars and fossil fuels. Sustainability asks for an inversion of the shrinking world due to innovations in transport "annihilating space trough time" (Harvey 1989, p. 241). From the world explored at 15 km/h by horse coaches and sailing ships to the revolution of steam power applied to transportation around the 1850s (100 km/h for steam locomotives, 50 km/h for steam ships) to the mass air transportation based in jet aircrafts (1000 km/h), the challenge of sustainable living needs to manage the schizophrenia between speed (Virilio 1997, 1993) and accessibility. Speed is more related with power and exclusivity, accessibility with inclusion, and justice.

What may appear as a purely technical and quantitative aspect assumes knowledge (and operational) significance because the choice of the size of the representation defines the portion of territory and the elements to be represented (Monmonier 2005). Size matters (De Blij 2009), as stressed by Bazrkar in Chap. 4 for water system modelling. Spatial size from local to regional (from the meters of soil profiles to thousands of km for catchment areas) and time period (from the second of flashflood events to the hundreds years of long-term representation) should be accurately selected, despite the calculation power of computers in order to build credible and understandable models and representations. Space grid size and time units are not neutral to the systems to be modelled and the different steps of the modelling process (data collection, conceptual modelling, mathematical modelling, calibration and validation). Up-scaling and downscaling are common processes which bridge the gap between modelling scale and organizational scale; however, these operations should take into account heterogeneity in space and time, and from a sustainability point of view, social implications and policy challenges. Size in modelling biodiversity is another key element in what can be considered "known" and "unknown": we seminally recall that Kelly Swing in Chap. 5 highlights how scientists know more about "organisms that exceed 10 cm or 10 g" and Sam Bonnett in Chap. 7 stresses how microbial community dynamics are ignored in many ecosystem-level models and "that less than 5 % of microbial species or less than 1 % of operational taxonomic units in soils are described".

Another important aspect related to scale is the level of analysis and the choice of level made by the researcher to organize the observational dimension of territorial dynamics (Zhang et al. 2014).

The level of analysis in sustainability research and policies can be influenced either by the hierarchy of political and economic meshes chosen (observation at district, province, country, continent level, or just local, national, global), or by ecological meshes (climate, ecosystem, river basin). Different types of analysis take predefined scale levels as a result of a scientific tradition or consolidated professional practices. Alternatively, analysis is made upon research customer requirements (Higgins et al. 2012) as is the case of an organism for biologists.

Sam Bonnett in Chap. 7 stresses the need to integrate scale and disciplines to grasp the different levels at which processes operate. The authors suggest involving ecologists, biochemists, and mathematicians in modelling wide ecosystems and not just small interactions. From one side, the arena of disciplines needs to be enlarged to social and territorial sciences and, from the other side, the multi-scale tools and the concept of mesocosm, suggested by the authors, can facilitate interdisciplinary and social dialogue. There are disciplines more at ease with microcosm and others more at ease with macrocosm, while inside disciplines the situation is almost scattered. However, a mesocosm located in the field with easy boundaries and with a manageable size can improve realism, scientific rigors, and social implementations. Mesocosm can bridge the different elements characterizing scale: size, level, and relations. Oceanic islands, so useful for ecosystem modelling (and to models of climate change impact for physical sciences notwithstanding policy and inclusive decision making), are interesting examples of mesocosms managing reasonable size and at the same time multiple levels of social and environmental relations, opening the door to the framework of nested systems (Maturana and Varela 1979, 1984; Koestler 1980; Roos and Oliver 1999).

The third element of scale is determined from a primary focus on relationships and the dynamics of the analyzed phenomena. The perspective of the relationship is typical of the geographical approach based on the vision of network taking a multi-scale look. This approach involves placing the study area in a multi-area context from local to biosphere (Collinge 2006; Marston 2000; Marston et al. 2005; Raven et al. 2012; Santos 2008). To forget a scale, or to avoid a scale of analysis, means mauling understanding of the complexities.

Sustainable development research should address the need to analyze different observational and organizational levels of territorial dynamics, adopting multiple representations at different sizes to account for the wholeness of the conceptual implications of territorial scales. We refer to the case highlighted in Chap. 7—middled out modelling. Top-down and bottom-up approaches, from scientific modelling to sustainability governance, have represented a battle field of theoretical and pragmatic points of view in framing knowledge and in implementing practices. Middle-out modelling can be either a starting point for observational and organizational scales or a meeting point from different paths of knowledge building and implementation of practices, with the awareness that sustainability challenges do not ask sustainability actors to defend positions, but to share visions and proposals from above, from below, from outside, and from the middle.

The goal is to produce tools (including maps) offering the view on different portions of the territorial reality and at the same time offering the story of the relationship between actors and places involved in the networking of scale's pluralities (Helfenbein 2010; Marston 2000; Montesuma Oliveira et al. 2011; Santos 2008). Now with the evolution of models, especially digital Earth and Information Modelling, we face this challenge by building good, descriptive, integrative, and explanatory models (see Chaps. 3 and 7).

14.4 Developing Regional Modelling: Case Study of a Heritage Site, Ajanta in India

Models envisaged by policy makers and sustainability scientists cooperatively produce tools with desired practical implications. We comparatively evaluate models presented in the volume, with existing policy methods for resource management and Global perspective for resource planning methods in developing nations per se. Models are discussed in the context of a case study of Rural South Asia (with special reference to India). Since 1947, under the Ministry of Rural Development (MRD) and the Ministry of Tribal Affairs (MTA), India's policies were made to create sustainable models for heritage, biodiversity, and natural resource management. Through the MRD and MTA, methods of organic farming, sustainable water, and energy management and collaborative approaches of human participation are essential in rural/tribal areas of India. In the given region, emphasis is laid on human support for resource management—historically, humans are guardians of natural resources. The expertise of a community in sustainable development of resource management increases if the said community owes a legacy of living in the region for historical times (Selin 1999; Agarwal 2011). The ownership and sharing of resources with community participation helps in alleviating various challenges which are posed in a developing economy of the likes of India (Sunderlin et al. 2008).

A prospective superpower, India is still grappling with risks that threaten to hamper its progress. These range from environmental threats caused by GM crops and pollution; dangers to health from HIV/AIDS and maternal–child morbidity and mortality; safety concerns about natural hazards, nuclear power, and industrial disasters; and challenges to livelihoods and values. Some of the issues that this volume explores are: what counts as an 'acceptable' risk, and who decides? How should divergent perceptions of risks be reconciled? Where is the line between science and politics? We attempt to breach the interface between policy formation and direction. Advocating a more multidimensional approach to managing risks, the authors challenge many of the dominant perspectives in India.

The field of risk research, which has emerged over the last 40 years in the West, has been relatively unexplored in India. To bridge this gap, this volume brings together Indian and scholars and practitioners across the fields of biological sciences and social sciences to work on a common strategy and framework as a solution for the challenges of the environmental issues. With inclusion of qualitative research and suggestions from biodiversity, forestry, and anthropology experts, it intends to frame the models of sustenance in various environmental areas of the developed and developing regions of the world (Sunderlin et al. 2008).

Studies in India/South Asia related to forest and community participation have given favorable results. These are substantiated through HEH correlations in forest management; Joint Forest Management (JFM) has given pro-environmental results substantiating involvement of the community (Agarwal 2011; Kjosavik et al. 2004; Behera 2006; Kumar 2002). Native communities' role in ecological sustenance, in prevention of deforestation and bio-diversity loss and conservation remains one of the most sought after solutions in the forest policy of India. Additionally, results of studies carried out by the Forest Departments in India (Agrawal and Chhatre 2006; Joshi 1999; Agrawal and Gupta 2005; Poffenberger 1996) indicate potential solution to the ecological crisis.

14.4.1 Suitability Policies of Ministry of Rural Development and Ministry of Tourism Administration: Evaluating 'Ajanta and Around'

Ajanta has a combination of UNESCO Buddhist-cave heritage sites located at the sanctuary 'Gautala-Outram', a prominent biodiversity zone of the central-Western India (Hidaka 2007). The region of the UNESCO Heritage site is found in the hub of the forest and the biodiversity zone (Singh and Anand 2013). The cave site of Ajanta and its associated environment of forest and mountain range, with its flora–fauna tract and rivulets, provide natural resources of water and vegetation for the local communities in Aurangabad, Jalna, and Jalgaon (Kshirsagar et al. 2012). Historically, the region has been a major source of natural resources and an area of conservation. Regard for the sustainable development of the entire region is dependent on Ajanta and its forest resources (Sengar 2016), since becoming a UNESCO heritage site in 1983. Around 1982, the region received special attention for its monuments and also for its biodiversity value. The Ministry of Rural Development (MRD), Ministry of Environment and Forestry (MoEF), and Maharashtra Tourist Development Corporation (MTDC) gave special emphasis for the governance and sustenance of the region. Special measures to maintain an environmental balance remain a major objective of the regions' environmental policy. The survival and sustenance of the Buddhist cave of Ajanta as an UNESCO heritage site is largely dependent on the well-being of the environment of the region (Doug 2014).

The main objective of this case study is to understand the inputs of the community and the State approach in maintaining an ecological balance. Ajanta is a fine example of human architectural heritage in excellence in the vicinity of a forest area, which is rich in biodiversity and local landscape. It has a fine balance of the State-Community participation in ecological sustenance for a heritage site, a region which is core to the biodiversity in river Godavari valley region in Central-Western India (Deshmukh 2008). The study undertaken elaborates an impact evaluation of MRD, MoEF, MTDC, and MTA policies and how these could be enhanced through global genres of resource planning and management. To understand resource sustainability, while enhancing resource utilization capacity of traditional practices; the tribal (indigenous communities) village—'Ajanta' (Western India)—is elaborated upon.

The broader region of Ajanta hills and forests comprises the Buddhist cave heritage sites with 30 rock cut caves constructed during second century BC to fifth century BC. The hills and forests in and around Ajanta are closely connected to the heritage and regional history of the cave heritage sites. The prominent villages of Lenapur, Savarkheda, Dutt wadi, and Ajanta are inhabited by the forest communities of Bhil, Koli, and Banjara clans. Since 1954, the villages in and around the Buddhist cave heritage sites have been governed by the MTA and Department of Forestry. Local government administration, policies protecting biodiversity, and the environs of the region were formulated (Sengar 2011). The composite region of Ajanta presents an integrated ecological site where existence and well-being of a heritage site are inversely proportional to the life and well-being of the ecology of the region (Bharti 2013; Deshmukh 2008).

Ajantas' rock cut caves are world famous for the tempera paintings which are carved on inner walls of each cave. The paintings of the Ajanta caves are made with locally available mineral resources and flora. History speaks through the technique utilized in the caves. The communities' and resources contribution in making this unique heritage site is unquestionable (Deshmukh 2008; Bharti 2013). However, changing times and increasing touristic pressure ignore the role of the local people and resources, leading to severe and rapid deterioration in the cave sites (Doug 2014; Deshmukh 2008). The caves, which were excavated in hill ranges, are composed of chaotic piles of irregular basalt flows. The lower parts of some of the flows have several joints. As a result, although the paintings on the flows have a smooth appearance, the statues and pillars along the joints have developed cracks. Seventeen panels in Ajanta and 117 in Ellora are deteriorating. The presence of chlorophaeite in the basalt rocks has also contributed to the degradation. Chlorophaeite absorbs moisture, resulting in the formation of thin scales on the rock surfaces. The scales fall off in summer, disfiguring the paintings; 13 panels in Ellora and 6 in Ajanta have deteriorated (Bharti 2013; Deshmukh 2008).

The proportion of harmful reactive chemicals in the atmosphere has increased substantially, due to uncontrolled industrialization in the Aurangabad and Jalgaon districts of the state (Deshmukh 2008). This affects the paintings in the caves and its associated environment, bringing an UNESCO heritage site under threat and increasing the vulnerability of a naturally rich biodiversity zone. Deshmukh suggests healing of the cracks in the rocks with epoxy resin and applying a thin chemical coat on the affected panels to constitute a humidity-proof transparent film over the paintings. The restoration of many of the components of the cave site of Ajanta requires a sustainable ecological environment of the region. This will be feasible if the native communities and the local environment can be revived and conserved to their original form. A further concern is sustenance of the water level and geological structures of the terrain (Bharti 2013).

Declining water resources and depletion of the forest cover are major regional challenges for the heritage and biodiversity sites of the region. The excessive intake of tourism depletes resources, increasing erosion from foot pressure of visitors, and results in less community participation in the ecological terrain. Native people of the region are threatened by the decreasing resources in terms of water, vegetation, and fauna; their survival and their threatened eviction will further challenge the maintenance of the heritage site (Sengar 2016; Doug 2014). There are colossal challenges for a heritage site in a biodiversity zone, facilitating HEH relationships and visualization of qualitative models for maintenance of human-created heritage and coexistence with the ecological models.

14.5 Good Living (*buen vivir*), Extractivism, and Yasunization: Suggestions from a Latin American Mega-Diverse Country

The Yasuní National Park in the Ecuadorian Amazon is a paradigmatic case study of the complexities of land management in environments with high biological and cultural diversity. Conflicts are structuring and structured to effect the simultaneous activation of territorial policies. Established as a park in 1979, recognized as a Biosphere Reserve by UNESCO in 1989, it is the region in the world with the highest biological diversity (of all five kingdoms per hectare is incomparable). The park is the home of the un-contacted tribes, the Waorani (people of recent contact), Kicwa, and Shuar people (Bass et al. 2010; Narvaez et al. 2013; Pappalardo 2013; Villaverde et al. 2005).

In 2007, Ecuador was the first country in the world to reserve an area of 7500 km^2 as an Intangible area, reserved for the rights to self-determination of uncontacted people Tagaeri Taromenane. This perpetually prohibits any industrial activity. A buffer zone of 10 km to ensuring further respect to the area was also set up. Intangible zone and buffer areas partially overlap Yasuní Biodiversity Reserve (YBR). However, the nomadic people have for centuries moved on an area of about 20,000 km^2 between the rivers Napo and Curaray (north/south) and between the first Andean hills and the confluence of Nashino with Curaray (west/east). Hence, the Intangible Zone is not aligned on territoriality of Tagaeri Taromenane (Pappalardo et al. 2013). Political anthropology and self-determination rights of societies' minorities in isolation should be integrated to design adequate territorial policies and the protection and maintenance of appropriate ecosystem conditions for survival (De Marchi et al. 2013).

The territory of YBR established for the conservation of biodiversity and the recognition of human rights is simultaneously overlapped by petroleum activities, illegal logging, and hunting. A geographical representation (a model of understanding) not distorting the territory and allowing the comprehension of the system of large projects around the four sides of the park (Via Auca, Napo River, the border with Peru and southern Curaray) to develop alternative proposals for sustainable territorial development is required (De Marchi 2013; De Marchi et al. 2015). Yasuni Biodiversity Reserve, its biological and cultural diversity, is at the core of a paradigmatic process of inclusive sustainable policy design in Latin America.

This case study attracted attention to the concept of *buen vivir*/good living; conversely, it is the theatre of the first proposal of living oil under soil.

In the last 20 years, many Latin American countries' endogenous declination of sustainability has been shaped around the perspective of *buen vivir*/good living (*Ally Kawsay*, *Sumak Kawsay*) combining quality of life, social justice, and rights of nature. However, good living (*buen vivir*) is not a monolithic corpus of common vision for sustainable development. It is possible to identify three main perspectives: a good living of indigenous people (good living from below), a good living of State (good living from above), and good living of social scientist (good living from theoretical reflection) (Altmann 2013a, b).

The origins of the good living concept are rooted in the indigenous movement starting in the 1980s, reclaiming a political role and plurinationality in Latin American countries. In Ecuador, a paradigmatic country for its mega diversity in both biological and cultural domains, the issue of plurinationality and ownership of territory and natural resources was framed with high intensity and structure elaborated due to the weight of indigenous people inside Ecuadorian society (De Marchi et al. 2010c; Hidalgo-Capitán et al. 2014). The defense of territory, the strict relation between plurinationality and territorial autonomy, saw the good living (*sumak kawsai*) as a vision of harmonic life behind political struggles for autonomous control over natural resources by indigenous nations (Altmann 2013a, b; CONAIE 2007, 2010). The fight of the Sarayaku people against oil operations in their territories frames the *sumak kawsay* as a way of strengthening people's identity. This constitutes a renovated pact among humans and nature, where management of resources by hunting the necessary, shared rules, harmony, regeneration time, life renewal, life cycle are key concepts of a local knowledge declination of a world quest for sustainability (Sarayaku 2003).

Critical intellectuals of ecological economics prescribe that good living should be based on a circular economy, a wise management of natural resources, and the abandonment of extraction economy based on oil and minerals with heavy environmental and social impacts (Alier 2011; Acosta 2010; Gudynas 2011).

The debate about *buen vivir* consolidates the new constitution of Ecuador of 2008, recognizing the right to nature and the *sumak kawsai* as a constitutional principle of Ecuador Plurinational State (Acosta 2010). However, between constitution principles and implemented policies, the State enacted a different concept of *buen vivir* in operational terms. The three approaches can be distinguished by the role assigned to ecosystems and natural resources (renewable and not renewable) in shaping the new alliance between people, State, and ecosystems summarized in the polarities of extractivism and Yasunization.

Extractivism bases social justice on natural resource extraction putting nature to the service of Nation, while civil society struggles for Yasunization. The latter is a neologism avoiding the extraction of fossil fuels and the simultaneous construction of territorial development alternatives, which combines the fight against climate change with community engagement, biodiversity conservation, and promotion and protection of human rights and environmental justice.

Yasunization is a social response to climate crisis. For the period 2011–2050, cumulative carbon dioxide emissions, projectively, must stay within the limits of 2 °C above the average global temperature of pre-industrial times, between 870 and 1240 Gt of CO_2. In contrast, the carbon stored in global reserves of fossil fuels is estimated at 11,000 Gt of carbon dioxide (Jakob and Hilaire 2015). This discrepancy represents a conflict between policies for maintaining carbon unexploited in order to protect life on Earth and the foolhardy options for carbon capturing and geo-engineering to protect a growing carbon economy. As climate scientists say "the continuation of high fossil fuel emissions, given current knowledge of the consequences, would be an act of extraordinary willing intergenerational injustice" (Hansen et al. 2013). Scientific communities confirm the need to maintain huge amount of carbon reserves buried: more than 80 % of coal, 50 % of gas, and 30 % of oil reserves must remain underground to save Earth (Meinshausen et al. 2009; McGlade and Ekins 2015).

Impacts on climate change ensuing decades are detailed by scientific literature and IPCC research; however; fossil fuels are key drivers not only for future climate change, but also for past and current social and environmental impacts. The ecological footprint of fossil fuel production is increasing as accessible reserves are depleted due to greater use of water, energy, and diluents limiting positive effect of energy efficiency and consumption reduction (Davidson and Andrews 2013; Jordaan et al. 2009). Impacts on health, water, and biodiversity in conventional and unconventional fossil fuel operations in different geographical contexts are widely reported in scientific literature (Allen et al. 2012; Cooley and Donnell 2014; Finer et al. 2015; Hansen et al. 2013; Kelly et al. 2010; Kurek et al. 2013; Narvaez et al. 2013; Osborn et al. 2011; Pappalardo et al. 2013; Schmidt 2011; Vidic et al. 2013). From a socio-economic point of view, petro-violence and Faustian Pacts (Watts 2001), the illusion of production (Coronil 1997), the paradox of plenty (Lynn 1997) and resource course (Stevens 2003; Sachs and Warner 2001), accumulation by dispossession (Harvey 2003) are dynamics related to fossil fuel economies.

Energy is a key topic of sustainability, and the way of looking to energy represents an important contribution to imagine, plan, and implement the energy transition toward a postcarbon society. Raphaël Fonteneau (Chap. 3) focuses on framing energy options through a reflection on Energy Return on Energy Investment (ERoEI) and the implementation of MODERN, a model visualizing the deployment of energy transition. Looking at energy through the lens of ERoEI exhibits the vested energy investment costs of producing energy, normally invisible in the market where prices do not internalize environmental and energy costs. The precipitation of ERoEI by oil extraction in the last century declined from 1000 to 5 (ERoEI is dimensionless) and oil import from 1990 to 2005 declined from 35 to 12. The crude mathematics of ERoEI and the use of MODERN supply important elements of reflection from one side to the energy, environmental and social costs of energy transition, and the risk to kill renewable sources to maintain linkage with fossil: the waste of energy to produce energy. The authors highlight the 'geographicalness' of energy deployment to avoid generalized solution and to find the right energy system for each territory.

Despite actual and future social, environmental, and climate impacts of fossil fuels and the need to maintain carbon reserves buried, there has been only one policy experiment in the world: the Yasuni ITT Initiative implemented in Ecuador in the period 2006–2013 avoids the exploitation of ITT oil block (Ishpingo, Tambococha, Tiputini), partially located into the Yasuni National Park. Launched in 2006 as a citizen's initiative, it was adopted in 2007 by the Ecuadorian Government proposing international commitment; in the framework of United Nations, there is a monetary compensation for the avoided CO_2 resulting from leaving the ITT block untapped, thus creating a sustainable scenario designed to improve life conditions in the Amazon (Espinosa 2013; Larrea and Warnars 2009; Narvaez et al. 2013; Rival 2010; Vallejo et al. 2015). In 2013, considering the scarce results in terms of accumulated funds, the government abandoned the initiative. However, civil society and the scientific community still support the ITT initiative, coining the neologism "Yasunization" to describe either past experience or the need to maintain oil underground to reduce carbon emissions, thereby granting climate justice, human rights, and biodiversity conservation, through citizen involvement.

As pointed by Kelly Swing (Chap. 5 in this book), Ecuador (may be the most mega-diverse country in the world with only 10 % of all leaving species catalogued) is particularly vulnerable to extractive operations and habitat fragmentation, considering the limited knowledge on biodiversity and the potential risk to lose suitable resources for human development. It is impossible to notice an extinction if the existence of species has never been confirmed. Swing quotes Ed Begley Jr. about the perspective in loss and gain: "When we destroy something created by man, we call it vandalism; when we destroy something created by nature, we call it progress" and Aldo Leopold about the representation of development for industrial society: "to those devoid of imagination, a blank place on the map is a useless waste; to others, the most valuable part".

Yasunization is the citizen commitment from below to conserve apparently empty spaces, where cultural and ecological diversity are not yet noticed to build supply side policies for climate change and territorial sustainability. It pledges a connection between local and global quality of life: at the moment the unique policy experiment with so ambitious objectives; a case study to be diffused of possible sustainable futures.

14.6   Dichotomy and Accord: Weaving the Main Threads

The quest for sustainability incorporates the basic challenge of humanity in producing territory: to build bi-modular systems combining society and ecosystems through the use of technology. Humans learn ecological niche (Colinvaux 1993) and supply culture and ingenuity to the task of living in different ecosystems. From pristine Earth life to the robotics revolution, techniques and technology accompany humans in building the territory as a spatial machine for social reproduction. Territory is the result of combination between technology and space (ecosystems) through the manipulation of social institutions in regulating technical application (Santos 1994, 1996).

Incorporating humans into Earth's fragile habitats means constructing a co-evolutionary process between society and ecosystems (Ostrom 1990), and among society, ecosystems and technology (the other way of speaking about territory is the historical application of human work and technology to space). It requires weaving social institutions, ecosystem dynamics, and evolutions of technology; in other words, to tailor the socio-logics, the bio-logics, the eco-logics, and the techno-logics (Raffestin 1980).

Chapter 13 of this book "Intelligent and Robust Path Planning and Control of Robotic Systems" is not only a review of technical challenges for robotics research, but offers an important contribution to weave the main threads running across the different chapters. Authors highlight that allowing robotic performance of complex tasks in structured and predictable spaces does not feedback only to manage the sophisticated mechanics of actuators. These new technological organisms should be able to perform predicted behaviors and to learn from environment. Robots spread especially in industry and car manufactories from welding to spray painting, avoiding human works in dangerous environments. Despite not appearing as humanoid with biped motion, many type of robots are widely used in medicine (from surgical application to rehabilitation), civil applications, and commonly distributed in consumer market as toys, home, and outdoor devices. Mobile robots for land displacement are legged or wheeled; flying robots, water, and underwater robots are regularly used in warfare and in other civil applications.

Among the main topics of robotics development, there is the control to perform tasks in different environments granting control of motion and control of contacts. Borrowing ideas from biological systems gave robotics the key improvement of intelligent control as emulation of biological intelligence, either for performing planned tasks or to solve problems. Safannah Raafat remarks that artificial intelligence allows robots to "learn from what is happening rather than from what is predicted". Different learning paradigms are commonly used to make robots self-adaptive and perform autonomous capabilities using neural networks, fuzzy logic, genetic algorithms, and hybrid techniques based on integration. Robot vision and visual servoing (introduced in Chap. 12) are other applications of intelligent controls, based on sensor control instead of model control, developing an iterative learning control; combination of visual control and positioning systems with GPS is widely used, for example, in flying robots and not only in photogrammetric operations. Intelligent controls, robot vision, servoing, and serving are important elements to manage obstacle avoidance and motion planning: one of the key tasks of autonomous robot systems.

A further evolution of artificial intelligence detailed is the swarm robotics as emulation of biological swarm organization of insects to implement the self-organization and interactions of a large number of distributed autonomous and decentralized robot systems. Swarmanoid clusters of foot-bots hand-bots and eye-bots perform complex tasks dismantling the model of humanoid biped robots and recombine motion, manipulation, and vision in flexible and suitable ways to perform in different environments.

In robotics, with artificial intelligence, humans are reproducing evolution not only for anatomy combination and recombination, but also for intelligence, cognition, and enaction. Robots operate at the local scale of factory, house, office, and hospital and occupy the wide scale conquering wide spaces of land, water, and sky; at the same time, robotic systems cross the different contexts of structured and predictable spaces to spread on unstructured, unpredictable open environments.

The change of scale and context and related social and ethical implications of robot applications are particularly visible in Unmanned Systems (not only aerial), widely known as the "drone revolution" (Fahlstrom and Gleason 2014; Nonami et al. 2010; Tsourdos et al. 2011; Valavanis and Vachtsevanos 2014). Developed by military needs for a clean, ethic, safe, and effective warfare, drones contributed in redefining the right to kill, the just war, the spatial distance of enemies' executions, and simultaneously consolidate war against civilians putting unprotected people under the drones (Bashir and Crews 2012; Chamayou 2015; Langewiesche 2011; Rae 2014; Yenne 2004). Conversely, drones enter everyday life in profession, home life, and spare time and are driving a parallel revolution in civil use of space and in the relation among humans, environment, and technology (Choi-Fitzpatrick et al. 2016).

Precision farming, for example, is a promising area of Unmanned Systems' development (Whelan and Taylor 2013; Xiang and Tian 2011), offering a bifurcation between using technology to implement regulated supply of external inputs to environment or a wise use of ecological regulation of agro-ecosystems. Precision farming robotics does not turn humanity away from its allelopathic option, as highlighted by Kelly Swing in Chap. 6. Allelopathy is the production of chemical substances limiting or eliminating competitors, allowing control of space and resources reducing the life possibility of other organisms. Humans allelopathic nature is a successful strategy to dominate other organisms and landscapes, winning the games humans "are encouraged to continue using the winning strategy, hoard resources, eliminate competitors, occupy more space". The lack of specialized organs producing chemical compounds, as other species, is supplied in humans by ingenuity developing a technological allopathy.

Technological opportunities put humans in front of the bifurcation, deepening the allelopathic strategy through territorial substitution; reestablishing a new alliance between society and ecosystems regulating flows and withdraws. The mis-match between technology, society and environment has social and environmental implications on environmental sustainability and social justice.

Unmanned vehicles are a well-known paradigmatic case study on technology and its social and environmental relations and implications. Flying robots are changing not only farm landscapes, but also urban sky, prospecting a new system of movement of people and goods and new options for delivering goods and services. Technological evolution has its own speed (Virilio 1997) compared to evolution of institution; despite twenty-first century technologies, societies are rooted in fifteenth century institutions (Beck et al. 1994) and the reflexive modernity of technology clashes with early modernity of institutions.

Robots developed occupying dangerous and unsafe working environments; however, this is not human achievement but an achievement of product relations; the luddite fight between workers and machines cannot be the solution to the freeing of humanity from working time. Sustainability is related with economy and institutions using technology to improve human quality of life. Technology can take many paths as clarified by the "Drones for good" approach (Choi-Fitzpatrick 2014) to find options consolidating human rights and avoiding geo-slavery.

Sustainability asks for a combined approach in weaving the threads of technology, scale, and modelling in ecosystems and institutions, putting technology into the framework of common pool resources (Ostrom and Hess 2007), and developing multiple co-evolutionary processes among technology as Global commons, ecosystems, and institutions together effect improvement of biological and cultural diversity (Ostrom 2005).

The challenge of human sustainable development highlighted in the 1994 Human development report opens into the framework of sustainable development objectives at 2030 and beyond.

This book allows the reader to tour theories, giving perspective on sustainability, places of implementation, scientific experimentation, and governance experiments between models and scales.

To close this chapter, we detail a case study from an European Alpine territory, which complements the tour of a sustainable development transition.

Trentino is an autonomous province in northern Italy and, together with the Autonomous Province of Bolzano, forms the Autonomous Region of Trentino Alto Adige/Südtirol. Trentino has an area of 6212 km^2 and around 530,000 people live in the province of Trento, distributed in 217 municipalities.

Trentino has one of the highest per-capita income and public expenditure at a national level. Thanks to the special situation of autonomy, in addition to the typical administrative functions of Italian provincial authorities, the Autonomous Province of Trento (PAT) has legislative powers in areas normally under state or regional jurisdiction. Health, education, training, employment, transport and roads, planning, energy, and environment have all been identified as particularly important competences. The special administrative grid is responsible for a special territory with high biodiversity from the Mediterranean ecosystem of the Lago di Garda to the tundra ecosystems of alpine cliffs. Forests cover more than 50 % of province surface and naturalistic forestry is widely applied to produce wood, protect territory, conserve biodiversity, and improve landscape quality for people and tourists.

Trentino is highly committed to an international quest for development and sustainability due to its historical background; additionally, a high rate of migration from Trentino in the late nineteenth and first half of the twentieth century, the widespread presence of social enterprises (cooperatives) commonly seen as the socio-economic mechanism, has lifted Trentino out of poverty in the last 50–60 years. Cooperatives in Trentino are active in different fields, particularly agriculture, credit, commercial distribution, and social services. Their presence in almost all economic sectors makes Trentino a "cooperative district" like few others in the world. Cooperatives are a cornerstone of a complex and diversified economic system based in quality food production (wine, cheese, meat, and fruits), community and industrial tourism (around 5,118,853 tourist arrivals and 29,668,503 tourist presences in 2013), industry, tertiary sector, research, and innovation.

On a national level, the Autonomous Province of Trento (PAT) has paved the way for territorial and environmental policies, adopting its first territorial plan in 1967 in Italy and its environmental impact assessment law in 1988. In 1999, Trentino prepared a first regional policy on sustainability and in 2010 prepared PASSO (*Patto per lo Sviluppo Sostenibile*—Pact for Sustainable Development).

PASSO was built in a dialogue among public administration experts, major groups, and civil society using face to face and online interactions. A common document and a set of indicators to monitor sustainability efforts to 2020 was developed around five areas of interest. To maintain an easy communicational approach, the five sustainability areas were simply represented with the first five letters of the alphabet: A, B, C, D, E (De Marchi et al. 2010a, b).

The first letter "A" represented the "Agenda setting" for sustainable development of Trentino as community looking for a multiple identity through sustainability. The commitment was around the development of a sustainable multiple citizenship, considering each inhabitant as alpine, Italian, European, and world citizens; the aim was to promote a declination of sustainability based on the sense of belonging and the responsibility to build Trentino as an alpine European cosmopolitan province. Indicators used for this topic were: Energy intensity and private transport (as a commitment for global climate change) and expenses in research and international cooperation as global commitment for innovation and sustainable development.

Letter B was related to Biodiversity as a corner stone for sustainability of life, supporting systems, and improving provincial ecosystem and landscapes for ethical, cultural, economical, and environmental reasons. If a landscape is an economical valuable asset for tourism and agriculture, the local community should not forget its importance for health, culture, and ethics, in a holistic reconnaissance of complex values of ecosystem services. Indicators for monitoring sustainability efforts in this area were: index of birds in agricultural habitats, soil consumption, area occupied by organic farming, water withdrawn, and tourist beds for 1000 inhabitants.

Closing the cycle of consumption and production (the letter C of the spelling book of Trentino sustainability), circular economy, and social metabolism, industrial ecology reproducing circulation in natural ecosystems was the declination of a topic of social responsibility for resource use. The need to create conditional innovations in social practices and adopt a different approach to consumption and production necessitates change both in the private and public sphere through the development of appropriate cultural, technological, regulatory, and economic models. Indicators for this component were: resources productivity, pro-capita waste production, certifications, and cars for 1000 inhabitants.

Democracy (letter D), information, participation, sustainability, and social innovation were declared as essential. Trentino participated in research of effectiveness of democracy, analyzing space for representative and deliberative democracy. PASSO intend to consolidate sustainable citizenship through responsible re-appropriation of the places, the development of an active territoriality, social inclusion, openness, transparency, and accountability of decision making. Indicators were participation to election, lifelong learning, e-government use, and environmental taxation.

Finally, "E" represents the "energy challenge" represented by the triangle: energy, transportation, and climate. PASSO attempts to define a perspective of sustainability and complete 'de-carbonisation' of the polarities moving and living in Trentino. The key issue is around sustainable mobility integrating territorial planning and transportation planning playing at the scales of accessibility and proximity. Indicators for this topics were: greenhouse gas emissions, greenhouse gas emission on transport sector, population exposed to particulate pollution, freight transport by roads, energy consumption from renewable sources.

PASSO defined strategies, objectives, and actions to be integrated into day-to-day policies to make sustainability "ordinary policy" of a consolidated direction for change.

The model of Trentino collated with the biodiversity and heritage site details of Ajanta in India meets the objectives of the environmental studies. The objectives of the Trentino model and detail of biodiversity and heritage of Ajanta help to develop a parallel construction of policies for sustainable ideals and possible implementation of models working on the practical grounds. The objective of sustenance could be achieved by bringing the developed and developing nation's models in a common platform. This certainly provides 'food for thought', which should be considered by international and developmental governance bodies such as the UN and smaller national or regional bodies, for formation and implementation of sustainability policies.

It is important to stress the contribution of different authors of this book in looking for the roots of a science of sustainability with more than two centuries of thickness before the framing of sustainability policies in international institutions starting from the seminal Bruntdtland report of 1987. This secular knowledge is another baggage of world cultural heritage we can combine with the wisdom of local knowledge to navigate the challenge of sustainability objectives at 2030 and beyond.

Linneus, Humboldt, Darwin, Muir, and many others contributed to complex thinking that subsequent authors have recalled, bringing us a culture not completely new, a scientific contribution to roots and long duration, and the discovery of the world cultural heritage of some exponents of the republic of letters.

The confusion and misunderstanding of using cosmos and universe as synonyms and interchangeable words do not help us in understanding the complexity of models and knowledge attached to the deep meaning of this word, resulting from the contribution of more than 25 centuries of formalized philosophy and thousands of years of *animus mundi* inside local knowledge.

As Hillman (1992) says cosmos is a Greek world meaning something of harmonious, beautiful, pleasant nature, offering something coherent and aesthetically appreciable, cosmos has more to do with cosmetic than with sidereal space. The cosmos is related to a Greek culture of harmony between space, organization, and policies. Universe is a Latin name, the uniform space of Roman Empire, Universus, and is a set of rules and structure commonly joining a unique empire with roads, language, and water distribution. Exchanging cosmos and universe is not useful in a questing for human sustainability. Cosmos enshrines beauty and diversity, from the cosmic scale of the city to the cosmic scale of the Earth and the nested hierarchy of complex systems. Kosmos (1845–1858) was the title of the work of Humboltd refunding geography and natural sciences in the nineteenth centuries. *Kosmos, Entwurf einer physischen Weltbeschreibung* in German can be translated as "Cosmos, project for a physical description of the world". *Entwurf* maintains all the dynamic characters of something not defined but under construction: a project, outline, blueprint, and sketch. *Beschreibung* is the description and *Welt* is the world; many translations use the word universe as synonymous with the world.

Advances in mathematic research are married to challenges of sustainability in this book, with both socioeconomic/human dimensions and enhanced sustenance of ecological systems. Seminally, De Marchi and colleagues established PASSO as an example of a refined system which may be used to detail sustainability categories in a grouped approach. The different chapters and contributions in this volume represent a global effort to further refine key areas which must be taken into consideration to facilitate ongoing sustainability efforts.

There is prediction potential both for past and future systems from the combined intuition of each chapter separately and also in holistic terms. Qualitative description and quantitatively refining elements of sustainability serve the purpose of enabling an accurate multi-criteria decision making process to be formed. If one considers each chapter as a 'block' of information within the wider concept of a sustainable Earth model, accurate Ariadne threads supply the information with which we can expand and contract models for further information value on specific areas and the wider management of the Earth system, as the human population increases into the future. Forming policies with use of the tools such as Digital Earth frameworks and GIS helps to harmonize interdisciplinary methods, with use of both social (including historic) and scientific approaches. Much synergy can result from this book. We highlight the use of specific computer-aided tools and thinking; however, it is important to retain understanding of the underlying systems in order that we can program and form coding for application to all developing technological platforms and hardware. This book is essential in helping to bridge the gap between policy formation, direction, and implementation; hence, the information and combinatorial approaches are pertinent for organizational units such as international organizations and other bodies, national and regional governments as well as the general public or 'man on the street' in both developed and developing contexts. Throughout this articulated volume, there are many paths challenging the quest for sustainability around the building of cosmos of knowledge and practices combining opportunity or fate to find a balanced combination among communities, institutions, ecosystems, and technology.

References

Acosta A (2010) El Buen Vivir en el camino del post-desarrollo. Una lectura desde la Constitución de Montecristi. Policy Paper 9. FES-ILDIS, Quito

Agarwal A (2001) Participatory exclusions, community forestry, and gender: an analysis for South Asia and a conceptual framework. World Dev 29(10):1623–1648

Agarwal B (2011) Food Crises and Gender Inequality. Economic and Social Affairs: DESA Working Paper No. 107

Agrawal A, Chhatre A (2006) Explaining success on the commons: community forest governance in the Indian Himalaya. World Dev 34(1):149–166

Agrawal A, Gupta K (2005) Decentralization and participation: the governance of common pool resources in Nepal's Terai. World Dev 33(7):1101–1114

Ajanta-Ellora Threatened: http://www.indiaenvironmentportal.org.in/content/25271/ajanta-ellora-threatened/. Accessed 1 Mar 2016

Alier JM (2011) El Ecologismo de los pobres: conflictos ambientales y lenguajes de valoración. Editorial Icaria, Barcelona

Allen L, Cohen MJ, Abelson D, Miller B (2012) Fossil fuels and water quality. World Water 7:73–96

Altmann P (2013a) Good life as a social movement proposal for natural resource use: the indigenous movement in Ecuador. Consilience J Sustain Dev 10(1):59–71

Altmann P (2013b) El Sumak Kawsay en el discurso del movimiento indígena ecuatoriano. Indiana 30:283–299

Andrienko G, Andrienko N, Bak P, Keim D, Wrobel S (2013) Visual analytics of movement. Springer, Berlin

Atzmanstorfer K, Blaschke T (2013) The geospatial web: a tool to support the empowerment of citizens through E-participation? In: Silva CN (ed) Citizen E-participation in urban governance: crowdsourcing and collaborative creativity. IGI Global, Hershey, pp 144–171

Bashir S, Crews RD (2012) Under the drones: modern lives in the Afghanistan-Pakistan borderlands. Harvard University Press, Cambridge

Bass MS, Finer M, Jenkins CN, Kreft H, Cisneros-Heredia DF (2010) Global conservation significance of Ecuador's Yasuní National Park. PLoS One 5:e8767

Beck U, Giddens A, Lash S (1994) Reflexive modernization. Politics, tradition and aesthetics in the modern social order. Stanford University Press, Stanford

Behera B, Stefanie E (2006) Institutional analysis of evolution of joint forest management in India: A new institutional economics approach, Forest Policy and Economics. 8(4): 350–362

Bharti G (2013) Ajanta caves: deterioration and conservation problems (a case study). Int J Sci Res Publ 3(11):392–395

Blanco FGB, Chen H (2014) The implementation of building information modelling in the United Kingdom by the transport industry. Procedia Soc Behav Sci 138:510–520

Boyd E, Nykvist B, Borgström S, Stacewicz I (2015) Anticipatory governance for social-ecological resilience. AMBIO 44(1):149–161

Briscoe D (ed) (2016) Beyond BIM: architecture information modeling. Taylor & Francis, New York

Butler D (2006) Virtual globes: the web-wide world. Nature 439:776–778

Chamayou G (2015) A theory of the drone. The New Press, New York

Choi-Fitzpatrick A (2014) Drones for good: technological innovation, social movements and the state. J Int Aff 68(1):1–18

Choi-Fitzpatrick A, Chavarria D, Cychosz E, Dingens JP, Duffey M, Koebel K, Siriphanh S, Tulen MY, Watanabe H (2016) A global estimate of non-military drone usage: 2009–2015. The Good Drone Lab, Kroc School of Peace Studies and Center for Media, Data, and Society, Central European University

Colinvaux PA (1993) Ecology II. Wiley, New York

Collinge C (2006) Flat ontology and the deconstruction of scale: a response to Marston, Jones and Woodward. Trans Inst Br Geogr 31:244–251

CONAIE (2007) Propuesta de la CONAIE frente a la Asamblea Constituyente. Principios y lineamientos para la nueva constitución del Ecuador. Por un Estado Plurinacional, Unitario, Soberano, Incluyente, Equitativo y Laico. CONAIE, Quito

CONAIE (2010) Declaración al pie de taita Imbabura y mama. Los Pueblos y Nacionalidades Indígenas del Ecuador frente a la Cumbre de los Presidentes del ALBA-TCP con 'autoridades indigenas y afrodecendientes'

Cooley H, Donnell K (2014) Hydraulic fracturing and water resources. What do we know and need to know? In: The world's water, vol 7. Island Press, Pacific Institute, Washington, pp 63–81

Coronil F (1997) The magical state. University of Chicago Press, Chicago

Craglia M, Ostermann F, Spinanti L (2012) Digital Earth from vision to practice: making sense of citizen-generated content. Int J Digital Earth 5(5):398–416

Craig JW, Harris T, Weiner D (2002) Community participation and geographic information systems. Taylor & Francis, London

Davidson DJ, Andrews J (2013) Not all about consumption. Science 339:1286–1287

De Blij H (2009) The power of place, geography destiny and globalization's rouge landscape. Oxford University Press, New York

De By RA, Georgiadou R (2014) Digital earth applications in the twenty-first century. Int J Digital Earth 7(7):511–515

De Marchi M (2013) Territorio y representaciones: geografías del Yasuní. In: Narvaez I, De Marchi M, Pappalardo SE (eds) Yasuní zona de sacrificio, Análisis de la iniciativa ITT y los derechos colectivos indígenas. FLACSO Ecuador, Quito, pp 242–275

De Marchi M, Dalla Libera L, Dalla Libera P, Dalla Libera S, Fracon C, Ropelato L (2010a) PA.S. SO., Patto per lo Sviluppo Sostenibile del Trentino, 2020. Provincia Autonoma di Trento

De Marchi M, Dalla Libera L, Dalla Libera P, Dalla Libera S, Fracon C, Ropelato L (2010b) Quadro Sinottico—PA.S.SO., Patto per lo Sviluppo Sostenibile del Trentino, 2020. Provincia Autonoma di Trento

De Marchi M, Natalicchio M, Ruffato M (2010c) I territori dei cittadini, il lavoro dell'OLCA (Observatorio Latinoamericano de Conflictos Ambientales). CLEUP, Padova

De Marchi M, Pappalardo SE, Ferrarese F (2013) Zona Intangible Tagaeri Taromenane (ZITT): ¿una, ninguna, cien mil? CLEUP, Padova

De Marchi M, Pappalardo SE, Codato D, Ferrarese F (2015) Zona intangible Tagaeri Taromenane y expansion de las Fronteras Hidrocarburifera. CLEUP, Padova

Deshmukh M (2008) The importance of geological studies in the conservation of cave tourism. In: Shalini S (ed) Profiles in Indian tourism. APH, Delhi

Di Biase D (1990) Visualization in the earth sciences. Earth Miner Sci 59(2):13–18

Doug A (2014) Impact of tourism on places of World Heritage. Scholedge Int J Multidiscip Allied Stud 1(1):10–20

Echeverria B (1995) Las ilusiones de la modernidad. UNAM/El equilibrista, México

Espinosa C (2013) The riddle of leaving the oil in the soil, Ecuador's Yasuní-ITT project, from a discourse perspective. Forest Policy Econ 36:27–36

Fahlstrom PG, Gleason TJ (2014) Introduction to UAV systems. Wiley, Chichester

Farinelli F (1998) (a cura), Quadri della natura/Alexander von Humboldt. La nuova Italia, Scandicci

Finer M, Babbitt B, Novoa S, Farrarese F, Pappalardo SE, De Marchi M, Saucedo M, Kumar A (2015) Future of oil and gas development in the western Amazon. Environ Res Lett 10 (2):024003. doi:10.1088/1748-9326/10/2/024003

Gill L, Kumar V, Lange E, Lerner D, Morgan E, Romano D, Shaw E (2010) An interactive visual decision support tool for sustainable urban river corridor management. In: Proceedings of iEMSs 2010, Ottawa, Canada, pp 1438–1445

Goodchild MF (2000) Cartographic futures on a digital earth. Cartographic Perspect 36:7–11

Goodchild MF (2012) The future of digital earth journal. Ann GIS 18(2):93–98

Gore A (1992) Earth in the balance. Houghton Mifflin, Boston

Gore A (1998) The digital earth: understanding our planet in the 21st century. http://www.isde5. org/al_gore_speech.htm. Accessed 1 Mar 2016

Grossner K, Clarke K (2007) Is Google Earth, "Digital Earth?": defining a vision. In: Proceedings of the fifth international symposium on digital earth, Berkeley

Grossner KE, Goodchild MF, Clarke KC (2008) Defining a digital earth system. Trans GIS 12 (1):145–160

Gudynas E (2011) Buen Vivir: today's tomorrow. Development 54(4):441–447

Hansen J, Kharecha P, Sato M, Masson-Delmotte V, Ackerman F, Beerling DJ, Hearty PJ, Hoegh-Guldberg O, Hsu S, Parmesan C, Rockstrom J, Rohling EJ, Sachs J, Smith P, Steffen K, Van Susteren L, von Schuckmann K, Zachos JC (2013) Assessing "Dangerous Climate Change": required reduction of carbon emissions to protect young people, future generations and nature. PLoS One 8(12):e81648

Harvey D (1989) The condition of postmodernity: an enquiry into the origins of cultural change. Blackwells, Cambridge

Harvey D (2003) The new imperialism. Oxford University Press, Oxford

Helfenbein R (2010) Thinking through scale: critical geography and curriculum spaces. In: Malewski E (ed) Curriculum studies handbook: the next moment. Routledge, New York, pp 304–317

Hesse-Biber SN (2010) Mixed methods research: merging theory with practice. Guilford Press, New York

Hidaka K (2007) Ajanta–Ellora Conservation and Tourism Development Project. http://www.jica. go.jp/english/our_work/evaluation/oda_loan/post/2007/pdf/project28_full.pdf

Hidalgo-Capitán AL, Guillén García A, Deleg Guazha A (eds) (2014) Sumak Kawsay Yuyay, Antología del Pensamiento Indigenista Ecuatoriano sobre Sumak Kawsay. Centro de Investigación en Migraciones (CIM), Universidad de Huelva, Programa Interdisciplinario de Población y Desarrollo Local Sustentable (PYDLOS), Universidad de Cuenca, Huelva y Cuenca

Higgins S, Mahon M, McDonagh J (2012) Interdisciplinary interpretations and applications of the concept of scale in landscape research. J Environ Manage 113:137–145

Hillman J (1992) The thought of the heart and the soul of the world. Spring, Dallas

Howitt R (1998) Scale as relation: musical metaphors of geographical scale. Area 30(49):58

Jalaei F, Jrade A (2015) Integrating building information modeling (BIM) and LEED system at the conceptual design stage of sustainable buildings. Sustain Cities Soc 18:95–107

Jakob M, Hilaire J (2015) JClimate science: Unburnable fossil-fuel reserves, Nature 517(7533): 150–152

Jameson F, Speaks M (1992) Envelopes and enclaves: the space of post-civil society (an architectural conversation). Assemblage 17:30–37

Johnston BR, Hiwasaki L (2011) Water, cultural diversity, and global environmental change: emerging trends, sustainable futures? Springer, Berlin

Jordaan SM, Keith DW, Stelfox B (2009) Quantifying land use of oil sands production: a life cycle perspective. Environ Res Lett 4:1–15

Joshi A (1999) Progressive bureaucracy: an Oxymoron? The Case of Joint Forest Management in India, Rural Development Forestry Network, Network Paper 24a, Winter 98/99

Kelly EN, Schindler DW, Hodson PV, Short JW, Radmanovich R (2010) Oil sands development contributes elements toxic at low concentrations to the Athabasca River and its tributaries. Proc Natl Acad Sci U S A 107:16178–16183

Kjosavik DJ, Shanmugaratnam N (2004) Integration or Exclusion? Locating Indigenous Peoples in the Development Process of Kerala, South India. Forum for Development Studies 31(2): 2–44

Koestler A (1980) Briks to babel. Random House, New York

Kshirsagar AA, Pawar SM, Patil NP, Mali VP (2012) Diversity of medicinal plants in Gautala sanctuary of Kannad, District Aurangabad (MS) India. Biosci Discov 3(3):355–361

Kumar S (2002) Does "Participation" in Common Pool Resource Management Help the Poor? A Social Cost–Benefit Analysis of Joint Forest Management in Jharkhand, India. World Development. 30(5):763–782

Kurek J, Kirk JL, Muir DCG, Wang X, Evans MS (2013) Legacy of a half century of Athabasca oil sands development recorded by lake ecosystems. Proc Natl Acad Sci USA 110(5):1761–1766. doi:10.1073/pnas.1217675110

Langewiesche W (2011) Esecuzioni a distanza. Adelphi, Milano

Leichenko RM, O'Brien KL (2002) The Dynamics of Rural Vulnerability to Global Change: The Case of southern Africa. Mitigation and Adaptation Strategies for Global Change 7(1):1–18

Larrea C, Warnars L (2009) Ecuador's Yasuni-ITT initiative: avoiding emissions by keeping petroleum underground. Energy Sustain Dev 13:219–223

Liu S, Meng X, Tam C (2015) Building information modeling based building design optimization for sustainability. Energy Build 105:139–153

Lynn KT (1997) The paradox of plenty. University of California Press, Berkeley

Masson V & et al. (2014) Adapting cities to climate change: A systemic modelling approach. Urban Climate 10:407–429

Mahdavi-Amiri A, Alderson T, Samavati F (2015) A survey of digital earth. Comput Graph 53 (B):95–117

Marston SA (2000) The social construction of scale. Prog Hum Geogr 24(219):242

Marston SA, Jones JP, Woodward K (2005) Human geography without scale. Trans Inst Br Geogr 30:416–432

Maturana H, Varela F (1979) Autopoiesis and cognition: the realization of the living. Boston studies in the philosophy of science. Reidel, Dordrecht

Maturana H, Varela F (1984) The tree of knowledge. Biological basis of human understanding. Shambhala, Boston

Mayr E (1997) This is biology. Belknap Press of Harvard University Press, Cambridge

McGlade C, Ekins P (2015) The geographical distribution of fossil fuels unused when limiting global warming to 2 °C. Nature 517:187–189

McHarg IL (1969) Design with nature. Natural History Press, New York

Meinshausen M, Meinshausen N, Hare W, Raper SCB, Frieler K, Knutti R, Frame DJ, Allen RM (2009) Greenhouse-gas emission targets for limiting global warming to 2 °C. Nature 458:1158–1163

Miettinen R, Paavola S (2014) Beyond the BIM utopia: approaches to the development and implementation of building information modeling. Autom Constr 43:84–91

Monmonier M (2005) Lying with maps. Stat Sci 20:215–222

Montesuma Oliveira I, Maziero Pinheiro Bini G, de Campos LE, Elke Debiasi R (2011) Escala e seus agentes em dissolução: Uma perspectiva transescalar. Rev Geogr Am Cent Semest 2011:1–10

Moor R, Gowda MVR (2014) India's risks: democratizing the management of threats to environment, health, and values. Oxford University Press, Oxford

Narvaez I, De Marchi M, Pappalardo SE (2013) Yasuní zona de sacrificio, Análisis de la iniciativa ITT y los derechos colectivos indígenas. FLACSO Ecuador, Quito

Naveh Z (1998) Ecological and cultural landscape restoration and the cultural evolution towards a post-industrial symbiosis between human society, and nature. Restor Ecol 6(2):135–143

Newell JP, Cousins JJ (2015) The boundaries of urban metabolism: towards a political industrial ecology. Prog Hum Geogr 39:702–728

Nir D (1990) Region as a socio-environmental system: an introduction to a systemic regional geography. Kluwer, Dordrecht

Nonami K, Kendoul F, Suzuki S, Wang W, Nakazawa D (2010) Autonomous flying robots: unmanned aerial vehicles and micro aerial vehicles. Springer, Berlin

Novotny V (2009) Notebooks from New Guinea: field notes of a tropical biologist. Oxford University Press, Oxford

Orr DW (1992) Ecological literacy: education and the transition to a postmodern world. SUNY Press, Albany

Osborn SG, Vengosh A, Warner NR, Jackson RB (2011) Methane contamination of drinking water accompanying gas-well drilling and hydraulic fracturing. Proc Natl Acad Sci U S A 108:8172–8176

Ostrom E (1990) Governing the commons: the evolution of institutions for collective action. Cambridge University Press, Cambridge

Ostrom E (2005) Understanding institutional diversity. Princeton University Press, Princeton

Ostrom E, Hess C (2007) Understanding knowledge as a commons: from theory to practice. MIT Press, Cambridge

Pappalardo SE (2013) Expansión de la frontera extractiva y conflictos ambientales en la Amazonia ecuatoriana: el caso Yasuni. Tesis de doctorado en Geografía humana y física, Universidad de Padova

Pappalardo SE, De Marchi M, Ferrarese F (2013) Uncontacted waorani in the Yasuní biosphere reserve: geographical validation of the zona intangible Tagaeri Taromenane (ZITT). PLoS One 8:e66293

Poffenberger M (1996) Communities and forest management: a report of the IUCN Working Group on Community Involvement in Forest Management with Recommendations to the Intergovernmental Panel on Forests. IUCN, Washington

Prigogine I, Stengers I (1997) The end of certainty: time, chaos and the new laws of nature. Free Press, New York

Rae JD (2014) Analyzing the drone debates: targeted killings, remote warfare, and military technology. Palgrave Macmillan, Basingstoke

Raffestin C (1980) Pour une géographie du pouvoir. Librairies Techniques, Paris

Raven R, Schot J, Berkhout F. (2012) "Space and scale in socio-technical transitions", Environmental Innovation and Societal Transitions 4, 63–78

Rival L (2010) Ecuador's Yasuní-ITT initiative: the old and new values of petroleum. Ecol Econ 70:358–365

Rokooei S (2015) Building information modeling in project management: necessities, challenges and outcomes. Procedia Soc Behav Sci 210:87–95

Roos J, Oliver D (1999) From fitness landscapes to knowledge landscapes. Syst Pract Action Res 12(3):279–293

Sachs J, Warner A (2001) The curse of natural resources. Eur Econ Rev 45(4–6):827–838

Santos M (1994) Técnica, espaço, tempo. Editora Hucitec, São Paulo

Santos M (1996) A Natureza do espaço. Técnica e tempo, razão e emoção. Editora Hucitec, São Paulo

Santos M (2008) Por uma Geografia Nova. EDUSP, São Paulo

Sarayaku (2003) El libro de la vida de Sarayaku para defender nuestro futuro

Sayre NF (2005) Ecological and geographical scale: parallels and potential for integration. Prog Hum Geogr 29(276):290

Sengar B (2011) Lovers admist Caves and Frescoes of Ajanta, Il Giornale Di Giulietta 57:26–29

Schmidt CW (2011) Blind Rush? Shale gas boom proceeds amid human health questions. Environ Health Perspect 119:348–353

Scolozzi R, Poli R (2015) System dynamics education: becoming part of anticipatory systems. On Horiz 23(2):107–118

Selin S (1999) Developing a typology of sustainable tourism partnerships. J Sustain Tour 7 (3–4):260–273

Sengar B (2016) Colonial landscape in a princely state: British land policies in rural spaces of Ajanta. Conference Souvenir: old and new worlds: the global challenges of rural history, international conference, Lisbon 27–30 Jan 2016. https://lisbon2016rh.files.wordpress.com/2015/12/onw-0118.pdf

Simao A, Densham PJ, Haklay MM (2009) Web-based GIS for collaborative planning and public participation: an application to the strategic planning of wind farm sites. J Environ Manage 90 (6):2027–2040

Singh RB, Anand S (2013) Geodiversity, geographical heritage and geoparks in India. Int J Geoherit 1(1):10–27

Stevens P (2003) Resource impact: curse or blessing? A literature survey. J Energ Lit 9(1):3–42

Sunderlin WD, Dewi S, Puntodewo A, Müller D, Angelsen A, Epprecht M (2008) Why forests are important for global poverty alleviation: a spatial explanation. Ecology and Society 13(2)

Tashakkori A, Teddlie C (2010) SAGE handbook of mixed methods in social & behavioral research. Sage, Thousands Oaks

Torrance H (2012) Triangulation, respondent validation, and democratic participation in mixed methods research. J Mixed Methods Res 6:111–123

Tsourdos A, White B, Shanmugavel M (2011) Cooperative path. Planning of unmanned aerial vehicles. Wiley, New York

Turri E (1998) Il paesaggio come teatro. Marsilio, Venezia

Valavanis KP, Vachtsevanos JG (eds) (2014) Handbook of unmanned aerial vehicles. Springer, Berlin

Vallejo MC, Burbano R, Falconí F, Larrea C (2015) Leaving oil underground in Ecuador: the Yasuní-ITT initiative from a multi-criteria perspective. Ecol Econ 109:175–185

Vidic RD, Brantley SL, Vandenbossche JM, Yoxtheimer D, Abad JD (2013) Impact of shale gas development on regional water quality. Science 340:1288

Villaverde X, Ormaza F, Marcial V, Jorgenson JP (2005) Parque Nacional y Reserva de Biosfera Yasuni: historias, problemas y perspectivas. Imprefepp, Quito

Virilio P (1997) Speed and politics: an essay on dromology. Semiotext(e), New York

Virilio P (1993) O espaço crítico e as perspectivas do tempo real. Editora 34, Rio de Janeiro

Watts M (2001) Petro-violence: community, extraction, and political ecology of a mythic commodity. In: Peluso NL, Watts M (eds) Violent environments. Cornell University Press, Ithaca, pp 189–212

Whelan B, Taylor J (2013) Precision agriculture for grain production systems. CSIRO, Clayton

Wu W, Kaushik I (2015) Design for sustainable aging: improving design communication through building information modeling and game engine integration. Procedia Eng 118:926–933

Xiang H, Tian L (2011) Development of a low-cost agricultural remote sensing system based on an autonomous unmanned aerial vehicle (UAV). Biosyst Eng 108:174–190

Yenne B (2004) Attack of the drones: a history of unmanned aerial combat. MBI, Minneapolis

Zajíčková V, Achten H (2013) Landscape information modeling, plants as the components for information modelling. In: Proceedings of the eCAADe conference, 2013, computation and performance, building information modelling, vol 2, Delft University of Technology, Delft, pp 515–524, 18–20 Sep 2013

Zhang C, Lin H, Chen M, Li R, Zeng Z (2014) Scale compatibility analysis in geographic process research: a case study of a meteorological simulation in Hong Kong. Appl Geogr 52:135–143

Printed in the United States
By Bookmasters